Materials for Carbon Capture

Materials for Carbon Capture

Edited by

De-en Jiang
Department of Chemistry, University of California, Riverside, CA, USA

Shannon M. Mahurin
Chemical Sciences Division, Oak Ridge National Laboratory, Oak Ridge, TN, USA

Sheng Dai
Chemical Sciences Division, Oak Ridge National Laboratory, Oak Ridge, TN, USA
Department of Chemistry, University of Tennessee, Knoxville, TN, USA

Registered Offices
John Wiley & Sons, Inc., 111 River Street, Hoboken, NJ 07030, USA
John Wiley & Sons Ltd, The Atrium, Southern Gate, Chichester, West Sussex, PO19 8SQ, UK

Editorial Office
The Atrium, Southern Gate, Chichester, West Sussex, PO19 8SQ, UK

For details of our global editorial offices, customer services, and more information about Wiley products visit us at www.wiley.com.

Library of Congress Cataloging-in-Publication Data applied for

Hardback ISBN: 9781119091172

Cover Design: Wiley
Cover Image: Courtesy of De-en Jiang

Set in 9.5/12.5pt STIXTwoText by SPi Global, Chennai, India

Printed and bound by CPI Group (UK) Ltd, Croydon, CR0 4YY

10 9 8 7 6 5 4 3 2 1

Contents

List of Contributors

Ravichandar Babarao
Manufacturing Flagship
Commonwealth Scientific and Industrial
Research Organisation
Clayton, Victoria
Australia

and

School of Science
RMIT University
Melbourne, Victoria
Australia

Tae-Hyun Bae
Singapore Membrane Technology Centre
Nanyang Technological University
Singapore

and

School of Chemical and Biomedical
Engineering
Nanyang Technological University
Singapore

Teng Ben
Department of Chemistry
Jilin University
Changchun
China

Jason E. Bara
Department of Chemical & Biological
Engineering
University of Alabama
Tuscaloosa, AL
USA

Chong Yang Chuah
School of Chemical and Biomedical
Engineering
Nanyang Technological University
Singapore

Sheng Dai
Chemical Sciences Division
Oak Ridge National Laboratory
TN
USA

and

Department of Chemistry
University of Tennessee
Knoxville, TN
USA

Chi-Linh Do-Thanh
Department of Chemistry
University of Tennessee
Knoxville, TN
USA

Xueying Ge
Department of Chemistry
University of South Florida
Tampa, FL
USA

W. Jeffrey Horne
Department of Chemical & Biological
Engineering
University of Alabama
Tuscaloosa, AL
USA

Kuan Huang
Key Laboratory of Poyang Lake
Environmental and Resources Utilization
of Ministry of Education
School of Resources Environmental and
Chemical Engineering
Nanchang University
Jiangxi
China

Aman Jain
Manufacturing Flagship
Commonwealth Scientific and Industrial
Research Organisation
Clayton, Victoria
Australia

and

Indian Institute of Technology
Kanpur Uttar Pradesh
India

De-en Jiang
Department of Chemistry
University of California
Riverside, CA
USA

Siew Siang Lee
Singapore Membrane Technology Centre
Nanyang Technological University
Singapore

An-Hui Lu
School of Chemical Engineering
State Key Laboratory of Fine Chemicals
Dalian University of Technology
China

Shengqian Ma
Department of Chemistry
University of South Florida
Tampa, FL
USA

Shannon M. Mahurin
Chemical Sciences Division
Oak Ridge National Laboratory
TN
USA

Mingguang Pan
Department of Chemistry
ZJU-NHU United R&D Center
Zhejiang University
Hangzhou
China

Shilun Qiu
State Key Laboratory of Inorganic
Synthesis and Preparative Chemistry
Jilin University
Changchun
China

Jennifer Schott
Department of Chemistry
University of Tennessee
Knoxville, TN
USA

Ziqi Tian
Department of Chemistry
University of California
Riverside, CA
USA

Ikuo Taniguchi
International Institute for Carbon-Neutral
Energy Research (WPI-I^{2}CNER)
Kyushu University
Fukuoka
Japan

Aaron W. Thornton
Manufacturing Flagship
Commonwealth Scientific and Industrial
Research Organisation
Clayton, Victoria
Australia

Congmin Wang
Department of Chemistry
ZJU-NHU United R&D Center
Zhejiang University
Hangzhou
China

and

Key Laboratory of Biomass Chemical
Engineering of Ministry of Education
Zhejiang University
Hangzhou
China

Rong Wang
Singapore Membrane Technology Centre
Nanyang Technological University
Singapore

and

School of Civil and Environmental
Engineering
Nanyang Technological University
Singapore

Song Wang
Department of Chemistry
University of California
Riverside, CA
USA

Sunee Wongchitphimon
Singapore Membrane Technology Centre
Nanyang Technological University
Singapore

Xiang-Qian Zhang
School of Chemical Engineering
State Key Laboratory of Fine Chemicals
Dalian University of Technology
China

Preface

Fossil fuels are a relatively inexpensive source of energy, and the combustion of these fuels has enabled significant technological advances, has fostered prosperity, and largely powers the global economy of today. Fossil fuels are used over a broad range of sectors including transportation, the industrial sector, and the generation of electricity. The combustion of fossil fuels, however, results in the emission of carbon dioxide into the atmosphere, which leads to negative environmental impacts. Despite the development and growth of renewable energy sources, fossil fuels will continue to play a key part in the energy landscape for the foreseeable future as the global demand for energy continues to grow at an unprecedented rate. Carbon capture, which is a process where carbon dioxide is separated from power plant effluents or industrial processes, offers a technological solution to reduce carbon dioxide emissions while enabling the continued use of fossil fuels. Though carbon capture technologies currently exist, new materials and processes are needed to drive technological advances for more energy-efficient and cost-effective separation of carbon dioxide from a mixed gas stream. The importance of this topic is surely reflected in the heightened interest across many sectors including industry.

This book aims to highlight the current state of the art in materials for carbon capture, providing a comprehensive understanding of separations ranging from solid sorbents to liquid sorbents and membranes. The knowledge, expertise, and dedication of the diverse group of contributors have made this book a reality. We feel this book will be helpful to those new to the area of carbon capture, affording an overview of the novel materials currently being explored. Graduate students will find this book useful both as an introduction to the various materials that are on the cutting edge of separations and as a way to expand their fundamental understanding of the separations process. Hopefully, it will also inspire these graduate students and spark their imagination to go beyond the novel materials highlighted in this book and develop new materials with enhanced separations properties. Even experts in the field, experimentalists and theorists alike, will benefit from the diverse and unconventional topics covered in this book. The combined efforts of experts and those new to the field, experimentalists and theorists, scientists and engineers will foster discovery and innovation in carbon capture as well as storage and utilization. We hope that readers of all levels will enjoy this book and discover the wonder of this separations process while also being inspired to contribute their knowledge to a global challenge that affects us all.

Acknowledgments

We are supported by the US Department of Energy, Office of Science, Office of Basic Energy Sciences, Chemical Sciences, Geosciences, and Biosciences Division. We are grateful to Sarah Higginbotham and Emma Strickland of Wiley for working with us on the book from proposal to production and to Aruna Pragasam and Adalfin Jayasingh of Wiley for helping us deliver this book. Jianbo Xu helped index the book. We thank all the contributors to this book for their collaboration, time, and patience.

1

Introduction

De-en Jiang[1], Shannon M. Mahurin[2] and Sheng Dai[2,3]

[1]*Department of Chemistry, University of California, Riverside, CA, USA*
[2]*Chemical Sciences Division, Oak Ridge National Laboratory, Oak Ridge, TN, USA*
[3]*Department of Chemistry, University of Tennessee, Knoxville, TN, USA*

CHAPTER MENU

References, 3

Burning fossil fuels for electricity and transportation has led to steadily increasing CO_2 levels in the atmosphere, as recorded in the Keeling curve [1], and, consequently, global warming. This concern has become a major driving force for a larger share of renewable energy in power generation and for electrifying transportation. However, coal-fired and natural gas-fired power plants have a long lifetime, which makes post-combustion carbon capture necessary. In addition, pre-combustion carbon capture will be an important part of clean-coal technology. Removal of CO_2 from natural gas is also important, especially given the shale-gas boom. Moreover, direct air capture of CO_2 has also been explored by many, since there is already a large amount of emitted CO_2 in the air. Hence, carbon capture and storage (CCS) is important for mitigating global warming and climate change [2].

Novel materials hold the key to energy-efficient carbon capture. As a frontier research area, carbon capture has been a major driving force behind many materials technologies. This book aims to present an overview of the advances in materials research for carbon capture, beyond the commercial amine-based solvent-sorption technologies. Broadly speaking, carbon-capture materials can be divided into two categories: sorbents and membranes. Common sorbents are high-surface-area porous materials, such as zeolites, metal-organic frameworks (MOFs), covalent-organic frameworks (COFs), and amorphous porous carbonaceous materials. Membranes are mainly of the polymeric type, while inorganic, carbonaceous, and mixed-matrix membranes (MMMs) are being actively explored.

MOFs are promising large-capacity adsorbents for CO_2 due to their great chemical tunability in controlling the pore size, pore shape and topology, metal-site chemistry, and linker functional groups [3]. In Chapter 2, Ge and Ma present an overview of the MOF materials for carbon capture, focusing on the correlation between MOF structure and CO_2 uptake and tabulating the best-performing MOFs; they also briefly discuss pure MOF membranes and MOF-containing mixed-matrix-membranes.

One weakness limiting the application of many MOFs in capturing CO_2 from water-vapor-saturated flue gas is their sensitivity to moisture. Porous carbonaceous materials, on the other hand, are both chemically and thermally stable. They are usually made from pyrolysis of a carbon-atom-containing precursor that can be either a polymer or a small molecule [4]. At the high-temperature-treatment end (~900 °C or higher), the carbon content is high (>90 mol%), and the resulting materials are just called porous carbons. In Chapter 3, Zhang and Lu review the different approaches to make porous carbons, from the perspectives of templates and precursors, and their performances for carbon capture as adsorbents.

Ben, Qiu, and their workers have pioneered the design and synthesis of a different type of porous carbonaceous materials called porous aromatic frameworks (PAFs), which can be visualized by replacing all the C—C bonds in the diamond with groups such as the biphenyl, leading to a material with a huge surface area of over $5000 \, m^2 \, g^{-1}$ [5]. PAFs have generated a lot of interest as a material platform for gas storage and separation. In Chapter 4, Ben and Qiu review PAFs for carbon capture and strategies for their further improvement.

Computational modeling and virtual screening are playing an increasingly important role in materials discovery for catalysts, batteries, thermoelectrics, and topological phases, to name a few. So carbon capture is not an exception. In Chapter 5, Jain, Babarao, and Thornton comprehensively review the computational methods, candidate materials, and criteria for virtual screening of materials as membranes and sorbents for carbon capture. Moreover, they show the physical insights that can be gained from computational modeling in understanding the many factors that come into play.

In Chapter 6, Jiang and workers further summarize the advances in using computational modeling to guide the development of ultrathin membranes based on 2D materials such as graphene for gas separations. Interlayer-spacing tuning exhibits great potential in control of molecular and ionic transport in 2D membranes [6–8]. The field of 2D membranes for gas separations was to a large extent initiated by the original proof of concept of one-atom-thin membranes for gas separations by Jiang et al. [9]. In this chapter, they review the progress made both experimentally and computationally in this field since their original work in 2009, focusing on the computational aspects for guiding future experimental developments.

Polymeric membranes are commercially used for gas separations and water desalination [10]. Their performances are limited by a trade-off between selectivity and permeability called the Robeson upper bound [11]. In Chapter 7, Bara and Horne review the polymeric membranes for CO_2 separation for different types of polymers; they also briefly touch upon facilitated transport and membrane contactors. In Chapter 8, Huang and Dai present an overview of carbon-based membranes for CO_2 separation.

Increasing materials complexity has been a key driver in recent advances in membrane separations to leverage both interactions and transport via different components and building blocks of the composite materials [12]. The complexity built in the composite materials supplies a large space of imagination for use-inspired fundamental studies via mixing and matching of materials, a point emphasized in a special issue of *Science* (2 November 2018). In the context of gas separations, the best example is the MMMs [13]. Strategies in designing MMMs to overcome the upper bound include the use of nano-sized or nanosheet-shaped molecular sieving fillers with a polymer and the elimination of the interfacial gaps [10]. In Chapter 9, Bae and coworkers review composite materials for carbon capture in terms of both adsorbents and membranes. Dendrimers provide a different approach toward materials complexity. In Chapter 10, Taniguchi discusses how poly(amidoamine) dendrimers can be used for carbon capture.

Ionic liquids (ILs) as a nonvolatile but versatile medium have attracted great interest in the areas of separations, energy storage, and catalysis, among others. Advanced ionic systems, such as confined ionic liquids [14], poly(ionic liquid)s [15], and porous ionic polymers [16] offer many opportunities in utilizing the long-range Coulombic interaction to tune the structure and assembly of the molecular building blocks that impact molecular/ionic transport in either a sorbent [17] or a membrane setup [18, 19]. In Chapter 11, Pan and Wang summarize the recent advances in using ionic liquids for chemisorption of CO_2, while in Chapter 12, Mahurin and coworkers review IL-based membranes for CO_2 separation.

In sum, this book is aimed at presenting to the reader the latest advances in the materials aspect of carbon capture, drawing from the contributors' expertise. This field is still quickly advancing, driven by the urgent need to mitigate carbon emissions. Although there are still developments that are not covered in the following 11 chapters, we hope that they do present some of the most important classes of materials currently being pursued for carbon capture.

References

1 Scripps Institution of Oceanography. (2019).The Keeling Curve. https://scripps.ucsd.edu/programs/keelingcurve (accessed 27 May 2019).

2 Bui, M., Adjiman, C.S., Bardow, A. et al. (2018). Carbon capture and storage (CCS): the way forward. *Energy & Environmental Science* 11: 1062.

3 Yu, J.M., Xie, L.H., Li, J.R. et al. (2017). CO_2 capture and separations using MOFs: computational and experimental studies. *Chemical Reviews* 117: 9674.

4 Zhai, Y., Dou, Y., Zhao, D. et al. (2011). Carbon materials for chemical capacitive energy storage. *Advanced Materials* 23: 4828.

5 Ben, T., Ren, H., Ma, S. et al. (2009). Targeted synthesis of a porous aromatic framework with high stability and exceptionally high surface area. *Angewandte Chemie International Edition* 48: 9457.

6 Sun, P.Z., Wang, K.L., and Zhu, H.W. (2016). Recent developments in graphene-based membranes: structure, mass-transport mechanism and potential applications. *Advanced Materials* 28: 2287.

7 Wang, L.D., Boutilier, M.S.H., Kidambi, P.R. et al. (2017). Fundamental transport mechanisms, fabrication and potential applications of nanoporous atomically thin membranes. *Nature Nanotechnology* 12: 509.

8 Zheng, S.X., Tu, Q.S., Urban, J.J. et al. (2017). Swelling of graphene oxide membranes in aqueous solution: characterization of interlayer spacing and insight into water transport mechanisms. *ACS Nano* 11: 6440.

9 Jiang, D.E., Cooper, V.R., and Dai, S. (2009). Porous graphene as the ultimate membrane for gas separation. *Nano Letters* 9: 4019.

10 Park, H.B., Kamcev, J., Robeson, L.M. et al. (2017). Maximizing the right stuff: the trade-off between membrane permeability and selectivity. *Science* 356: eaab0530.

11 Robeson, L.M. (2008). The upper bound revisited. *Journal of Membrane Science* 320: 390.

12 Koros, W.J. and Zhang, C. (2017). Materials for next-generation molecularly selective synthetic membranes. *Nature Materials* 16: 289.

13 Dechnik, J., Gascon, J., Doonan, C.J. et al. (2017). Mixed-matrix membranes. *Angewandte Chemie International Edition* 56: 9292.

14 Zhang, S.G., Zhang, J.H., Zhang, Y., and Deng, Y.Q. (2017). Nanoconfined ionic liquids. *Chemical Reviews* 117: 6755.

15 Qian, W.J., Texter, J., and Yan, F. (2017). Frontiers in poly(ionic liquid)s: syntheses and applications. *Chemical Society Reviews* 46: 1124.

16 Xu, D., Guo, J.N., and Yan, F. (2018). Porous ionic polymers: design, synthesis, and applications. *Progress in Polymer Science* 79: 121.

17 Zeng, S.J., Zhang, X., Bai, L.P. et al. (2017). Ionic-liquid-based CO_2 capture systems: structure, interaction and process. *Chemical Reviews* 117: 9625.

18 Dai, Z.D., Noble, R.D., Gin, D.L. et al. (2016). Combination of ionic liquids with membrane technology: a new approach for CO_2 separation. *Journal of Membrane Science* 497: 1.

19 Tome, L.C. and Marrucho, I.M. (2016). Ionic liquid-based materials: a platform to design engineered CO_2 separation membranes. *Chemical Society Reviews* 45: 2785.

2

CO$_2$ Capture and Separation of Metal–Organic Frameworks

Xueying Ge and Shengqian Ma

Department of Chemistry, University of South Florida, Tampa, FL, USA

2.1 Introduction

The goal of this chapter is to briefly introduce CO$_2$ capture and separation related to metal–organic framework (MOF) materials. The concentration of CO$_2$ in the atmosphere has increased rapidly from 310 to 411 ppm over the period 1960–2019 and is expected to reach more than 500 ppm by 2050 (Figure 2.1). Burning of fossil fuels like coal, oil, and natural gas to support 85% of global energy demand is the mainly anthropogenic

Materials for Carbon Capture, First Edition. Edited by De-en Jiang, Shannon M. Mahurin and Sheng Dai.

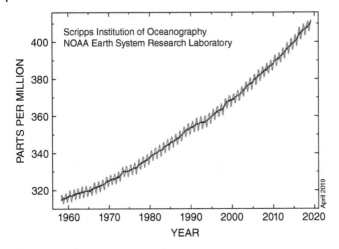

Figure 2.1 Atmospheric CO_2 concentration during 1958–2019 (at Mauna Loa Observatory), showing the increase of CO_2 in the atmosphere. Source: reproduced from Ref. [1] with permission from the NOAA Global Monitoring Division.

source of CO_2 emissions [2]. Due to the increase of the global population and industrial development, fossil fuels will continue playing an important role in the energy supply. And the emission of CO_2 will continue increasing, influencing the balance of incoming and outgoing energy in the atmosphere and raising the average surface temperature of Earth. CO_2 is often regarded as the primary anthropogenic greenhouse gas, which is the main contributor to global warming.

It is imperative to reduce CO_2 emissions from various industrial processes in order to minimize its influence on climate change [3]. Thus, carbon capture and storage (CCS) technologies need to be developed to reduce CO_2 emissions, such as capturing CO_2 from emission sources like power plants, improving energy efficiency, and shifting to renewable energy sources [4, 5]. Based on the fundamental chemical process involved in the combustion of fossil fuel, there are mainly three basic CO_2 separation and capture options: post-combustion, pre-combustion, and oxyfuel combustion [2].

Currently, aqueous alkanolamine solutions are the best technology to capture CO_2 from post-combustion flue gas [3]. However, it is hard to regenerate the sorbent (the amine solution) from the formation of C–N chemical bonds between amine functionalities and CO_2 in this process. Thus, it is highly demanding to develop the physical sorbent with a lower regeneration cost.

As an emerging new class of porous solids, MOFs, also known as *coordination networks* or *coordination polymers*, are novel materials constructed with metal or metal oxide connected by organic linkers (referred to as secondary building units [SBUs]) via a self-assembly process [6]. MOFs are robust enough to allow the removal of guest species for permanent porosity, and therefore are promising for adsorption-related applications. CO_2 capture is the most attractive research area in the application of MOFs. This chapter provides an overview of recent work in this area and considers the evaluation of the adsorption process and its selectivity using the isosteric heat of adsorption (Q_{st}) and Ideal Adsorbed Solution Theory (IAST). Following the introduction, Section 2.1 provides a summary of three capture technologies and MOFs as a promising candidate for CO_2 separation and capture. Section 2.2 outlines the mathematical methods to evaluate the performance of practical gas separation. Section 2.3 illustrates the ability of MOFs in CO_2 capture and separation. And Section 2.4 discusses CO_2 capture based on MOFs' ability and the relevant evaluation theory. A brief summary of CO_2 capture in MOF membranes is also provided in Section 2.5. Section 2.6 draws conclusions including the improvement of the MOF structure and future study for CO_2 capture. Our goal in this chapter, from basic math methods to the practice application, is to provide sufficient detail about CO_2 capture and separation by MOFs.

2.1.1 CO_2 Capture Process

Broadly, three lines of capturing technologies exist to reduce CO_2 emissions in combustion processes: post-combustion, pre-combustion, and oxyfuel combustion [2]. The operational conditions for each process are different in terms of temperature, pressure, and optimized materials. Post-combustion capture is the most feasible choice because it can be directly used in power plants by retrofitting processes [7]. This capture system can be deployed in both coal-fired power plants and natural-gas power plants [8, 9]. Combustion of impurities in coal produces unwanted by-products including sulfur dioxide, nitrogen oxides, and particulate matter (fly ash) [7], resulting in air pollution. Other trace species, such as mercury, are also formed during this process. All these unwanted products must be removed to meet applicable emission standards. The main drawback of the post-combustion technique is low CO_2 capture efficiency when streaming a large volume gas at 40–60 °C in low pressure. Usually, the flue gas in power plants contains 15–16% CO_2 and 73–77% N_2 by volume, as well as other components such as water, SO_2, and O_2 at ambient conditions [10]. The separation of CO_2 from N_2 is the central issue in post-combustion CO_2 capture.

The pre-combustion capture technique converts fossil fuels to syngas (a mixture of H_2 and CO) followed by water-gas shift reactions to form a mixture of H_2 and CO_2 at high partial pressures before combustion [11]. The significant advantage of pre-combustion capture is that the higher component concentrations and pressures reduce the energy capture penalty in the process to 10–16%, roughly half that of the post-combustion CO_2 capture. Separation of CO_2/H_2 is easier than CO_2/N_2, compared with post-combustion technology. However, high temperature requirements, low efficiency, and high cost make pre-combustion a hard choice [10].

The third alternative, oxyfuel combustion, has been attracting people's interest for power generation: it requires pure oxygen rather than air for the burning process to produce the flue gas, which is mainly CO_2 and water vapor (H_2O), as well as small amounts of NO_X and SO_X. CO_2 capture is not required in oxyfuel combustion as it is easy to condense water vapor from the flue gas. However, pure oxygen is produced by air separation processes, resulting in a high cost for this technology, which hinders its application [9]. Technical options for CO_2 capture are shown in Figure 2.2.

2.1.2 Introduction to MOFs for CO_2 Capture and Separation

MOFs are attractive to scientists because of their unique structural properties, including high thermal and chemical stability, robustness, unprecedented internal surface area, 3D structure incorporating uniform pore size, and a network channel [13–16]. The pore

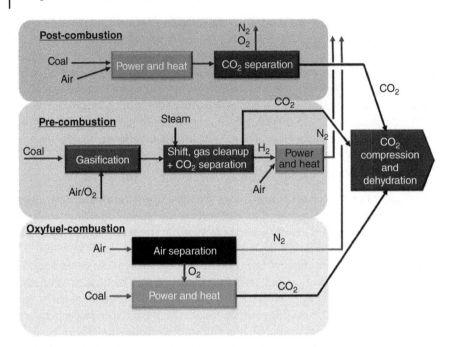

Figure 2.2 Technical option for CO_2 capture. Source: reproduced from Ref. [12] with permission from *Chemical Society Reviews*.

and channel can be maintained after removal of the guest molecules from the pores. The remaining void can adsorb other guest molecules, such as CO_2, H_2, N_2, and CH_4. Furthermore, due to the virtually limitless combinations of metals and ligands, incalculable number of MOFs can be designed and synthesized. Particularly, by the judicious choice of SBUs and linkers, MOFs can be designed and systematically tuned by predesign in synthesis and post-synthetic modification for specific applications, compared with zeolites and carbon-based adsorbents [17]. Therefore, MOFs are the ideal adsorbents or membrane materials for storage and separation of a gas like CO_2.

Compared with zeolites, the high surface-area-to-weight ratio of MOFs is a way to increase the capacity for CO_2 capture. Even though zeolites have higher storage capacities at pressures of less than 10 bar, some MOFs perform better for gas capture at pressures greater than 10 bar. Following are the capacities of porous materials with the amount of active area per unit weight at high pressure: frameworks of 1500–4500 m^2 g^{-1}, activated carbon of 400–1000 m^2 g^{-1}, and zeolites of up to 1500 m^2 g^{-1} [16].

2.2 Evaluation Theory

2.2.1 Isosteric Heat of Adsorption (Q_{st})

The adsorption process is exothermic, and we can quantify this process using the Clausius-Clapeyron relation [18]. The Q_{st} for gas on porous materials is calculated from the adsorption isotherms measured at two temperatures. Under ideal conditions, the bulk gas phase is considered ideal, and the adsorbed phase volume is neglected [19]. Therefore, the isosteric heat of adsorption is calculated from the adsorption isotherms measured at 77 K and 87 K, which are the temperatures of liquid nitrogen and liquid argon, respectively. However, this small temperature range usually causes a very high uncertainty in the heat of adsorption. To ensure that the isosteric heat of adsorption is determined with higher accuracy, it requires at a minimum three closely spaced adsorption isotherms (in temperature, e.g. 77 K, 87 K, and 97 K) [20]. The variation and magnitude from the function can quantitatively evaluate the interaction between adsorbates and adsorbents. The three models presented next have been primarily studied by interpolating isotherm pressure values at specific amounts adsorbed from the data, calculating the Q_{st}.

2.2.1.1 The Virial Method 1

$$\ln\left(\frac{n}{p}\right) = A_0 + A_1 n + A_2 n^2 + \ldots \tag{2.1}$$

where n is the amount adsorbed, p is the pressure, and A_0, A_1, and A_2 are constants. The isosteric enthalpy of adsorption at specific surface coverages is determined by the van't Hoff isochore. The isosteric enthalpies of adsorption at zero surface coverage are obtained from the A_0 values [20].

2.2.1.2 The Virial Method 2

$$\ln(p) = \ln(n) + \frac{1}{T}\sum_{i=0}^{m} a_i n^i + \sum_{j=0}^{m} b_j n^j \tag{2.2}$$

Where p is pressure in torr, n is the amount adsorbed in mmol g^{-1}, T is the temperature in K, and a_i and b_j are temperature-independent empirical parameters. This method provides the equilibrium pressure p and concentration n for any temperature T when a_i and b_j have been determined from a fit to a set of experimental isotherms [21]. This is essentially the Clausius-Clapeyron linear relationship between $\ln(p)$ and $\frac{1}{T}$, where the slope (second term in Eq. (2.2)) is the isosteric enthalpy of adsorption (Q_{st}) from the following equation:

$$Q_{st} = -R\sum_{i=0}^{m} a_i n^i \tag{2.3}$$

where R is the gas constant. This is the most common way to calculate isosteric heat of adsorption [21].

2.2.1.3 The Langmuir–Freundlich Equation

$$\frac{Q}{Q_m} = \frac{B \times P^{1/T}}{1 + B \times P^{1/T}} \tag{2.4}$$

where Q is the moles adsorbed, Q_m is the maximum moles adsorbed, P is the pressure, and B and T are the fitting constants [22].

Isotherm data can be fitting by using one of these three relationships between p and n to develop isosteric $\ln(p)$ versus $1/T$ plots. From these plots, the isosteric heat of adsorption can by determined by the Clausius-Clapeyron equation.

2.2.2 Ideal Adsorbed Solution Theory (IAST)

IAST is used to predict binary mixture adsorption from an experimental pure-gas isotherm. Single-component isotherms should be fitted by a proper model. However, the choice of which methods to use to fit the adsorption isotherm unrestricted. For example, several isotherm models were tested to fit the experimental pure isotherm for CH_4 and CO_2, with the materials at 273 K. The dual-site Langmuir–Freundlich model is fitted by the following equation

$$q = q_A + q_B = \frac{q_{sat,A} b_A p}{1 + b_A p} + \frac{q_{sat,B} b_B p}{1 + b_B p} \tag{2.5}$$

where there are two distinct adsorption sites A and B, q is the amount of adsorbed gas, p is the gas phase pressure, q_{sat} is the saturation amount of absorbed gas, and b is the Langmuir–Freundlich parameter. Then, q_{sat} and b can be obtained from the fitting Langmuir–Freundlich equation [23]. Selectivity factors of the mixture of gas for component 1 and 2 are shown in Eq. (2.6) [24], where q_i is the uptake amount and p_i is the partial pressure of component i.

$$S_{abs} = \frac{q_1/q_2}{p_1/p_2} \tag{2.6}$$

2.3 CO₂ Capture Ability in MOFs

2.3.1 Open Metal Site

Metal atoms in most MOFs are saturated by framework components via coordination bonds. However, some metal atoms in certain MOFs are partially coordinated by the guest solvent molecules. The coordinatively open metal sites are created within the MOFs when removing the guest solvent molecules. These open metal sites could be regarded as Lewis acid sites that strongly combine with gas molecules like CO_2 by electrostatic interaction for improving CO_2 capture ability [25–28]. The Snurr group [29] reported the selectivities of CO_2/CH_4 compared with carborane-based MOFs with and without metal open sites, and the results showed that open metal sites in MOFs have high selectivities for CO_2 over CH_4.

The isostructural series of MOFs typed $M_2(dobdc)$ ($dobdc^{4-}$ = 2,5-dioxido-1,4-benzenedicarboxylate; M = Mg, Fe, Co, Mn, Ni, Cu, Zn) with open metal sites show high CO_2 uptake capacity and selectivity. These MOFs have high CO_2 adsorption especially at low pressure, which fits for the flue gas separation pressure. And a high Q_{st} value for CO_2 adsorption demonstrates favorable adsorption of CO_2 on the metal sites [30]. Up to now, Mg-MOF-74 [31] represents a benchmark for a solid adsorbent with the highest CO_2 adsorption capacities at low to moderate CO_2 partial pressure according to CO_2 capture from flue gas. Li et al. [32] reported a "single-molecule trap" (SMT) with the desired size and properties suitable for trapping target CO_2 molecules. As molecular building blocks, these SMTs could be linked to build 3D MOFs to transport CO_2 in and out of the SMTs efficiently. In particular, the distances between two opposite copper sites in each paddlewheel tetragonal cage are 2.4 and 6.7 Å, which will accommodate one CO_2 molecule between them through electrostatic interaction. As a result, SMT-1 shows the

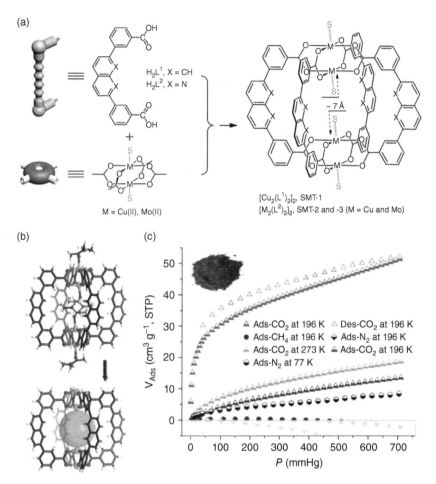

Figure 2.3 (a) Schematic representation of the construction of SMT-1–3: S is the coordinated solvent molecule, which can be removed through sample activation to yield an empty molecular cage for gas adsorption. (b) The molecular structure of SMT-1 with (top) and without (bottom) coordinated solvent molecules. The color schemes: Cu, cyan; O, red; N, blue; C, brown; and H, light gray. The green sphere represents the free space inside the molecular cage. (c) Gas adsorption isotherms of SMT-1, showing selective adsorption of CO$_2$ over CH$_4$ and N$_2$ (inset: a picture of an activated SMT-1 sample). Source: reproduced from Ref. [32] with permission from *Nature Communications*. (*See color plate section for color representation of this figure*).

uptake of CO$_2$ is 2.32 mmol g^{-1} at 196 K and 1 atm, which indicates around four CO$_2$ molecules absorbed per paddlewheel tetragonal cage. At 273 K, SMT-2 also absorbed CO$_2$ about to 1.11 mmol g^{-1}. Both show significantly low uptake of N$_2$ and CH$_4$. These precisely designed SMTs are clearly efficient for the CO$_2$ selectivities, as shown in Figure 2.3.

2.3.2 Pore Size

Pore size is one of the most important factor for gas adsorption and selectivity. According to the molecule steric effect, when the pore size of the materials is between the kinetic

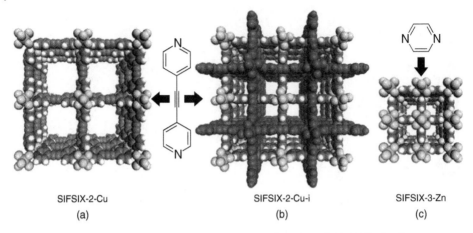

SIFSIX-2-Cu   SIFSIX-2-Cu-i   SIFSIX-3-Zn

(a)   (b)   (c)

Figure 2.4 (a) SIFSIX-2-Cu; BET apparent surface area (N_2 adsorption) 3140 m² g⁻¹. (b) SIFSIX-2-Cu-i: BET apparent surface area (N_2 adsorption) 735 m² g⁻¹. (c) SIFSIX-3-Zn; apparent surface area (determined from CO_2 adsorption isotherm) 250 m² g⁻¹. Color code: C (gray), N (blue), Si (yellow), F (light blue), H (white). All guest molecules are omitted for clarity. Note that the green net represents the interpenetrated net in SIFSIX-2-Cu-i. The nitrogen-containing linker present in SIFSIX-2-Cu and SIFSIX-2-Cu-i is 4,4′-dipyridylacetylene (dpa), whereas that in SIFSIX-3-Zn is pyrazine (pyr). Source: reproduced from Ref. [33] with permission from *Nature*. (*See color plate section for color representation of this figure*).

diameter of two gas molecules (e.g. N_2, 3.64 Å; CH_4, 3.80 Å), one could separate a gas from this mixture. Nugent et al. [33] reported that a series of isoreticular MOFs with periodically arrayed hexafluorosilicate (SIFSIX) pillars shows a significant impact of pore size for CO_2 adsorption capacity, as shown in Figure 2.4. The pore sizes of SIFSIX-2-Cu, SIFSIX-2-Cu-i, and SIFSIX-3-Zn are 13.05, 5.15, and 3.84 Å. At 0.1 bar, close to the partial pressure of CO_2 in flue gas, SIFSIX-3-Zn uptakes more CO_2 than SIFSIX-2-Cu-i (2.4 mmol g⁻¹ versus 1.7 mmol g⁻¹). At 0.4 mbar, corresponding to the average partial pressure of CO_2 in the atmosphere, SIFSIX-3-Cu (1.24 mmol g⁻¹) has much higher uptake of CO_2 than SIFSIX-3-Zn (0.13 mmol g⁻¹) or SIFSIX-2-Cu-i (0.0684 mmol g⁻¹). The increment of CO_2 adsorption capacity of SIFSIX-3 Cu compared to SIFSIX-3-Zn or SIFSIX-2-Cu-i may be attributed to the smaller pore size of SIFSIX-3-Cu, which results in a relative enhancement of the charge density surrounding the adsorbed CO_2 molecules.

Chen et al. [34] reported that five supramolecular isomerism with the same chemical composition, but different topology – Qc-5-M-dia (M = Co, Ni, Zn, and Cu, dia = twofold, 3D diamondoid network) and Qc-5-Cu-sql-a (sql = 2D square lattice network) – can fine-tune the pore size for CO_2 molecular sieving. Qc-5-Cu-dia, Qc-5-Cu-sql-α, and Qc-5-Cu-sql-β exhibit 1D channels with diameters of 4.8, 3.8, and 3.3 Å, respectively as shown in Figure 2.5. In addition, Qc-5-Cu-sql-α undergoes an irreversible phase change upon desolvation to Qc-5-Cu-sql-β. At 293 k and 1 bar, Qc-5-Ni-dia and Qc-5-Cu-dia exhibit greater N_2 and CH_4 uptake than Qc-5-Cu-sql-β. However, Qc-5-Cu-sql-β exhibits a higher CO_2 uptake than Qc-5-Cu-dia. To further understand the sorption behavior, they also performed molecular simulation, and Qc-5-Cu-sql-β was found to exhibit a better close-fitting interaction between CO_2 molecules and the pore walls.

Figure 2.5 Pore size for Qc-5-Cu-dia, Qc-5-Cu-sql-α, and Qc-5-Cu-sql-β polymorphs: C (gray), Cu (maroon), O (red), N (blue), H (white). Source: reproduced from Ref. [34] with permission from *Angewandte Chemie International Edition*. (*See color plate section for color representation of this figure*).

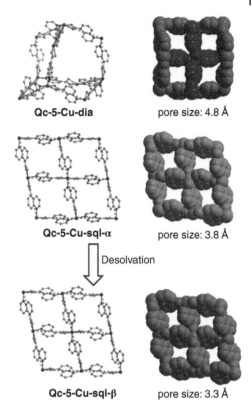

Qc-5-Cu-dia   pore size: 4.8 Å

Qc-5-Cu-sql-α   pore size: 3.8 Å

Desolvation

Qc-5-Cu-sql-β   pore size: 3.3 Å

2.3.3 Polar Functional Group

Since polar functional groups have strong interactions with CO_2 in the pore surfaces of MOFs, grafting polar functional groups onto an MOF's structure through direct synthesis or post-synthesis modification provides a huge opportunity to enhance the adsorption capacity and selectivity of CO_2. [17].

Until now, a large number of MOFs with varied polar functional groups have shown enhanced CO_2 adsorption capacity and selectivity. Banerjee and his co-workers [35] described an isoreticular series of eight zeolitic imidazole frameworks (ZIFs) with the desired gmelinite topology (zeolite code GME). Specifically, by changing the imidazole linker from polar (-NO_2, ZIF-78; -CN, ZIF-82; -Br, ZIF-8; -Cl, ZIF-69) to nonpolar (-CH_3, ZIF-79; -C_6H_6, ZIF-68), they found that ZIF-78 is the most effective CO_2 absorbent and shows the highest selectivity for capture of CO_2, since the -NO_2 group has greater dipole moments than the other functional groups.

Aromatic amine groups (-NH_2) are a promising functional group to enhance the CO_2 adsorption capacity and selectivity of MOFs, since it can form a strong interaction with CO_2. Lots of MOFs with aromatic amine-functionalized ligands have been designed, and the adsorption capacity and selectivity of CO_2 have been studied among these MOFs. As an example, $Zn_2(C_2O_4)(C_2N_4H_3)_2 \cdot (H_2O)_{0.5}$ shows high CO_2 uptake, while the heat of CO_2 adsorption is moderate. This Zn-aminotriazolato-oxalate exhibits the CO_2 uptake of 4.35 mmol^{-1} g^{-1} at 273 K and 1.2 bar [36]. At zero loading, indicating the interaction of

CO_2 with the most energetically favored sites in the framework, ΔH_{abs} was determined to be 40.8 kJ mol⁻¹. At max loading, ΔH_{abs} was still high at 38.6 kJ mol⁻¹. The combination of CO_2 with aromatic -NH_2 group in Zn-aminotriazolato-oxalate causes high ΔH_{abs}.

Recently, Zr-based MOFs have attracted people's attention thanks to their high stability and good adsorption ability for some small molecules, such as CO_2. Several studies have demonstrated that introducing a polar group to Zr-based MOFs could enhance CO_2 adsorption and separation in experiments and theoretical calculations [37–39]. Among these MOFs, Wang and co-workers [40] showed that two UiO-67 analogues, $[Zr_6O_4(OH)_4(FDCA)_6]$ (BUT-10) and $[Zr_6O_4(OH)_4(DTDAO)_6]$ (BUT-11), are functionalized with two groups of carbonyl and sulfone, respectively, in the ligand. Both Zr-based MOFs showed enhanced CO_2 adsorption and separation selectivity over N_2 and CH_4. At 298 K and 1 atm, the maximum CO_2 uptakes of UiO-66, BUT-10, and BUT-11 are 22.9, 50.6, and 53.5 cm³ g⁻¹, respectively. The selectivities of CO_2/CH_4 and CO_2/N_2 are over 1.9 and 2.0 times in BUT-10 and 3.3 and 3.4 times in BUT-11, respectively, compared to UiO-67, demonstrating that CO_2 molecules are located around the sulfone groups in pore surfaces of BUT-11 and verifying that sulfone groups significantly increase the affinity toward CO_2 molecules of the framework at the molecular level.

2.3.4 Incorporation

Incorporating lithium ions into MOFs has attracted interest due to the high Q_{st} values for H_2 [41]. MOFs containing extraframework ions should be promising for CO_2 separation processes. Babarao and Jiang [42] reported a molecular simulation study for separation of gas mixtures (CO_2/H_2, CO_2/CH_4, and CO_2/N_2) in a rho zeolite-like metal–organic framework (rho-ZMOF). Since rho-ZMOF contains a wide-open anionic framework and charge-balancing extraframework Na^+ ions, CO_2 is adsorbed predominantly over other gases due to strong electrostatic interactions.

The incorporation of sites that bind CO_2 into porous materials is a promising strategy to realize highly selectivity CO_2 capture. McDonald and co-workers [43] report a series of diamine-appended M_2(dobpdc) (M = Mg, Mn, Fe, Co, Ni, Zn) frameworks with an expanded structure of the well-known M-MOF-74, demonstrating an exceptionally high CO_2 adsorption capacity at low pressure except Ni_2(dobpdc) frameworks due to inserting CO_2 into metal-amine bonds. At 0.39 bar and 25 °C, Mg-MOF-74 uptakes 2.0 mmol g⁻¹ CO_2 and at 0.15 bar and 40°C which is standard flue gas adsorption conditions, it uptakes 3.14 mmol g⁻¹ CO_2, as shown in Figure 2.6.

2.4 MOFs in CO₂ Capture in Practice

2.4.1 Single-Component CO₂ Capture Capacity

The single-component CO_2 capture capacity depends on the pure CO_2 sorption isotherms at a given temperature, with pressures ranging from very low to atmospheric in most case, to high pressure in a few cases. Measurements at room temperature and low pressure are relevant for practical CO_2 capture, since low-CO_2 components need to be separated from gas streams. There are two different capacities: gravimetric capacity (mg g⁻¹, ww⁻¹) and

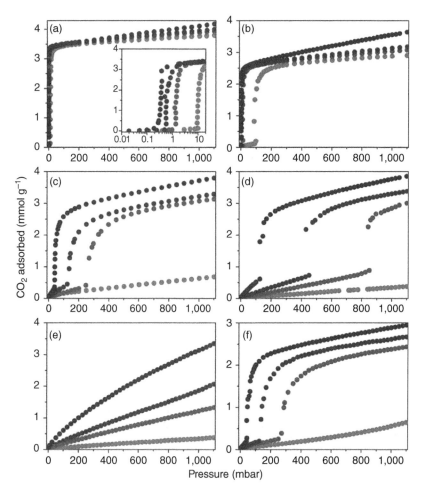

Figure 2.6 Carbon dioxide adsorption isotherms at 25 °C (blue), 40 °C (blue-violet), 50 °C (red-violet), and 75 °C (red) for (a) mmen-Mg₂(dobpdc); (b) mmen-Mn₂(dobpdc); (c) mmen-Fe₂(dobpdc); (d) mmen-Co₂(dobpdc); (e) mmen-Ni₂(dobpdc); and (f) mmen-Zn₂(dobpdc). Source: reproduced from Ref. [43] with permission from *Nature*. (*See color plate section for color representation of this figure*).

volumetric capacity (cc cc^{-1} or vv^{-1}) [44]. In general, a higher Brunauer–Emmett–Teller (BET) surface area indicates a higher CO_2 adsorption capacity without considering other strong adsorbate-adsorbent interactions, but not a higher selectivity over other gases [45]. Some MOFs listed in Table 2.1 have high surface areas: nearly 1000 m^2 g^{-1}, with pore volumes around 0.5 cm^{3}g^{-1}. However, the high surface areas don't guarantee a high CO_2 adsorption capacity of MOFs at 298 K and 1 bar. Indeed, the CO_2 adsorption capacities of these highly porous MOFs are low. As an example, MOF-177 at 298 K and 1 bar shows a high BET surface area of 4690 m^2 g^{-1} with a CO_2 uptake of 0.8 mmol g^{-1} and a pore volume of 1.59 cm^3 g^{-1} [23]. NU-100 also shows a high BET surface area 6143 m^2 g^{-1} with a CO_2 uptake of 2.7 mmol g^{-1} and a pore volume of 2.82 cm^3 g^{-1} [46]. Those highly porous MOFs

Table 2.1 Summary of single-component CO_2 uptake capacity in selected metal–organic frameworks (MOFs) at ambient temperature.

MOF	BET surface area (m² g⁻¹)	Pore volume (cm³ g⁻¹)	Pore size (Å)	CO_2 uptake (mmol g⁻¹)			Ref
				1 bar	0.15 bar	0.1 bar	
Mg-MOF-74	1174	0.648	10.2	8.6	–	–	[49]
	1495 (296 K)	–	11	8.0	6.1	5.4	[31, 50]
Co-MOF-74	957	0.498	–	7.5		2.8	[51]
MAF-X25ox	–	0.46	–	7.1	4.1	–	[52]
Ni-MOF-74	936	0.495		7.1		4.1	[51]
HP-e	1210	0.45	9.1	7.0			[53]
SIFSIX-1-Cu	1468	0.56	–	5.3	–	–	[54]
SIFSIX-2-Cu-i	734	0.26	5.2	5.4	1.7	–	[33, 55]
mmen-Mg₂(dobpdc)	3270	1.25	18.4	6.4	4.9		[43, 56, 57]
dmen-Mg₂(dobpdc)	675	–	–	5.0	3.8	–	[58]
Mg₂(dobdc)-(N₂H₄)₁.₈	1012	0.3	–	6.49	6.10	–	[59]
Opt-UiO-66(Zr)-(OH)₂	1230	0.56	3.93	5.63	2.50	–	[60]
[Cu(Me-4py-trz-ia)]	1473		0.586	6.1			[61]
Cu-TDPAT	1938	0.93		5.9	1.4		[62]
CuTPBTM	3160	1.268	–	5.3	–	–	[48]

"–" = not available.

are not suitable for CO_2 capture at low pressure, such as post-combustion CO_2 capture for flue gas. However, they are potentially relevant for storage of CO_2 at high pressure [47, 48].

Millward and Yaghi reported pioneering work about CO_2 isotherms up to 42 bars [63]. Bourrelly et al. [64] reported that the hydroxyl groups in MIL-53 (Al) (MIL is the Material Institute Lavoisier) is interacted with CO_2 molecules, leading to the shrinkage of the structure. When increasing the CO_2 pressure, the pore structure of MIL-53 (Al) will reopen, causing an adsorption capacity of CO_2 of 10.4 mol kg⁻¹ at 30 bar and 304 K, which is well above conventional zeolites and comparable with microporous carbons [65]. Recently, our group [66] reported a flexible Cd-MOF with fourfold interpenetration and diamondoid (dia) topology, which can show "gating effect" and "breathing" behavior upon CO_2 sorption, leading to the first time the reversible closed/open end states of the structure induced by CO_2 molecules has been directly visualized. As shown in Figure 2.7, two structural transformations of the Cd-MOF are **1a** and **1b**: **1a** was activated by exchanging N.N-dimethylmethanamide (DMF) molecules with dry dichloromethane under vacuum, and **1b** is the desolvated sample. At 298 K, **1b** exhibited very low uptake of CO_2 below 8 bars; then there was a steep increase at approximately 8 bar, which meant gate opening. The total uptake of CO_2 is 117 cm³ g⁻¹ at 30 bar and 298 K.

2.4.2 Binary CO_2 Capture Capacity and Selectivity

Even though single-component CO_2 uptake capacity is high, MOFs are not necessary to perform well in binary CO_2/N_2 and CO_2/CH_4 mixture tests. CCS-related gas

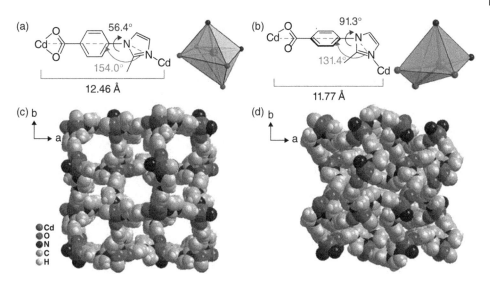

Figure 2.7 Cd⋯Cd distances, bond angles, and dihedral angles of the miba ligand and the coordination environment of the Cd(II) atoms of (a) 1a and (b) 1b. 3D framework of 1a (c) and 1b (d) [66]. Source: adapted by permission of the American Chemical Society. (*See color plate section for color representation of this figure*).

separation includes CO_2/N_2 separation in post-combustion capture, CO_2/H_2 separation in pre-combustion capture, O_2/N_2 and CO_2/CO separation in oxy-combustion, and CO_2/CH_4 separation in the purification of natural gas [67]. Adsorption capacity and selectivity are the major concerns in selective adsorption. In most situations, selective adsorption uses single-component isotherms and the ISAT to calculate the materials' selectivity factor [67], as shown in Table 2.2 for CO_2/N_2 separation and Table 2.3 for CO_2/CH_4 separation.

Lin et al. [68] reported a series of polyethyleneimine (PEI)-modified MIL-101 for highly efficient CO_2 capture. Due to the weight of PEI, the surface area and pore volume of MIL-101 decrease dramatically. At 100 wt% PEI loading, the CO_2 adsorption capacity at 0.15 bar and 25 °C reached 4.2 mmol g⁻¹, and it reached 3.4 mmol g⁻¹ at 50 °C, which is over 10 times MIL-101 (0.33 and 0.2 mmol g⁻¹) at 0.15 bar, at 25 °C and 50 °C, respectively. Furthermore, in the designed flue gas with 0.15 bar CO_2 and 0.75 bar N_2, the selectivity of CO_2 over N_2 reaches 770 and 1200 at 25 °C and 50 °C: this demonstrates that poly alkylamines are also promising to enhance the adsorption capacity of CO_2, since they occupy more active amine groups. Wu et al. [69] simulated a Li-modified IRMOF-1, and CHEM-4Li MOF shows exceptional CO_2 capture capability due to the strong electrostatic interactions between the framework atoms and gas molecules with the presence of lithium: the CO_2/N_2 selectivity (CO_2: N_2 = 15.6 : 84.4) of CHEM-4Li MOF is 395.

Xiang and co-workers [50] constructed a series of MOFs with different surface areas, pore-surface functionalities, and pore surfaces, such as UTSA-20a, UTSA-15a, UTSA-33a, UTSA-25a, UTSA-34b, Yb(BPT), $Zn_5(BTA)_6(TDA)_2$, Cu(BDC-OH), $Zn_4(OH)_2(1,2,4-BTC)_2$, and UTSA-16, for the capture of CO_2. Compared with those MOFs, UTSA-16 has the best performance for CO_2/CH_4 selectivity. The breakthrough experiment also supports the affinity of CO_2 with the UTSA-16 by showing the retention of CO_2.

Table 2.2 Selective adsorption of CO_2 over N_2 in selected metal–organic frameworks (MOFs).

MOF	Uptake CO₂ vs N₂ (conditions)	Selectivity (conditions)	Ref
Qc-5-Cu-sql-β	15/85 gas mixture (293 K and 1.0 bar)	40 000	[34]
SIFSIX-3-Cu	15/85 gas mixture (298 K and 1.0 bar)	10 500	[55]
SIFSIX-3-Zn	10/90 gas mixture (298 K and 1.0 bar)	1818	[33]
PEI-MIL-101-125	15/75 gas mixture (323 K)	1200	[68]
PEI-MIL-101-100	15/75 gas mixture (323 K)	750	[68]
IRMOF-1-4Li (chem-4Li MOF)	15.6/84.4 gas mixture (298 K and 1 bar)	395 (simulated)	[69]
mmen-CuBTTri	15/75 gas mixture (298 K)	327	[70]
UTSA-16	15/85 gas mixture (296 K and 1.0 bar)	315	[50]
Zn₂(bpdc)₂(bpee)(DMF)₂	- (298 K and 0.16 bar)	294	[71]
MAF-66	15/75 gas mixture (298 K)	225	[72]
mmen-Mg₂(dobpdc)	15/75 gas mixture (298 K)	200	[56]
Mg₂(dobdc)	15/75 gas mixture (313 K)	182	[23]
dmpn-Mg₂(dobpdc)	15/85 gas mixture (298 K and 1.0 bar)	153	[73]
bio-MOF-11	6.0 vs 0.43 mmol g⁻¹ (273 K, 1 bar)	81	[74]
SIFSIX-2-Cu-i	15/85 gas mixture (298 K and 1.0 bar)	72	[33]
ZIF-78	51 vs 4.2 cm³ g⁻¹ (298 K, 1 atm)	50	[75]
ZIF-82	54 vs 3.9 cm³ g⁻¹ (298 K, 1 atm)	35.5	[75]

Table 2.3 Selective adsorption of CO_2 over CH_4 in selected metal–organic frameworks (MOFs).

MOF	Uptake CO₂ vs N₂ (conditions)	Selectivity	Ref
Qc-5-Cu-sql-β	50/50 gas mixture (293 K and 1.0 bar)	3300	[34]
Zn₂(bpdc)₂(bpee)(DMF)	298 K and 0.16 bar	257	[71]
SIFSIX-3-Zn	50/50 gas mixture (303 K and 1.0 bar)	231	[33]
rho-ZMO	50 CO₂ concentration% (298 K and 1 bar)	80	[42]
ZIF-78	51 vs 4.2 cm³ g⁻¹ (298 K, 1 atm)	50	[75]
SIFSIX-2-Cu-i	50/50 gas mixture (298 K and 1.0 bar)	33	[33]
UTSA-16	50/50 gas mixture (298 K and 200 kPa)	29.8	[49]
ZIF-79	34 vs 2.9 cm³ g⁻¹ (298 K, 1 atm)	22.5	[75]
BUT-11	10/90 gas mixture (298 K and 1.0 bar)	9.0	[40]
MAF-66	- (298 K)	5.8	[72]

Acetylene (C_2H_2) is a chemical product of vinyl compounds, acrylic acid derivatives, and α-ethynyl alcohols via an addition reaction in the chemical industry. The production of high-purity acetylene is very important. Since the size, shape, and boiling point (189.3 and 194.7 K, respectively) of C_2H_2 and CO_2 are identical, an efficient way to separate C_2H_2 and CO_2 is in high demand. In 2005, Kitagawa et al. first [76] reported selective adsorption of C_2H_2 over CO_2 by using a MOF with a one-dimensional channel

$(4 \text{ Å} \times 6 \text{ Å})$ due to strong hydrogen bonding. In particular, Luo, Chen, and co-workers [77] constructed a new MOF-74 isomer $[Zn_2(dobdc)(H_2O)] \cdot 0.5H_2O$ (Zn-UTSA-74, H_4dobdc = 2,5-dioxido-1,4-benzenedicarboxylic acid) with two binding sites per metal and one-dimensional open channel of about 8.0 Å, leading to a moderately high amount of C_2H_2 adsorption (145 cm^3 cm^{-3}). Very recently, Lin and co-workers [78] reported a novel micro-porous material $[Zn(dps)_2(SiF_6)]$ (UTSA-300, dps = 4,4′-dipyridylsulfide) with multiple potential binding sites that can completely exclude CO_2 from C_2H_2 at ambient temperature.

2.4.3 Other Related Gas-Selective Adsorption

Since flue gas usually contains 5–15% water, CO_2 capture performance in MOFs needs to be re-evaluated in the presence of water vapor [79]. For most MOFs, especially with vacant Lewis acid sites, H_2O is more competitive for adsorption over CO_2 due to the decrease of CO_2 uptake under wet conditions. Considerable effort must be dedicated to removing H_2O from MOFs. Nguyen et al. [80] reported a hydrophobic chabazite-type ZIF-300 that can keep a constant CO_2 uptake under conditions with 80% humidity. Other than hydropho-bic functional groups, another promising approach is to modify multiple functional groups into MOFs to increase CO_2 affinity and water repellence. Hu et al. incorporated multi-functional ligands into the parental UiO-66 framework, which is UiO-66(Zr)-NH$_2$-F$_4$-0.53, leading to a minor loss of 30% CO_2 uptake capacity under wet CO_2/N_2 (15/85) mixture conditions [81].

Direct air capture (DAC) is another area of research for multicomponent CO_2 capture under atmospheric conditions (400 ppm, 298 K, 1 bar) [11]. However, it doesn't need to deal with the high concentration of contaminants in flue gas. The highest CO_2 uptake capacity for DAC is relate to amine-doped MOFs, since there are strong chemisorption sites. For example, $Mg_2(dobdc)(N_2H_4)_{1.8}$ and mmen-Mg_2(dobpdc) show excellent CO_2 uptake capacities of 4.22 mmol g^{-1} (dry CO_2/N_2 mixture and 313 K) and 2 mmol g^{-1} (50% relative humidity, 298 K) [82], respectively, which are comparable to that of amine-doped TEPA-SBA-15 (mesoporous silica) (3.59 mmol g^{-1}) [83].

2.5 Membrane for CO₂ Capture

Membrane separation technology is also a promising route for gas separation, due to envi-ronmental friendliness, high energy efficiency with low cost, high surface area, and ease of processing and maintenance [84, 85]. The feed stream is driven by the different pressure between the two sides of the membrane across one side of the membrane under steady-state conditions, leaving the non-permeating molecules in the retentate stream. In this case, membranes with well-defined pores can dominate the performance of separation. Nowa-days, polymer membranes are most commonly developed since they are easy to fabricate and show a fundamental trade-off relationship between various gas pairs like CO_2/CH_4 and CO_2/N_2, which is illustrated by the Robeson upper bound [86, 87]. As a novel class of hybrid nanoporous solid, MOFs have attracted scientific interest more recently due to the tunabil-ity of pore size, surface chemistry in a broad range, and better compatibility with membrane polymer chains [12, 88–90].

2.5.1 Pure MOF Membrane for CO$_2$ Capture

A variety of membranes were developed using ZIFs constructed from tetrahedral metal ions bridged by imidazolates, due to numerous advantages such as designable structures and high thermal and chemical stability. Peng and co-workers reported that an isomer of ZIF-7, Zn$_2$(bim)$_4$ is fabricated on α-Al$_2$O$_3$ support with a 1-nm-thick molecular sieve nanosheet due to the large lateral area and high crystallinity from a layered MOF, achieving a H$_2$ permeance that varies from 760–3760 GPU (1 GPU = 10^{-6} cm^3 cm^{-2} s^{-1} cm Hg^{-1} at standard temperature and pressure (STP) with a H$_2$/CO$_2$ selectivity from 53–291. To solve the damage of the MOF structure caused by the conventional physical exfoliation, Zn$_2$(bim)$_4$ crystals were first wet ball-milled at very low speed (60 rpm) and ultrasonicated in volatile solvent to exfoliate. The obtained Zn$_2$(bim)$_4$ nanosheet has a side length of ~600 nm and uniform thickness of 1.12 nm. The Zn$_2$(bim)$_4$ membrane shows thermal and hydrothermal stability under different conditions for more than 400 hours. The authors claim the performance of this membrane exceeds the latest Robeson's upper bound for H$_2$/CO$_2$ gas separation and is higher than that of the molecular sieve membrane reported to date [91].

2.5.2 MOF-Based Mixed Matrix Membranes for CO$_2$ Capture

In addition to the pure MOF membrane, the MOF-based hybrid membrane or mixed matrix membranes (MMMS) have also attracted attention for CO$_2$ capture since they combine the advantages of mixing polymers and MOF particles, some of which show high separation performance for CO$_2$ capture. MIL-53(Al), which is a member of the MILs family, has good chemical stability, high porosity, and a special breathing character given varying pressures of CO$_2$. It is a good filler for fabricated MMMS for CO$_2$ capture and separation. Apart from pore size and functionality, morphologies of MOFs also affect the performance of MMMS for CO$_2$ capture and separation. Sabetghadam et al. reported that the particle morphology of MOFs has an impact on permeation studies of MMMS. They synthesized three different crystal morphologies (nanoparticles, nanorods [NRs], and microneedles [MNs]) of MOFs (Figure 2.8) functionalized with an amino group as a filler, which has a good interaction with polymer matrices, the polyimides Matrimid and 6FDA-DAM. The nanoparticles incorporating 8 wt% NH$_2$-MIL-53(Al) loading achieved the largest improvement of CO$_2$/CH$_4$

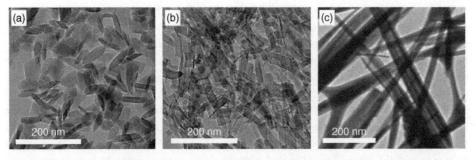

Figure 2.8 TEM micrographs of (a) NH$_2$-MIL-53(Al) nanoparticles, (b) NH$_2$-MIL-53(Al) nanorods, and (c) NH$_2$-MIL-53(Al) microneedles. Source: reproduced from Ref. [92] with permission from *Advanced Functional Materials*.

separation performance at 3 bar and 25 °C. Furthermore, due to the high free volume and higher adsorption capacity of permeable polyimide 6FDA-DAM compared with Matrimid, the permeability was increased up to 85% upon NH_2-MIL-53(Al) nanoparticles [92].

2.6 Conclusion and Perspectives

With significant advances in recent years, MOFs are promising novel adsorbents for CO_2 capture due to their large pore volumes, high surface area, and easily controllable pore structure [93]. Many researchers point out that the CO_2 capacities or selectivities of MOFs can be improved by reducing the pore size via interpenetration [94] and catenation, modifying the organic linkers by introducing the polar functional group [95]. Also, increasing the interaction strength between CO_2 molecules and MOFs, such as introducing unsaturated metal centers, can help increase MOFs' CO_2 adsorption capacities. In addition, the post-synthetic method becomes a popular and effective way to change the current properties of MOFs to meet specific needs. However, some important issues need to be addressed before applying MOFs in practical applications of CO_2 capture, such as reasonable cost for bulk MOFs synthesis and the stability of MOFs toward water vapor, heat, and acid gas. Researchers spare no effort to develop new synthesis and activation procedures to reduce the usage of expensive organic solvents, such as using raw materials in petroleum as organic linkers and modifying the MOFs' surface to reduce water effects on CO_2 adsorption. Recently, MOF membranes offer a number of benefits for gas separation, such as removing the phase change compared to other gas separation technologies and overcoming the limitations of polymeric and inorganic membranes. Computational screening of membrane materials has been successfully reported on the basis of selectivity and permeability properties, and MOF-based membranes hold great promise for application in gas separation. However, further tests that include full membrane operation have barely been explored. In the future, we expect to address the materials and membrane synthesis and processing aspects to a large extent by computational methods to develop a complete understanding of the synthesis steps and changes in chemical and mechanical properties in the CO_2 capture process.

Acknowledgments

We acknowledge NSF (DMR-1352065) and the University of South Florida for financial support of this work.

References

1 Earth System Research Laboratory (ESRL). 2019. Trends in atmospheric Carbon Dioxide. https://www.esrl.noaa.gov/gmd/ccgg/trends.
2 Rackley, S.A. (2009). *Carbon Capture and Storage*. Elsevier.
3 Rochelle, G.T. (2009). Amine scrubbing for CO_2 capture. *Science* 325: 1652.

4 IPCC (2005). *IPCC Special Report on Carbon Dioxide Capture and Storage.* Cambridge University Press.

5 Haszeldine, R.S. (2009). Carbon capture and storage: how green can black be? *Science* 325: 1647.

6 Rosi, N.L., Kim, J., Eddaoudi, M. et al. (2005). Rod packings and metal–organic frameworks constructed from rod-shaped secondary building units. *J. Am. Chem. Soc.* 127: 1504–1518.

7 Khurana, M. and Farooq, S. (2016). Adsorbent screening for postcombustion CO_2 capture: a method relating equilibrium isotherm characteristics to an optimum vacuum swing adsorption process performance. *Ind. Eng. Chem. Res.* 55: 2447–2460.

8 Brandl, P., Soltani, S.M., Fennell, P.S., and Dowell, N.M. (2017). Evaluation of cooling requirements of post-combustion CO_2 capture applied to coal-fired power plants. *Chem. Eng. Res. Des.* 122: 1–10.

9 Stanger, R., Wall, T., Spörl, R. et al. (2015). Oxyfuel combustion for CO_2 capture in power plants. *Int. J. Greenh. Gas. Con.* 40: 55–125.

10 Yu, J., Xie, L.-H., Li, J.-R. et al. (2017). CO_2 capture and separations using MOFs: computational and experimental studies. *Chem. Rev.* 117: 9674–9754.

11 Sanz-Pérez, E.S., Murdock, C.R., Didas, S.A., and Jones, C.W. (2016). Direct capture of CO_2 from ambient air. *Chem. Rev.* 116: 11840–11876.

12 Seoane, B., Coronas, J., Gascon, I. et al. (2015). Metal–organic framework based mixed matrix membranes: a solution for highly efficient CO_2 capture? *Chem. Soc. Rev.* 44: 2421–2454.

13 Gao, W.-Y., Wu, H., Leng, K. et al. (2016). Inserting CO_2 into aryl C–H bonds of metal–organic frameworks: CO_2 utilization for direct heterogeneous C–H activation. *Chem. Int. Ed.* 55: 5472–5476.

14 Gao, W.-Y., Chen, Y., Niu, Y. et al. (2014). Crystal engineering of an nbo topology metal–organic framework for chemical fixation of CO_2 under ambient conditions. *Chem. Int. Ed.* 53: 2615–2619.

15 Lu, W., Wei, Z., Gu, Z.-Y. et al. (2014). Tuning the structure and function of metal–organic frameworks via linker design. *Chem. Soc. Rev.* 43: 5561–5593.

16 D'Alessandro, D.M., Smit, B., and Long, J.R. (2010). Carbon dioxide capture: prospects for new materials. *Chem. Int. Ed.* 49: 6058–6082.

17 Wang, Z. and Cohen, S.M. (2009). Postsynthetic modification of metal–organic frameworks. *Chem. Soc. Rev.* 38: 1315–1329.

18 Sircar, S., Mohr, R., Ristic, C., and Rao, M.B. (1999). Isosteric heat of adsorption: theory and experiment. *J. Phys. Chem. B* 103: 6539–6546.

19 Pan, H., Ritter, J.A., and Balbuena, P.B. (1998). Examination of the approximations used in determining the isosteric heat of adsorption from the Clausius–Clapeyron equation. *Langmuir* 14: 6323–6327.

20 Chen, B., Zhao, X., Putkham, A. et al. (2008). Surface interactions and quantum kinetic molecular sieving for H_2 and D_2 adsorption on a mixed metal–organic framework material. *J. Am. Chem. Soc.* 130: 6411–6423.

21 Dincă, M., Dailly, A., Liu, Y. et al. (2006). Hydrogen storage in a microporous metal–organic framework with exposed Mn^{2+} coordination sites. *J. Am. Chem. Soc.* 128: 16876–16883.

22 Ma, S. and Zhou, H.-C. (2006). A metal–organic framework with entatic metal centers exhibiting high gas adsorption affinity. *J. Am. Chem. Soc.* 128: 11734–11735.

23 Mason, J.A., Sumida, K., Herm, Z.R. et al. (2011). Evaluating metal–organic frameworks for post-combustion carbon dioxide capture via temperature swing adsorption. *Energy Environ. Sci.* 4: 3030–3040.

24 Walton, K.S. and Sholl, D.S. (2015). Predicting multicomponent adsorption: 50 years of the ideal adsorbed solution theory. *AIChE J.* 61: 2757–2762.

25 Wang, Q., Bai, J., Lu, Z. et al. (2016). Finely tuning MOFs towards high-performance post-combustion CO_2 capture materials. *Chem. Commun.* 52: 443–452.

26 Gao, W.-Y., Palakurty, S., Wojtas, L. et al. (2015). Open metal sites dangled on cobalt trigonal prismatic clusters within porous MOF for CO_2 capture. *Inorg. Chem. Front.* 2: 369–372.

27 Zhang, Z., Yao, Z.-Z., Xiang, S., and Chen, B. (2014). Perspective of microporous metal–organic frameworks for CO_2 capture and separation. *Energy Environ. Sci.* 7: 2868–2899.

28 Wang, X.-S., Chrzanowski, M., Kim, C. et al. (2012). Quest for highly porous metal–metalloporphyrin framework based upon a custom-designed octatopic porphyrin ligand. *Chem. Commun.* 48: 7173–7175.

29 Bae, Y.-S., Farha, O.K., Spokoyny, A.M. et al. (2008). Carborane-based metal–organic frameworks as highly selective sorbents for CO_2 over methane. *Chem. Commun.*: 4135–4137.

30 Queen, W.L., Hudson, M.R., Bloch, E.D. et al. (2014). Comprehensive study of carbon dioxide adsorption in the metal–organic frameworks $M_2(dobdc)$ (M = Mg, Mn, Fe, Co, Ni, Cu, Zn). *Chem. Sci.* 5: 4569–4581.

31 Caskey, S.R., Wong-Foy, A.G., and Matzger, A.J. (2008). Dramatic tuning of carbon dioxide uptake via metal substitution in a coordination polymer with cylindrical pores. *J. Am. Chem. Soc.* 130: 10870–10871.

32 Li, J.-R., Yu, J., Lu, W. et al. (2013). Porous materials with pre-designed single-molecule traps for CO_2 selective adsorption. *Nat. Commun.* 4: 1538.

33 Nugent, P., Belmabkhout, Y., Burd, S.D. et al. (2013). Porous materials with optimal adsorption thermodynamics and kinetics for CO_2 separation. *Nature* 495: 80.

34 Chen, K.J., Madden, D.G., Pham, T. et al. (2016). Tuning pore size in square-lattice coordination networks for size-selective sieving of CO_2. *Chem. Int. Ed.* 55: 10268–10272.

35 Banerjee, R., Furukawa, H., Britt, D. et al. (2009). Control of pore size and functionality in isoreticular zeolitic imidazolate frameworks and their carbon dioxide selective capture properties. *J. Am. Chem. Soc.* 131: 3875–3877.

36 Vaidhyanathan, R., Iremonger, S.S., Dawson, K.W., and Shimizu, G.K.H. (2009). An amine-functionalized metal organic framework for preferential CO_2 adsorption at low pressures. *Chem. Commun.*: 5230–5232.

37 Yang, Q., Vaesen, S., Ragon, F. et al. (2013). A water stable metal–organic framework with optimal features for CO_2 capture. *Chem. Int. Ed.* 52: 10316–10320.

38 Wu, D., Maurin, G., Yang, Q. et al. (2014). Computational exploration of a Zr-carboxylate based metal–organic framework as a membrane material for CO_2 capture. *J. Mater. Chem. A* 2: 1657–1661.

39 GYang, Q., Wiersum, A.D., Llewellyn, P.L. et al. (2011). Functionalizing porous zirconium terephthalate UiO-66(Zr) for natural gas upgrading: a computational exploration. *Chem. Commun.* 47: 9603–9605.

40 Wang, B., Huang, H., Lv, X.-L. et al. (2014). Tuning CO_2 selective adsorption over N_2 and CH_4 in UiO-67 analogues through ligand functionalization. *Inorg. Chem.* 53: 9254–9259.

41 Rao, D., Lu, R., Xiao, C. et al. (2011). Lithium-doped MOF impregnated with lithium-coated fullerenes: a hydrogen storage route for high gravimetric and volumetric uptakes at ambient temperatures. *Chem. Commun.* 47: 7698–7700.

42 Babarao, R. and Jiang, J. (2009). Unprecedentedly high selective adsorption of gas mixtures in rho zeolite-like metal–organic framework: a molecular simulation study. *J. Am. Chem. Soc.* 131: 11417–11425.

43 McDonald, T.M., Mason, J.A., Kong, X. et al. (2015). Cooperative insertion of CO_2 in diamine-appended metal-organic frameworks. *Nature* 519: 303.

44 Hu, Z., Wang, Y., Shah, B.B., and Zhao, D. (2019). CO_2 capture in metal–organic framework adsorbents: an engineering perspective. *Adv. Sustain. Syst.* 3: 1800080.

45 Liang, W., Coghlan, C.J., Ragon, F. et al. (2016). Defect engineering of UiO-66 for CO_2 and H_2O uptake – a combined experimental and simulation study. *Dalton Trans.* 45: 4496–4500.

46 Farha, O.K., Özgür Yazaydin, A., Eryazici, I. et al. (2010). De novo synthesis of a metal–organic framework material featuring ultrahigh surface area and gas storage capacities. *Nat. Chem.* 2: 944.

47 Alezi, D., Belmabkhout, Y., Suyetin, M. et al. (2015). MOF crystal chemistry paving the way to gas storage needs: aluminum-based soc-MOF for CH_4, O_2, and CO_2 storage. *J. Am. Chem. Soc.* 137: 13308–13318.

48 Zheng, B., Bai, J., Duan, J. et al. (2011). Enhanced CO_2 binding affinity of a high-uptake rht-type metal–organic framework decorated with acylamide groups. *J. Am. Chem. Soc.* 133: 748–751.

49 Bao, Z., Yu, L., Ren, Q. et al. (2011). Adsorption of CO_2 and CH_4 on a magnesium-based metal organic framework. *J. Colloid Interface Sci.* 353: 549–556.

50 Xiang, S., He, Y., Zhang, Z. et al. (2012). Microporous metal-organic framework with potential for carbon dioxide capture at ambient conditions. *Nat. Commun.* 3: 954.

51 Yazaydın, A.Ö., Snurr, R.Q., Park, T.-H. et al. (2009). Screening of metal–organic frameworks for carbon dioxide capture from flue gas using a combined experimental and modeling approach. *J. Am. Chem. Soc.* 131: 18198–18199.

52 Liao, P.-Q., Chen, H., Zhou, D.-D. et al. (2015). Monodentate hydroxide as a super strong yet reversible active site for CO_2 capture from high-humidity flue gas. *Energy Environ. Sci.* 8: 1011–1016.

53 Jeong, S., Kim, D., Shin, S. et al. (2014). Combinational synthetic approaches for isoreticular and polymorphic metal–organic frameworks with tuned pore geometries and surface properties. *Chem. Mater.* 26: 1711–1719.

54 Burd, S.D., Ma, S., Perman, J.A. et al. (2012). Highly selective carbon dioxide uptake by [Cu(bpy-n)₂(SiF₆)] (bpy-1 = 4,4'-Bipyridine; bpy-2 = 1,2-Bis(4-pyridyl)ethene). *J. Am. Chem. Soc.* 134: 3663–3666.

55 Shekhah, O., Belmabkhout, Y., Chen, Z. et al. (2014). Made-to-order metal-organic frameworks for trace carbon dioxide removal and air capture. *Nat. Commun.* 5: 4228.

56 McDonald, T.M., Lee, W.R., Mason, J.A. et al. (2012). Capture of carbon dioxide from air and flue gas in the alkylamine-appended metal–organic framework mmen-Mg$_2$(dobpdc). *J. Am. Chem. Soc.* 134: 7056–7065.

57 Mason, J.A., McDonald, T.M., Bae, T.-H. et al. (2015). Application of a high-throughput analyzer in evaluating solid adsorbents for post-combustion carbon capture via multi-component adsorption of CO$_2$, N$_2$, and H$_2$O. *J. Am. Chem. Soc.* 137: 4787–4803.

58 Lee, W.R., Jo, H., Yang, L.-M. et al. (2015). Exceptional CO$_2$ working capacity in a heterodiamine-grafted metal–organic framework. *Chem. Sci.* 6: 3697–3705.

59 Liao, P.-Q., Chen, X.-W., Liu, S.-Y. et al. (2016). Putting an ultrahigh concentration of amine groups into a metal–organic framework for CO2 capture at low pressures. *Chem. Sci.* 7: 6528–6533.

60 Hu, Z., Wang, Y., Farooq, S., and Zhao, D. (2017). A highly stable metal-organic framework with optimum aperture size for CO$_2$ capture. *AIChE J.* 63: 4103–4114.

61 Lässig, D., Lincke, J., Moellmer, J. et al. (2011). A microporous copper metal–organic framework with high H$_2$ and CO$_2$ adsorption capacity at ambient pressure. *Chem. Int. Ed.* 50: 10344–10348.

62 Li, B., Zhang, Z., Li, Y. et al. (2012). Enhanced binding affinity, remarkable selectivity, and high capacity of CO$_2$ by dual functionalization of a rht-type metal–organic framework. *Chem. Int. Ed.* 51: 1412–1415.

63 Millward, A.R. and Yaghi, O.M. (2005). Metal–organic frameworks with exceptionally high capacity for storage of carbon dioxide at room temperature. *J. Am. Chem. Soc.* 127: 17998–17999.

64 Bourrelly, S., Llewellyn, P.L., Serre, C. et al. (2005). Different adsorption behaviors of methane and carbon dioxide in the isotypic nanoporous metal terephthalates MIL-53 and MIL-42. *J. Am. Chem. Soc.* 127: 13519–13521.

65 Loiseau, T., Serre, C., Huguenard, C. et al. (2004). A rationale for the large breathing of the porous aluminum terephthalate (MIL-53) upon hydration. *Chem-Eur. J.* 10: 1373–1382.

66 Yang, H., Guo, F., Lama, P. et al. (2018). Visualizing structural transformation and guest binding in a flexible metal–organic framework under high pressure and room temperature. *ACS. Cent. Sci.* 4: 1194–1200.

67 Myers, A.L. and Prausnitz, J.M. (1965). Thermodynamics of mixed-gas adsorption. *AIChE J.* 11: 121–127.

68 Lin, Y., Yan, Q., Kong, C., and Chen, L. (2013). Polyethyleneimine incorporated metal-organic frameworks adsorbent for highly selective CO$_2$ capture. *Sci. Rep.* 3: 1859.

69 Wu, D., Xu, Q., Liu, D., and Zhong, C. (2010). Exceptional CO$_2$ capture capability and molecular-level segregation in a Li-modified metal–organic framework. *J. Phys. Chem. C* 114: 16611–16617.

70 McDonald, T.M., D'Alessandro, D.M., Krishna, R., and Long, J.R. (2011). Enhanced carbon dioxide capture upon incorporation of N,N'-dimethylethylenediamine in the metal–organic framework CuBTTri. *Chem. Sci.* 2: 2022–2028.

71 Wu, H., Reali, R.S., Smith, D.A. et al. (2010). Highly selective CO_2 capture by a flexible microporous metal–organic framework (MMOF) material. *Chem-Eur. J.* 16: 13951–13954.

72 Lin, R.-B., Chen, D., Lin, Y.-Y. et al. (2012). A zeolite-like zinc triazolate framework with high gas adsorption and separation performance. *Inorg. Chem.* 51: 9950–9955.

73 Milner, P.J., Siegelman, R.L., Forse, A.C. et al. (2017). A diaminopropane-appended metal–organic framework enabling efficient CO_2 capture from coal flue gas via a mixed adsorption mechanism. *J. Am. Chem. Soc.* 139: 13541–13553.

74 An, J., Geib, S.J., and Rosi, N.L. (2010). High and selective CO_2 uptake in a cobalt adeninate metal–organic framework exhibiting pyrimidine- and amino-decorated pores. *J. Am. Chem. Soc.* 132: 38–39.

75 Phan, A., Doonan, C.J., Uribe-Romo, F.J. et al. (2010). Synthesis, structure, and carbon dioxide capture properties of zeolitic imidazolate frameworks. *Acc. Chem. Res.* 43: 58–67.

76 Matsuda, R., Kitaura, R., Kitagawa, S. et al. (2005). Highly controlled acetylene accommodation in a metal–organic microporous material. *Nature* 436: 238–241.

77 Luo, F., Yan, C., Dang, L. et al. (2016). UTSA-74: a MOF-74 isomer with two accessible binding sites per metal center for highly selective gas separation. *J. Am. Chem. Soc.* 138: 5678–5684.

78 Lin, R.-B., Li, L., Wu, H. et al. (2017). Optimized separation of acetylene from carbon dioxide and ethylene in a microporous material. *J. Am. Chem. Soc.* 139: 8022–8028.

79 Sumida, K., Rogow, D.L., Mason, J.A. et al. (2012). Carbon dioxide capture in metal–organic frameworks. *Chem. Rev.* 112: 724–781.

80 Nguyen, N.T.T., Furukawa, H., Gándara, F. et al. (2014). Selective capture of carbon dioxide under humid conditions by hydrophobic chabazite-type zeolitic imidazolate frameworks. *Chem. Int. Ed.* 53: 10645–10648.

81 Hu, Z., Gami, A., Wang, Y., and Zhao, D. (2017). A triphasic modulated hydrothermal approach for the synthesis of multivariate metal–organic frameworks with hydrophobic moieties for highly efficient moisture-resistant CO_2 capture. *Adv. Sustain. Syst.* 1: 1700092.

82 Liao, P.-Q., Chen, X.-W., Liu, S.-Y. et al. (2016). Putting an ultrahigh concentration of amine groups into a metal–organic framework for CO_2 capture at low pressures. *Chem. Sci.* 7: 6528–6533.

83 Kumar, A., Madden, D.G., Lusi, M. et al. (2015). Direct air capture of CO_2 by physisorbent materials. *Chem. Int. Ed.* 54: 14372–14377.

84 Sridhar, S., Smitha, B., and Aminabhavi, T.M. (2007). Separation of carbon dioxide from natural gas mixtures through polymeric membranes – a review. *Sep. Purif. Rev.* 36: 113–174.

85 Yampolskii, Y. (2012). Polymeric gas separation membranes. *Macromolecules* 45: 3298–3311.

86 Robeson, L.M. (1991). Correlation of separation factor versus permeability for polymeric membranes. *J. Membr. Sci.* 62: 165–185.

87 Robeson, L.M. (2008). The upper bound revisited. *J. Membr. Sci.* 320: 390–400.

88 Yao, J. and Wang, H. (2014). Zeolitic imidazolate framework composite membranes and thin films: synthesis and applications. *Chem. Soc. Rev.* 43: 4470–4493.

89 Shah, M., McCarthy, M.C., Sachdeva, S. et al. (2012). Current status of metal–organic framework membranes for gas separations: promises and challenges. *Ind. Eng. Chem. Res.* 51: 2179–2199.

90 Denny, M.S. Jr.,, Moreton, J.C., Benz, L., and Cohen, S.M. (2016). Metal–organic frameworks for membrane-based separations. *Nat. Rev. Mater.* 1: 16078.

91 Peng, Y., Li, Y., Ban, Y. et al. (2014). Metal-organic framework nanosheets as building blocks for molecular sieving membranes. *Science* 346: 1356.

92 Sabetghadam, A., Seoane, B., Keskin, D. et al. (2016). Metal organic framework crystals in mixed-matrix membranes: impact of the filler morphology on the gas separation performance. *Adv. Funct. Mater.* 26: 3154–3163.

93 Chen, C.-X., Wei, Z., Jiang, J.-J. et al. (2016). Precise modulation of the breathing behavior and pore surface in Zr-MOFs by reversible post-synthetic variable-spacer installation to fine-tune the expansion magnitude and sorption properties. *Chem. Int. Ed.* 55: 9932–9936.

94 Verma, G., Kumar, S., Pham, T. et al. (2017). Partially interpenetrated NbO topology metal–organic framework exhibiting selective gas adsorption. *Cryst. Growth Des.* 17: 2711–2717.

95 Chen, C.-X., Zheng, S.-P., Wei, Z.-W. et al. (2017). A robust metal–organic framework combining open metal sites and polar groups for methane purification and CO_2/fluorocarbon capture. *Chem-Eur J.* 23: 4060–4064.

3

Porous Carbon Materials

Designed Synthesis and CO_2 Capture

Xiang-Qian Zhang and An-Hui Lu

School of Chemical Engineering, State Key Laboratory of Fine Chemicals, Dalian University of Technology, Dalian, China

3.1 Introduction

The development of novel materials and new technologies for CO_2 capture and storage (CCS) has gained considerable attention in recent decades. The necessity of reducing the concentration of CO_2 in the atmosphere is a very important issue, since CO_2 is considered

Materials for Carbon Capture, First Edition. Edited by De-en Jiang, Shannon M. Mahurin and Sheng Dai.
© 2020 John Wiley & Sons Ltd. Published 2020 by John Wiley & Sons Ltd.

a greenhouse gas that causes global warming. In addition, higher concentrations of CO_2 are toxic for humans, especially in space-limited chambers like submarines and spaceships. Currently, for gases containing high concentrations of CO_2, chemical/physical sorption and cryodistillation processes are used in industry. For dilute sources such as natural gas, syngas, and biogas, physical adsorption is more effective because of its multiple advantages such as fast kinetics, easy regeneration, less corrosion of equipment, superior cycling capability, etc. For low-concentration CO2 sources, adsorbent materials with a high CO_2 adsorption capacity and excellent selectivity of CO_2 over other gases are essential for CCS [1–4].

Among these materials, porous carbons have been proven to be competitive candidates by virtue of their high specific surface area, moderate heat of adsorption, low-cost preparation, relatively easy regeneration, and lower sensitivity to the humidity than other CO_2-philic materials. Most importantly, nowadays researchers are able to synthesize carbon materials with defined nanostructure and morphology, and tunable surface area and pore size. This, in turn, leads to a high selectivity of CO_2/balance gas (in most cases, N_2). Activated carbons (ACs) derived from coal, petroleum, and coconut shells have been the most commonly used form of porous carbons for a long time. However, the surface chemistry and pore size have been uncontrollable due to uncertain structures of various precursors. Thus, their CO_2 capture efficiency is badly affected due to the low adsorption selectivity.

Recently, the designed synthesis of porous carbon adsorbents by using self-assembly of precursor molecules, controlled hydrothermal treatment, and pyrolysis has attracted significant attention due to the precise control and facile functionalization of the final carbons. This chapter reviews the recent literature on novel porous carbon materials for CO_2 capture. The aim is to describe the key advances, predominantly focusing on high-performance CO_2 capture materials with tailored macro- and micro-structures and targeted surface properties. In this chapter, the focus is on both an in-depth study of porous carbon materials and a high-level evaluation of their performance in CO_2 capture. Herein, diverse porous carbons are mainly classified into the following categories: synthetic polymer-based porous carbons, newly developed ionic liquids (ILs) and metal–organic framework (MOF)-derived carbons, sustainable resource–derived porous carbons, etc. In each category, synthesis principles and pore structures, macroscopic and microscopic morphologies, as well as their CO_2 capture behavior are discussed. After that, some critical design principles of fabricating porous carbon adsorbents that significantly influence the CO_2 capture capability are presented and highlighted. Finally, we present a brief summary and discuss future perspectives of porous carbons for CO_2 capture.

3.2 Designed Synthesis of Polymer-Based Porous Carbons as CO_2 Adsorbents

Porous carbon materials are ubiquitous and indispensable in many modern-day scientific applications. They are used extensively as sorbents for separation processes and gas storage. Conventional porous carbon materials, such as AC and carbon molecular sieves, are synthesized by pyrolysis and physical or chemical activation of organic precursors, such as coal, wood, fruit shells, and polymers, at elevated temperatures [5, 6]. These carbon materials normally have relatively broad pore-size distributions in both the micropore and mesopore ranges. Compared with conventional ACs and biomass-derived carbons, using synthetic polymers as porous carbon precursors enables better chemical composition control, easily tailored precise morphology, tunable pore systems, and targeted surface chemistry [7]. This can be achieved by a designed synthesis methodology.

Generally, CO$_2$ capture performance strongly depends on the microstructure of the adsorbents. Meanwhile, an understanding of the host-guest interactions between the adsorbent and CO$_2$ is a prerequisite for any great advance in CO$_2$ capture. Thus, a precisely controlled synthesis of carbon structures can provide a promising opportunity to authentically understand the physical and chemical properties of carbon materials at a molecular level and, in turn, efficiently guide practical applications. Thus, synthetic polymers-based porous carbons are extensively investigated nowadays. For clarity, in this part of the chapter, porous carbon materials derived from rationally designed synthetic polymers are classified into three groups according to synthesis methods: hard-template method, soft-template approach, and template-free synthesis.

3.2.1 Hard-Template Method

In the past, a nanocasting pathway (from pre-formed hard templates) has been demonstrated as a controllable method to prepare carbon monoliths with tailorable pore size over several length scales. *Nanocasting* is a process in which a mold (often referred to as a *hard template*, or *scaffold*) over nanometer scale is filled with a precursor; after processing, the initial mold is removed [8–13]. In this way, the space once occupied by the host mold is transferred into the pores of the final carbon products, and the carbon in the original template pores is released as a continuous carbon framework. The key is to rely on preparing templates with accessible porosity and thermally stable carbon precursors such as phenolic resin, sucrose, furfuryl alcohol (FA), acrylonitrile, acetonitrile, mesophase pitch, etc. In the following, we discuss the detailed synthesis principle based on several representative examples.

3.2.1.1 Porous Carbons Replicated from Porous Silica

Porous carbons replicated from porous silica have been extensively investigated [14]. Nanocasting for fabricating porous carbons usually involves the following steps, as shown in Figure 3.1: (a) preparation of a porous template with controlled porosity; (b) introduction of a suitable carbon precursor into the template pores through techniques such as wet impregnation, chemical vapor deposition (CVD), or a combination; (c) polymerization and carbonization of the carbon precursor to generate an organic–inorganic composite; and (d) removal of the inorganic template.

Using this strategy, numerous porous carbons with well-defined porosity have been obtained. For example, Lindén and co-workers prepared hierarchical porous monolithic carbons containing wormhole-like mesopores and macropores [15–17]. In a similar way, Hu et al. synthesized hierarchically porous carbons with a higher graphite-like ordered carbon structure by using meso−/macroporous silica as a template and using mesophase pitch as a precursor [18]. Considering the advantages of nanocasted porous carbons, the Yin group also reported the preparation of mesoporous, nitrogen-doped carbon (N-MC) with highly ordered two-dimensional hexagonal structures using diaminobenzene (DAB) as carbon

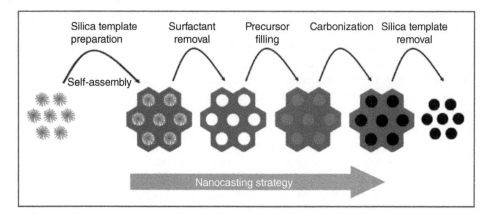

Figure 3.1 Typical method for the preparation of ordered mesoporous carbon (OMC) materials: the nanocasting strategy from mesoporous silica hard templates. Source: reproduced from Ref. [14] with permission from the Royal Society of Chemistry. (*See color plate section for color representation of this figure*).

and nitrogen sources, ammonium peroxydisulfate (APDS) as an oxidant, and SBA-15 as a hard template [19]. By adjusting the synthesis temperatures in the range of 70–100 °C, the pore diameter of the as-synthesized materials can be tuned from 3.4 to 4.2 nm, while the specific surface area of the N-MC with a nitrogen content of 26.5 wt% can be tuned from 281.8–535.2 m^2 g^{-1}. The C/N molar ratio of the samples can be tuned in a range of 3.25–3.65 by adjusting the mole ratio of DAB/APDS precursors at a synthesis temperature of 80 °C, while the pore diameter of the N-MC can be tuned in a range of 4.1–3.7 nm.

Recently, Suárez-García et al. synthesized ordered mesoporous carbons (OMCs) with high surface areas and pore volumes through polymerization of a polyamide precursor (3-aminobenzoic acid) inside the SBA-15 template porosity (Figure 3.2) [20]. The synthesis was accomplished in phosphoric acid medium in order to achieve additional porosity development while keeping the ordered arrangement as unchanged as possible. Furthermore, the authors also provide a systematic analysis of the influence of porous texture and heteroatoms on interactions between the carbon surface and CO$_2$ molecules.

To incorporate more nitrogen atoms in the carbon nanostructure, carbon nitride (CN), a well-known and fascinating material attracting worldwide attention, has been synthesized via such a nanocasting approach. Vinu prepared two-dimensional mesoporous carbon nitride (MCN) with tunable pore diameters using SBA-15 materials with different pore diameters as templates through a simple polymerization reaction between ethylenediamine (EDA) and carbon tetrachloride (CTC) using a nano hard-templating approach [21]. High-resolution transmission electron microscopy (HRTEM) was used to further examine the structural order and morphology of the mesoporous CN materials with different pore diameters. As seen in Figure 3.3, the pore diameter of the MCN materials can be easily tuned from 4.2–6.4 nm without affecting the structural order. The carbon to nitrogen ratio of the MCN decreases from 4.3 to 3.3 while increasing the EDA to CTC weight ratio from 0.3 to 0.9. Meanwhile, the specific surface area and the specific pore volume of MCN materials can be adjusted in the range between 505 and 830 m^2 g^{-1} and 0.55–1.25 cm^3 g^{-1}, respectively.

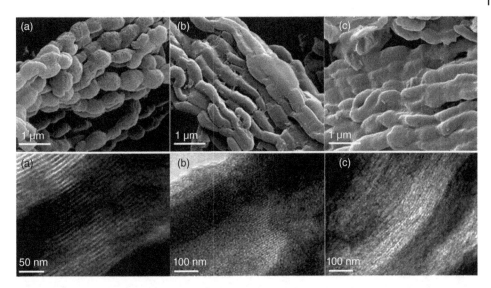

Figure 3.2 SEM micrographs (top) and TEM micrographs (down) of carbon materials obtained at 800 °C in the presence of (a) 5, (b) 50, and (c) 150 wt % H_3PO_4. Source: reproduced from Ref. [20] with permission from the American Chemical Society.

In a similar way, the Zhao group reported the preparation of porous CN spheres with partially crystalline frameworks and high nitrogen content (17.8 wt%) by using spherical mesoporous cellular silica foams (MCFs) as a hard template, and EDA and CTC as precursors [22]. The resulting CN spheres have a mesostructure with small and large mesopores with pore diameters centered respectively at ca. 4.0 and 43 nm, a relatively high Brunauer–Emmett–Teller (BET) surface area of ~550 m² g⁻¹, and a pore volume of 0.90 cm³ g⁻¹. The adsorption isotherms of the mesoporous CN spheres (Figure 3.4a) show that after adsorption for 150 minutes, CO_2 uptake reaches 2.90 mmol g⁻¹ at 25 °C. When the temperature was increased to 75 °C, the uptake greatly decreased to 0.97 mmol g⁻¹. The CO_2 uptake (2.50 mmol g⁻¹) of the pristine carbon material (Figure 3.4b) was similar to that of the mesoporous CN sample at 25 °C. However, when the temperature was increased to 75 °C, its CO_2 uptake decreased dramatically to 0.30 mmol g⁻¹, which is much lower than the corresponding value for the mesoporous CN spheres, suggesting that there is a weak interaction between the carbon pore walls and the CO_2 molecules [22].

The previously mentioned nanocasting method was very successful in preparing porous carbon with nicely controlled structures, but the multiple steps and long synthesis period involved are time consuming. Researchers have made considerable efforts to simplify these tedious procedures. Han and coworkers developed a one-step nanocasting technique to synthesize micro–/mesoporous carbon monoliths with very high BET surface area (ca. 1970 m² g⁻¹) and ca. 2 nm mesopores by the co-condensation of β-cyclodextrin with tetramethylorthosilicate [23].

3.2.1.2 Porous Carbons Replicated from Crystalline Microporous Materials
Using mesoporous silica as the hard template can generate OMCs with the nanocasting strategy. Efforts to construct molecular-sieve type porous carbon often require crystalline

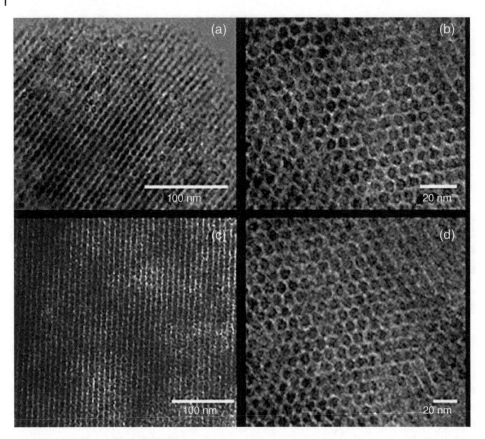

Figure 3.3 HRTEM images of (a, b) MCN-1-130 and (c, d) MCN-1-150. Source: reproduced from Ref. [21] with permission from John Wiley & Sons.

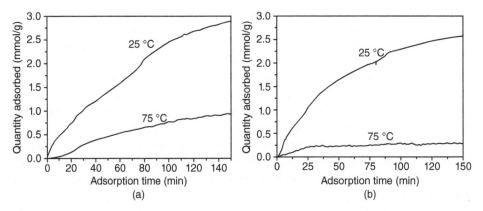

Figure 3.4 CO_2-capture capacities of (a) mesoporous carbon nitride (CN) materials and (b) pristine carbon spheres (CSs) at 25 and 75 °C. Source: reproduced from Ref. [22] with kind permission from Springer Science+Business Media.

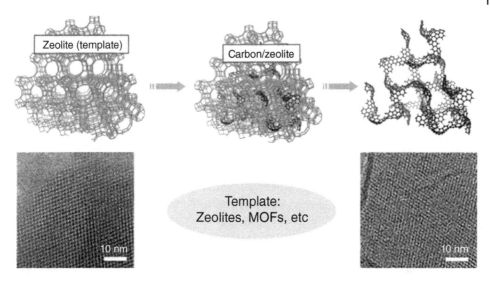

Figure 3.5 The synthesis of molecular sieve carbon (MSC) through a nanocasting pathway: the sacrificial templates include the crystalline zeolites and metal organic frameworks (MOFs). Source: reproduced from Ref. [24] with permission from John Wiley & Sons.

zeolite or MOF-related materials (zeolitic imidazolate frameworks [ZIFs], metal–organic coordination polymers [MOCPs]) as sacrificial templates. As shown in Figure 3.5, the nanoporous carbons (NPCs) produced in this way show remarkably high surface areas (up to 4000 m^2 g^{-1}) and precisely controlled microporous structures (0.5–1.5 nm) that are well suited for CO_2 capture [24]. Banerjee et al. reported the synthesis and gas adsorption properties of porous carbons by using isoreticular zeolitic imidazolate frameworks (IRZIFs) as a template and FA as a carbon source [25]. Similarly, Liang and co-workers synthesized a series of porous carbons from nonporous MOCPs, using in situ polymerized phenol resin as a carbon precursor [26]. The textural properties of highly microporous ZIF-templated carbon derived from commercially available ZIFs (Basolite Z1200) can be further improved via mild chemical activation, as demonstrated by Mokaya and co-workers [27].

3.2.1.3 Porous Carbons Replicated from Colloidal Crystals

Colloidal crystals are self-assembled periodic structures consisting of close-packed uniform particles. Replication of a colloidal crystal (colloidal silica/polymer sphere) in most cases leads to a high degree of periodicity in three dimensions. Removal of the crystal template leads to a replica with 3D ordered macroporous (3DOM) structures. The groups of Stein, Velev, and Lenhoff have independently achieved many great results in the field of colloidal crystals and related areas. Here we limit our discussion to a small aspect of the templating of colloidal crystals that serves as an effective path to obtain porous carbons with highly ordered macroporosity. For example, Lee et al. synthesized 3DOM carbons via a resorcinol-formaldehyde sol–gel process using a poly(methyl methacrylate) colloidal-crystal as a template [28]. Similarly, Adelhelm et al. also synthesized a hierarchical meso- and macroporous carbon using mesophase pitch as precursor and PS or PMMA as template through spinodal decomposition [29].

By carbonizing a thin layer of phenolic resin on suitable templates, Gierszal et al. reported the synthesis of one type of uniform carbon film with large pore volumes (6 cm g^{-1} [3] for 24 nm silica colloids), uniform pore sizes, and controlled thickness [30]. This synthesis involves the formation of a uniform polymeric film on the silica pore walls of silica colloidal crystals or colloidal aggregates and its carbonization and template removal. After proper pretreatment of the silica template and under controlled experimental conditions, the mixture of resorcinol and crotonaldehyde copolymerizes on the silica surface and forms a uniform film.

Recently, Zhang and co-workers presented a one-pot method to synthesize hierarchically bimodal-ordered porous carbons with interconnected macropores and mesopores, via in situ self-assembly of colloidal polymer (280, 370, and 475 nm) and silica spheres (50 nm) using sucrose as the carbon source. Compared with the classical nanocasting procedure, this approach is relatively simple; neither pre-synthesis of crystal templates nor additional infiltration is needed, and the self-assembly of polymer spheres into the crystal template and the infiltration are finished simultaneously in the same system [31]. Similarly, a hierarchically porous carbon with multi-modal (macropore and mesopore) porosity has also been prepared by using dual-template (PS [or PMMA]/colloidal silica), where PS (or PMMA) is for creating 3D ordered macropores, and colloidal silica is responsible for creating spherical mesopores [32, 33]. The unique hierarchical structures of 3DOM carbons, i.e. the open larger mesopores located in the ordered macropores, may enhance the kinetics greatly when used as CO$_2$ sorbents.

3.2.1.4 Porous Carbons Replicated from MgO Nanoparticles

Compared with conventional templates, the MgO-template method has the following advantages: (i) the MgO template is easily removed by a diluted noncorrosive acid; (ii) MgO can be recycled for the present method; and (iii) the size and volume of the mesopores are tunable by changing the MgO size and carbon precursor [34]. The MgO-template method can form micro/mesoporous carbons with high specific surface area and pore volumes by pyrolysis of organic polymers, such as poly(vinyl alcohol) (PVA), poly(ethylene terephthalate) (PET), hydroxy propylcellulose, poly(vinyl pyrrolidone), polyacrylamide (PAA), trimethylolmelamine (TMM), and others [34–36].

For example, Park et al. reported a series of porous carbons with well-developed pore structures, which were directly prepared from a weak acid cation exchange resin (CER) by the carbonization of a mixture with Mg acetate in different ratios [37]. By dissolving the MgO template, the porous carbons exhibited high specific surface areas (326–1276 m^2 g^{-1}) and high pore volumes (0.258–0.687 cm^3 g^{-1}). The CO$_2$ adsorption capacities of the porous carbons were enhanced to 164.4 mg g^{-1} at 1 bar and 1045 mg g^{-1} at 30 bar by increasing the Mg acetate to CER ratio. This result indicated that CER was a carbon precursor capable of producing the porous structure as well as improving the CO$_2$ adsorption capacities of the carbon species.

The Jang group reported a time-saving synthesis in which ordered mesoporous carbon-supported MgO (Mg-OMC) materials were synthesized by the carbonization of sulfuric-acid-treated silica/triblock copolymer/sucrose/Mg(NO$_3$)$_2$ composites. In the current approach shown in Figure 3.6, the triblock copolymer P123 and sucrose were employed as both structure-directing agents for the self-assembly of rice husk ash silica solution and carbon precursor. Sulfuric acid was used to cross-link P123 and sucrose in

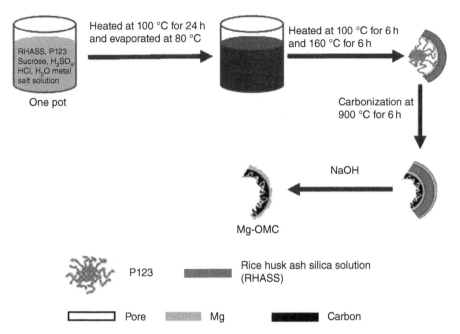

Figure 3.6 One-pot synthesis of Mg-ordered mesoporous carbon (OMC). Source: reproduced from Ref. [38] with permission from Elsevier.

the as-synthesized composites in order to improve the carbon yield. The CO_2 adsorption capacity of Mg-OMC-1 was observed to be $92\,\mathrm{mg\,g^{-1}}$ of sorbent, which is comparable to that of the well-established CO_2 sorbents [38].

Przepiórski et al. reported the competitive uptake of SO_2 and CO_2 on porous carbon material containing CaO and MgO, prepared by carbonization of PET mixed with a natural dolomite. The as-prepared porous carbon was examined as a sorbent material for simultaneous removal of CO_2 and SO_2 from air in dry conditions and in the presence of humidity, at temperatures ranging from 20–70 °C. The results clearly confirmed the importance of water on the amount of gases removed from air streams and on the removal mechanisms. The breakthrough curves registered clearly revealed overshoot in concentration of the gas during the first minutes of removal tests [39].

Recently, Park et al. reported the preparation of MgO-templated porous carbon (MPC) by carbonization of (styrene-divinylbenzene)-based ion exchange resin and activation by KOH. MPCs were prepared from a (styrene-divinylbenzene)-based ion exchange resin by the carbonization of a mixture with Mg gluconate at 900 °C. The prepared MPCs were subsequently treated with KOH at KOH/MPC ratios ranging from 0.5–4 at 800 °C. Low KOH/MPC ratios (KOH/MPC ratio = 1, MCK-1) tended to favor the formation of micropores, whereas higher KOH/MPC (KOH/MPC ratio = 4, MCK-4) led to the formation of mesopores. The treated MPCs with a KOH/MPC ratio = 1 exhibited the best CO_2 adsorption value of $266\,\mathrm{mg\,g^{-1}}$ at 1 bar. However, the treated MPCs with a KOH/MPC ratio = 3 (MCK-3) exhibited the highest CO_2 adsorption value of $1385\,\mathrm{mg\,g^{-1}}$ at 30 bar. This result indicates that the CO_2 adsorption capacity of MPCs can be attributed to the mesopore volume fraction at higher pressure [40]. (Figure 3.7)

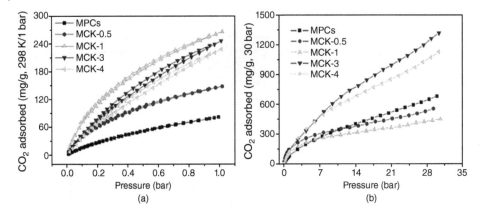

Figure 3.7 CO_2 adsorption isotherms of the pristine and chemical activated MPCs measured at (a) 1 bar and (b) 30 bar. Source: reproduced from Ref. [40] with permission from Elsevier.

3.2.2 Soft-Template Method

The soft-templating procedure is garnering a great deal of momentum for synthesizing porous carbons using the self-assembly of soft templates (e.g. amphiphilic block copolymers) and carbon precursors (e.g. phenolic resins). For soft-templating synthesis of mesoporous carbon, the interactions between template and organic precursor and, secondly, condensation/polymerization of precursor species, should be especially taken into consideration. Generally, phenolic resins and polyethylene oxide (PEO)-containing copolymers as the soft template are an appropriate pair to create the porous structure by direct self-assembly of phenolic resins and PEO segments of block copolymer through hydrogen-bonding interactions. In this section, the synthesis of porous carbons based on the soft-templating method is summarized according to the morphology and shape of the final materials.

3.2.2.1 Carbon Monolith

Porous carbons are versatile materials that possess a wide range of morphologies not only on the microscopic level but also on the macroscopic level. Macroscopically, a monolith generally shows a wide flexibility of operation in contrast to its powder counterparts [41]. Microscopically, the monolithic structure is characterized by its 3D bicontinuous hierarchical porosity, which usually leads to several distinct advantages such as low pressure drop, fast heat and mass transfer, high contacting efficiency, and straightforward handling [42, 43]. Thus, the monolithic carbons are well suited to gas sorption and separation, including CO_2 capture [44].

The synthesis of monolithic carbons generally relies on the means, including the sol–gel method and self-assembly approach [45, 46]. In recent years, much effort has been devoted to creating new types of carbon monoliths with enhanced functionality, i.e. developing new polymerization systems (solvents and/or precursors), precise pore engineering to obtain multi-modal porosities, and targeted surface/bulk functionalization for high-performance CO_2 capture [47–50].

Great progress has also been achieved in the synthesis of monolithic carbons with ordered mesoporosity by the self-assembly of a copolymer molecular template and carbon precursors [51]. However, it remains a great challenge to achieve highly developed mesoporosity while maintaining good monolith morphology, due to the following requirements. (i) A perfect matching interaction between the carbon-yielding precursors and the pore-forming component is required, which allows self-assembly of a stable micelle nanostructure. (ii) The micelle structures should be stable during the temperature required for curing a carbon-yielding component, but be readily decomposed during carbonization. (iii) The carbon-yielding component should be able to form a highly cross-linked polymeric material that can retain the nanostructure during the decomposition or extraction of the pore-forming component. In order to achieve a monolithic carbon with well-developed mesoporosity, not a single one of these conditions can be dispensed with.

The Dai group first synthesized highly OMC through a solvent annealing accelerated self-assembly method using polystyrene-block-poly (4-vinylpyridine) (PS-P4VP) as a soft template and N, N-Dimethylformamide (DMF) as the solvent [52]. However, the samples are in film form. Since then, using the self-assembly method to prepare porous carbons has been extensively investigated. At present, the products are mostly in the form of a powder or film. To overcome this issue, Valkama et al. reported a soft-template method to achieve carbon products in any desired shape, and the porosity can be tuned from mesoporous to hierarchically micro-/mesoporous simply by varying pyrolysis conditions for the cured block-copolymer phenolic resin complexes [53].

Based on the soft-templating principle, Dai's group reported a versatile synthesis of porous carbons (monolith, film, fiber, and particle) by using phenol-/resorcinol-/phloroglucinol-based phenolic resins as carbon precursors and triblock copolymer (F127) as the soft template. They found that due to the enhanced hydrogen bonding interaction with triblock copolymers, phloroglucinol with three hydroxyl groups is an excellent precursor for the synthesis of mesoporous carbons with a well-organized mesostructure [54]. This type of mesoporous carbon monolith shows good performance in gas capture [55]. At 800 Torr and 298 K, the adsorption equilibrium capacity of the OMC for CO_2 is $1.49 \, \text{mmol g}^{-1}$. Significantly higher adsorption uptake was observed for CO_2: $3.26 \, \text{mmol g}^{-1}$ at 100 bar and 298 K. More interestingly, the diffusion time constant of CO_2 decreased with adsorbate pressures due to the obvious mesoscale pore system.

Later, they prepared carbons with ordered mesopores based on the self-assembly approach of resorcinol-formaldehyde (RF) polymer and block copolymers under strong acidic conditions, and by subsequent centrifugation and shaping techniques. The Coulombic interaction (i.e. $I^+X^-S^+$ mechanism) and hydrogen bonding are believed to be the driving force for self-assembly between the RF resol and F127 template [56]. The polymerization-induced spinodal decomposition in glycolic solutions of phloroglucinol/formaldehyde polymers and block copolymers also leads to successful formation of the bimodal meso-/macroporous carbon monoliths [57].

Alternatively, Zhao's group developed a hydrothermal synthesis by using F127 and P123 as double soft templates and phenol/formaldehyde as carbon precursor (molar ratio between phenol and surfactant about 46 : 1), followed by hydrothermal aging at 100 °C for 10 hours [58]. Recently, the same group reported a controllable one-pot method to synthesize N-doped OMC with a high N content by using dicyandiamide as a nitrogen source via an evaporation-induced self-assembly (EISA) process [59]. In this synthesis, resol molecules can bridge the Pluronic F127 template and dicyandiamide via hydrogen bonding and electrostatic interactions. During thermosetting at 100 °C for formation of rigid phenolic resin and subsequent pyrolysis at 600 °C for carbonization, dicyandiamide provides closed N species, while resol can form a stable framework, thus ensuring the successful synthesis of ordered N-doped mesoporous carbon (Figure 3.8). Such N-doped OMCs possess tunable mesostructures (p6m and Im3m symmetry) and pore size (3.1–17.6 nm), high surface area (494–586 $m^2 g^{-1}$), and high N content (up to 13.1 wt%). Ascribed to the

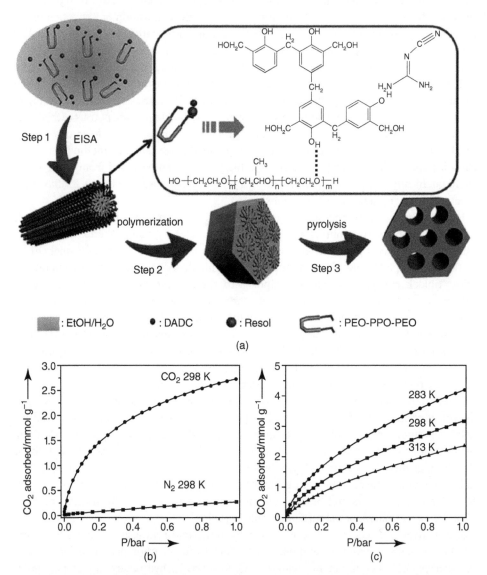

Figure 3.8 (a) The formation process of ordered N-doped mesoporous carbon from a one-pot assembly method using dicyandiamide (DCDA) as a nitrogen source. (b) CO_2 and N_2 adsorption isotherms at 25 °C for the N-doped mesoporous carbon H-NMC-2.5. (c) CO_2 adsorption isotherms at different temperature for A-NMC after activation by KOH. Source: reproduced from Ref. [59] with permission from John Wiley & Sons.

unique features of large surface area and high N contents, the materials showed high CO_2 capture of 2.8–3.2 mmol g^{-1} at 25 °C and 1.0 bar (Figure 3.8).

Similarly, Xiao et al. also reported a hydrothermal synthesis at even higher temperatures and longer time (i.e. 260 °C for more than 17 hours) to prepare carbon monoliths with well-ordered hexagonal or cubic mesopore systems [60]. Meanwhile, Gutiérrez et al. synthesized a low-density monolithic carbon exhibiting a 3D continuous micro- and macroporous structure, which derived from a PPO$_{15}$-PEO$_{22}$-PPO$_{15}$ block copolymer-assisted RF polymerization [61]. Zhang's group reported an organic–organic aqueous self-assembly approach to prepare B-/P-doped OMCs using boric acid and/or phosphoric acid as B- or P-heteroatom source, RF resin as the carbon precursor, and triblock copolymer Pluronic F127 as the mesoporous structure template [62]. The Lu group established a rapid and scalable synthesis of a crack-free and N-MC monolith with fully interconnected macropores and an ordered mesostructure using the soft-templating method. The monolith is achieved by using organic base lysine as a polymerization agent and mesostructure assembly promotor, through rapid sol–gel process at 90 °C [63]. As shown in Figure 3.9, lysine molecules could form intra-molecule salts, so the deprotonated carboxyl group and the protonated NH^{3+} group could form hydrogen bonds with the –OH group of resorcinol and the hydrophilic

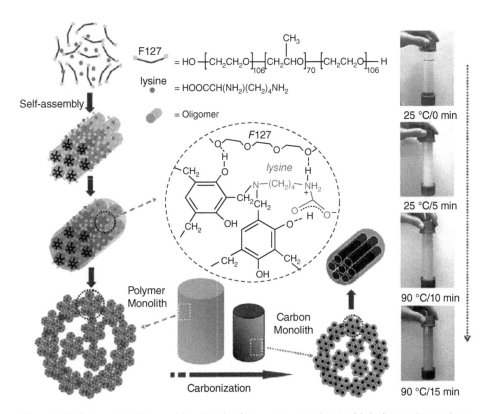

Figure 3.9 Schematic of the rapid synthesis of mesoporous carbons with lysine as the catalyst. Source: reproduced from Ref. [63] with permission from the Royal Society of Chemistry. (*See color plate section for color representation of this figure*).

EO (N–H···O) segment of F127. Thus the lysine molecule enhanced the assembly of the mesostructure between F127 and phenolic resins. Moreover, the basic lysine provided a basic environment for the polymerization of resorcinol and formaldehyde.

Later, the same group reported a new type of porous carbon monolith, which was synthesized through a self-assembly approach based on benzoxazine chemistry [64]. The resulting carbon monoliths show crack-free macro-morphology, well-defined multi-length scale pore structure, a nitrogen-containing framework, and high mechanical strength (Figure 3.10). With such a designed structure, the carbon monoliths show outstanding CO_2 capture and separation capacities, high selectivity, and facile regeneration at room temperature. At ~1 bar, the equilibrium capacities of the monoliths are in the range of 3.3–4.9 mmol g^{-1} at 0 °C, and of 2.6–3.3 mmol g^{-1} at 25 °C; while the dynamic capacities are in the range of 2.7–4.1 wt% at 25 °C using 14% (v/v) CO_2 in N_2 (Figure 3.11). The carbon monoliths exhibit high selectivity for the capture of CO_2 over N_2 from a CO_2/N_2 mixture, with a separation factor ranging from 13–28. Meanwhile, they undergo facile CO_2 release in an argon stream at 25 °C, indicating a good regeneration capacity. Similarly, Bao et al. prepared a new kind of N-doped hierarchical carbon that exhibits even higher Henry's law CO_2/N_2 selectivity. Their synthesis strategy was based on the rational design of a modified pyrrole molecule that can co-assemble with the soft Pluronic template via hydrogen bonding and electrostatic interactions to give rise to mesopores followed by carbonization. After low-temperature carbonization and activation processes, the ultrasmall pores (d < 0.5 nm) and preservation of nitrogen moieties allow for enhanced CO_2 affinity [65].

Due to the high precision in pore engineering by the nanocasting pathway and the great variety of micelle nanostructures derived from soft-templating, many researchers try to combine both techniques into an interdependent and interactive module with the aim of achieving porous carbons with controlled pore structure in a cost-effective manner.

Wang et al. prepared 3D ordered macro-/mesoporous porous carbons by combining colloidal crystal templating with surfactant templating in a gas-phase process [66]. In a vapor-phase infiltration, the wall thickness and window sizes of carbons are controllable through the variation of the infiltration time. Hierarchically ordered macro-/mesoporous carbon was prepared by dual templating with a hard template (silica colloidal crystal) and a soft template (Pluronic F127), using phenol-formaldehyde precursors dissolved in ethanol [67]. Zhao's group reported a mass preparation of hierarchical carbon-silica composite monoliths with ordered mesopores by using polyurethane (PU) foam as a sacrificial scaffold. The macroporous PU foam provides a large, 3D, interconnecting interface for EISA of the coated phenolic resin-silica-block copolymer composites, thus endowing composite monoliths with a diversity of macroporous architectures [68]. Recently, the same group reported a direct synthesis of transparent ordered mesostructured resin-silica composite monoliths with uniform rectangular shape through the EISA process by co-polymerization of tetraethyl orthosilicate and resol in the presence of triblock copolymer Pluronic F127 as a template [69]. The key factor of this synthesis is the good interoperability and compatibility of the plastic organic resin polymers and the rigid silica skeleton. As a result, multiple choices of the products (OMC or silica monoliths with integrated macroscopic morphologies similar to the original composite monoliths) can be realized by either removal of silica in HF solution or elimination of carbon by simple combustion.

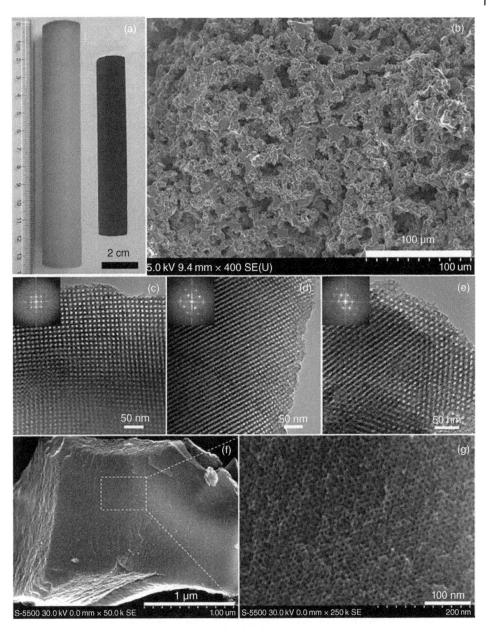

Figure 3.10 Photograph of (a) the synthesized polymer and (b) carbon monolith; (c–e) TEM images (c, d, and e: images viewed in the [100], [110], and [111] direction; the insets are the corresponding fast Fourier transform [FFT] diffractograms) and (f, g) HR-SEM images of the carbon monolith HCM-DAH-1. Source: reproduced from Ref. [64] with permission from the American Chemical Society.

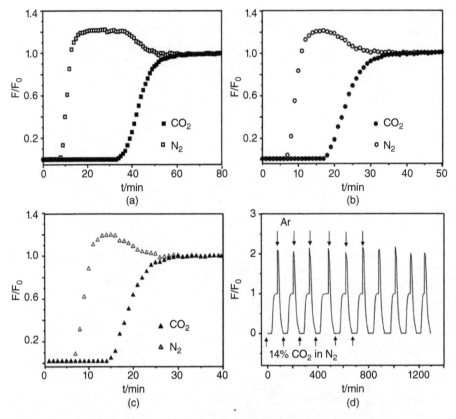

Figure 3.11 (a–c) Breakthrough curves: a 14% mixture of CO_2 in N_2 is fed into a bed of (a) HCM-DAH-1, (b) HCM-DAH-1-900-1, and (c) HCM-DAH-1-900-3. (d) Recycle runs of CO_2 adsorption–desorption on HCM-DAH-1 at 25 °C, using a stream of 14% (v/v) CO_2 in N_2, followed a regeneration by Ar flow. Source: reproduced from Ref. [64] with permission from the American Chemical Society.

To date, hydrogen-bonding interactions have been extensively explored as the self-assembly driving force between block copolymer surfactants and carbon precursors. As viewed from the current research, the success of hydrogen-bonding induced self-assembly is only in a small mesopore range (3–10 nm). Groundbreaking achievements in creating well-ordered porosity in either the micropore scale (<2 nm) or larger mesopore range (10–50 nm) remain a grand challenge. Moreover, the common features of most current syntheses are that they usually take a day, or even longer, and use inorganic catalysts (HCl or NaOH) for the polymerization and self-assembly. Hence, to explore new polymerization systems (new carbon precursors, organic catalysts) that are more time-effective is an exciting research area. In addition, hierarchically structured monolithic carbons with multimodal porosity would be more suitable for application in gas capture and separation. More desirably, the influence of multi-length scale pores on the sorption kinetics and storage capability should be elucidated through experimental and computational work.

3.2.2.2 Carbon Films and Sheets

In addition to the investigation of monolithic carbons with well-ordered mesoporosity as CO$_2$ sorbents, the synthesis of OMC films for CO$_2$ capture (membrane gas separation) has also attracted significant interest. Noticeably, highly OMC with cubic *Im3m* symmetry has been synthesized successfully via a direct carbonization of self-assembled F108 (EO$_{132}$PO$_{50}$EO$_{132}$) and RF composites obtained in a basic medium of non-aqueous solution [70]. Dai et al. demonstrated a stepwise self-assembly approach to the preparation of large-scale, highly ordered NPC films (Figure 3.12) [52].

The synthesis of well-defined porous carbon films involves four steps: (i) monomer–block copolymer film casting, (ii) structure refining through solvent annealing, (iii) polymerization of the carbon precursor, and (iv) carbonization. Zhao et al. reported the fabrication of freestanding mesoporous carbon thin films with highly ordered pore architecture via a simple coating-etching method [71]. The mesoporous carbon films were first synthesized by coating a resol precursor/Pluronic copolymer solution on a preoxidized silicon wafer, forming highly ordered polymeric mesostructures based on organic–organic self-assembly, followed by carbonization at 600 °C, and finally etching of the native oxide layer between the carbon film and the silicon substrate. Mild reacting conditions and wide composition

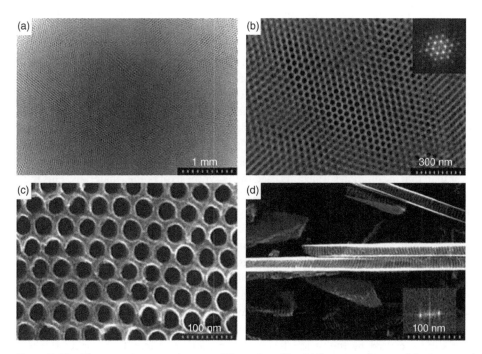

Figure 3.12 Electron microscopy images of the carbon film. (a) Z-contrast image of the large-scale homogeneous carbon film in a 4 × 3 mm area. (b) Z-contrast image showing details of the highly ordered carbon structure. (c) HRSEM image of the surface of the carbon film with a uniform hexagonal-pore array. The pore size is 33.7 ± 2.5 nm, and the wall thickness is 9.0 ± 1.1 nm. (d) SEM image of the film cross section, which exhibits all parallel straight channels perpendicular to the film surface. Source: reproduced from Ref. [52] with permission from John Wiley & Sons.

ranges are the obvious advantages of this method over the techniques previously reported [72–74].

Based on mesoporous RF-derived carbon films, Yoshimune et al. investigated the permeation properties of different gases, including H_2, He, N_2, CF_4, and condensable gases of CO_2 and CH_4 [75]. As expected, these membranes exhibited relatively high permeances due to their well-developed mesoporous structure. By comparing the six gas permeances of the RF carbon films at 303 K, the authors found that the gas permeances exhibited a sublinear dependence on the inverse square of the molecular weights, which are predominantly governed by the Knudsen mechanism. Interestingly, the permeance of condensable gases of CO_2 and CH_4 was accelerated. This is because the RF carbon films have both mesopores and a small number of micropores, and the enhanced permeance is attributed to the strong adsorption of condensable gases in the narrower micropores due to the surface diffusion mechanism. However, the effect of molecular sieving, which allows smaller gas molecules such as those of He to diffuse faster than larger gas molecules such as those of CF_4, is negligible since the gas permeances of the RF carbon membranes are independent of the kinetic diameter of the permeating molecules. Consequently, the transport mechanism of these membranes is mainly governed by Knudsen diffusion, and the existence of a small number of micropores might have induced the surface diffusion.

Recently, Lu et al. reported a wet-chemistry synthesis of a new type of porous carbon nanosheet whose thickness can be precisely controlled over the nanometer length scale (Figure 3.13) [76, 77]. This feature is distinct from conventional porous carbons that are composed of micron-sized or larger skeletons, and whose structure is less controlled. The synthesis cleverly uses graphene oxide (GO) as the shape-directing agent and asparagine as bridging molecules that connect the GO and in situ grown polymers by electrostatic interaction between the molecules. The assembly of the nanosheets can produce macroscopic structures, i.e. hierarchically porous carbon monoliths that have a mechanical strength up to 28.9 MPa, the highest reported for the analogues. The synthesis provides precise control of porous carbons over both microscopic and macroscopic structures at the same time. In all syntheses, the graphene content used was in the range of 0.5–2.6 wt%, which is significantly lower than that of common surfactants used in the synthesis of porous materials. This indicates the strong shape-directing function of GO. In addition, the overall thickness of the nanosheets can be tuned from 20–200 nm according to a fitted linear correlation between the carbon precursor/GO mass ratio and the coating thickness.

The porous carbon nanosheets show impressive CO_2 adsorption capacity under equilibrium, good separation ability of CO_2 from N_2 under dynamic conditions, and easy regeneration. The highest CO_2 adsorption capacities can reach 5.67 and 3.54 CO_2 molecules per nm^3 pore volume and per nm^2 surface area at 25 °C and ~1 bar (Figure 3.14). The probable reason is that the interaction of porous carbon nanosheets with CO_2 molecules is strong due to (i) the large amount of microporosity with pore size of ca. 8 Å and (ii) the polar surface caused by the residual heteroatom-containing (e.g. O, N) species. These dynamic data provide clear evidence that PCN-17 is extremely selective for adsorbing CO_2 over N_2, which represents a significant step forward in rationally designing a material for dilute CO_2 separation in humid conditions. These values are ideally consistent with the pure CO_2 adsorption data at partial pressures of 0.14, 0.09, and 0.04 bar, indicating its extraordinary moisture resistance.

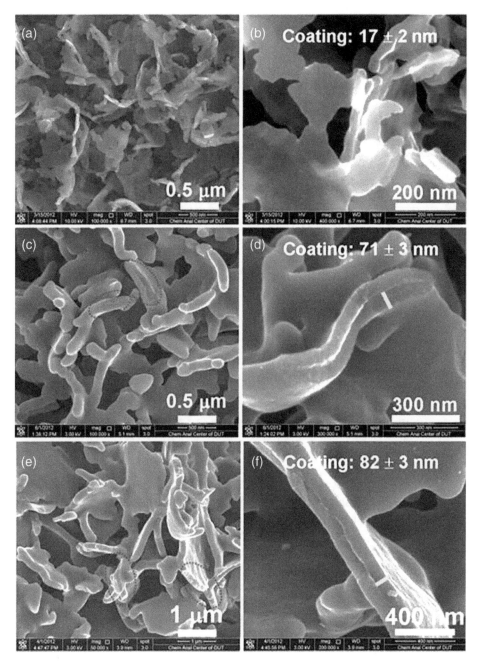

Figure 3.13 FE-SEM images with low and high magnification of the obtained porous carbon nanosheets with different thicknesses: (a, b) PCN-17, (c, d) PCN-71, (e, f) PCN-82. Source: reproduced from Ref. [76] with permission from the Royal Society of Chemistry.

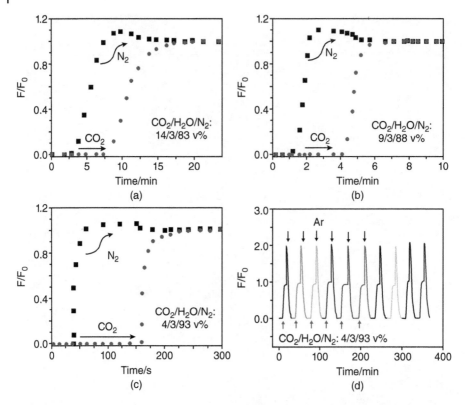

Figure 3.14 CO_2 separation evaluation of PCN-17 in dynamic breakthrough tests. (a–c) Breakthrough curves, using $CO_2/H_2O/N_2$ of (a) 14/3/83, (b) 9/3/88, and (c) 4/3/93 v%, respectively. (d) Cycling of CO_2 separation from a stream of $CO_2/H_2O/N_2$ of 4/3/93 v% at 25 °C, following a regeneration by an Ar purge at 50 °C. Source: reproduced from Ref. [76] with permission from the Royal Society of Chemistry.

Though these carbon sheets show promising CO_2 capture performances, their synthesis still requires a template, and the uniform ultramicropores have not been fully tunable. In order to address these issues, Lu et al. have developed a thermoregulated phase-transition method for the selective synthesis of flat carbon nanoplates with uniform and accessible ultramicropores in the absence of templates [78, 79]. Under this protocol, the 2D structures have more than 80% sp^2 carbon in composition, which allows a unimodal ultramicropore and a controllable thickness. Such 2D features allow quick kinetic gas transfer. When used for CO_2/CH_4 separation from, such carbon nanoplates exhibit a high selectivity of 7 under dynamic flow conditions [80].

3.2.2.3 Carbon Spheres

Porous carbon can also be made as spherical particles that are usually prepared by carbonization of polymer analogues. In this case, polymer precursors are required to be thermally stable and able to form carbon residue after a high-temperature pyrolysis. Phenolic resins derived from the polymerization of phenols (e.g. phenol, resorcinol, phloroglucinol) with aldehyde (e.g. formaldehyde, furfuraldehyde, hexamethylenetetraamine) are attractive due to their excellent performance characteristics such as high temperature resistance, thermal abrasiveness, and high yield of carbon conversion. As a result, varieties of chemical syntheses have been reported for the preparation and CO$_2$ adsorption performance of carbon spheres (CSs) [81–85]. The synthesis of porous carbon nanospheres for CO$_2$ capture is another important topic due to their combined features such as shortened dimensions, increased surface area, easy surface or bulk functionalization, etc. [86, 87].

For example, Jaroniec et al. reported the preparation of a series of CSs by carbonization of phenolic resin spheres via the modified Stöber method. All samples showed spherical morphology with an average diameter of 570, 420, 370, and 200 nm. The particle size decreased with increasing activation time, indicating deterioration of the outer surface of the CS with activation time. The resulting activated CSs have high surface area (730–2930 m^2 g^{-1}), narrow micropores (<1 nm), and, importantly, a high volume of these micropores (0.28–1.12 cm^3 g^{-1}), obtained by CO$_2$ activation of the aforementioned CSs. The remarkably high CO$_2$ adsorption capacities, 4.55 and 8.05 mmol g^{-1}, are measured on these carbon nanospheres at 1 bar and two temperatures, 25 and 0 °C, respectively [88].

Furthermore, it has been known that monodisperse colloidal spheres can self-assemble into three-dimensional periodic colloidal crystals only when their size distributions are less than 5% [89]. That is a particular challenge to establish a new and facile synthesis toward truly monodispersed carbon nanospheres [90]. Wang et al. have established a new synthesis of highly uniform carbon nanospheres with precisely tailored sizes and high monodispersity on the basis of the benzoxazine chemistry [91]. The polybenzoxazine-based spheres can be carbonized with little shrinkage to produce monodisperse CSs with abundant porosity and intrinsic nitrogen-containing groups that make them more useful for CO$_2$ adsorption [92]. The CO$_2$ adsorption capacity can reach 11.03 mmol g^{-1} (i.e. 485 mg g^{-1}) at −50 °C and 1 bar, which is highly desirable for CO$_2$ separation from natural gas feeds during the cryogenic process to produce liquefied natural gas. Moreover, the prepared CSs show the highest adsorption capacity for CO$_2$ per cm^3 micropore volume, when compared with recently reported carbon adsorbents with high CO$_2$ capture capacities at low temperature. The authors also used the CSs as models to investigate the influence of porous structure and surface chemistry on CO$_2$ adsorption behavior. The results indicated that porosity plays an essential role in achieving high CO$_2$ adsorption capacity at ambient pressure, while the nitrogen content of the carbon adsorbent can boost CO$_2$ adsorption capacity at low pressures. This finding may be beneficial for designing sorbents for the separation of dilute CO$_2$-containing gas streams in practical applications.

3.2.3 Template-Free Synthesis

The hard-template and soft-template synthetic strategies have achieved great success in preparing various porous carbons with precise microstructures. But they also have some inevitable limitations: for example, the procedure is tedious due to fabrication of some templates with a special nanostructure or molecular structure, removal of hard templates, and/or post-activation treatment; and many expensive templates are necessarily used. These limitations impart to porous carbons an uncompetitive price-to-performance ratio, as compared with other materials for any given application, and thus limit their commercial viability. Therefore, exploring new template-free preparation methods is urgently needed in the study of porous carbons [93–95].

The template-free synthesis of porous carbon generally involves the transformation of molecular precursors into highly cross-linked organic gels based on sol–gel chemistry [96]. Since the pioneering work of Pekala [97], the polymer-based monolithic carbons have scored remarkable achievements in the new polymerization system and further surface/bulk functionalization. Fairén-Jiménez and coworkers synthesized carbon aerogels with monolith density ranging from 0.37 to 0.87 g cm^{-3} by carbonization of organic aerogels deriving from RF polymers prepared in various solvents such as water, methanol, ethanol, tetrahydrofuran, and acetone solution [98]. They found that samples with a density higher than 0.61 g cm^{-3} had micropores and mesopores but no macropores.

Fu and Wu have studied the template-free fabrication of porous carbons by constructing carbonyl ($-CO-$) crosslinking bridges and $-C_6H_4-$ crosslinking bridges between polystyrene chains [93, 94]. As shown in Figure 3.15, the synthesis adopted linear polystyrene resin as raw material, anhydrous aluminum chloride as Friedel–Crafts catalyst, and CTC as crosslinker and solvent [93]. The $-CO-$ crosslinking bridges have been proven to play a decisive role in achieving good texture inheritability in the process of carbonization. The as-prepared hierarchically porous carbon (HPC) has a unique hierarchical porous texture: 10–30 nm of crosslinking polystyrene-based carbon nanoparticles contain numerous network micropores (<2 nm); their compact and loose aggregation leads to mesopores (2–50 nm) and macropores (50–400 nm), respectively; and these micro-, meso-, and macropores are interconnected to each other and account for 32, 56, and 12%, respectively.

Alternatively, copolymerization and/or cooperative assembly between carbon precursors and one or more additional modifiers (i.e. heteroatom-containing components) can be used to directly synthesize functional carbons with enhanced CO_2 adsorption capability [99]. Sepehri et al. synthesized a series of nitrogen-boron co-doped carbon cryogels by homogenous dispersion of ammonia borane in RF hydrogel during solvent exchange, followed by freeze-drying and pyrolysis. The nitrogen-boron co-doping results in a big improvement of the porous structure, and thus accelerates molecule/ions transport properties as compared to the non-modified carbons [100].

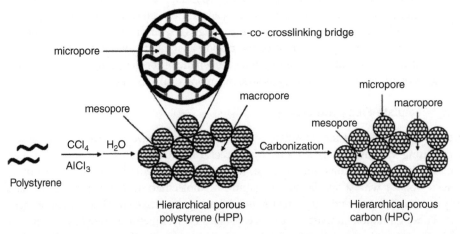

Figure 3.15 Schematic illustration of template-free synthesis of hierarchical porous polystyrene and carbon materials. Source: reproduced from Ref. [93] with permission from the Royal Society of Chemistry.

Lu's group reported a time-saving synthesis toward to a new type N-MC monolith through a sol–gel co-polymerization of resorcinol, formaldehyde, and L-lysine [101]. Based on N_2 sorption, TEM, and SEM results (Figure 3.16), it is clear that this carbon monolith possesses a hierarchical porous structure, i.e. contains both macropores and micropores. This should be advantageous for a CO_2 sorption process, since the macropores provide low-resistance pathways for the diffusion of CO_2 molecules, while the micropores are most

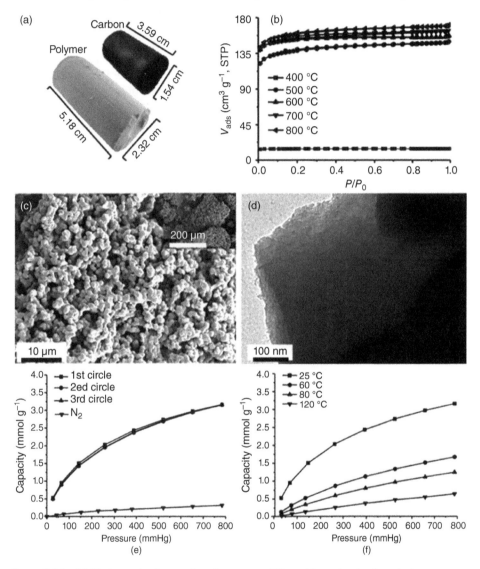

Figure 3.16 (a) Photograph of as-made polymer monolith and its carbonized product. (b) N_2-sorption isotherms of the obtained carbon monolithic pyrolyzed at different temperatures (P/P_0 is the relative pressure). (c) SEM image and (d) TEM image of sample carbonized at 500 °C (the inset in (c) shows an overview of the macroscopic structure). (e) CO_2 multi-circle sorption isotherms and N_2 sorption isotherm for sample at 25 °C. (f) Temperature-dependent CO_2 sorption isotherms. Source: reproduced from Ref. [101] with permission from John Wiley & Sons.

suitable for trapping of CO_2. As expected, such monolithic carbon performs very well in CO_2 capture with a capacity of 3.13 mmol g^{-1} at 25 °C (Figure 3.16e and f). With an increase in adsorption temperature, the adsorption capacities decrease from 3.13 to 1.64, 1.22 to 0.62 mmol g^{-1}, at the corresponding temperatures of 60, 80, and 120 °C, but still are at a high level as compared to its non-doped analogous carbon monoliths. Also note that the decrease of the CO_2 adsorption capacity at high temperature is the common effect of porous carbons. The high uptake at the initial stage of the isotherm may be attributed to the affinity between basic nitrogen groups and CO_2 via acid–base interaction. Interestingly, this series of samples undergo an easy regeneration process, i.e. by argon purge at room temperature.

Gu et al. have developed a template-free synthesis for new type of porous CSs, which show good performance in CO_2 capture. In their synthesis, the azide–alkyne 1,3-dipolar Huisgen cycloaddition reaction was employed for the condensation of 1,4-bis(azidomethyl)benzene and 1,3,5-ethynylbenzene (Figure 3.17) [102]. Because the resulting solid product contains periodically arranged aromatic 1,2,3-triazole rings in the polymer backbone, such carbon precursors contain a large percentage of nitrogen atom sources for the preparation of N-doped carbon materials. More importantly, it may be possible to control the N-doping

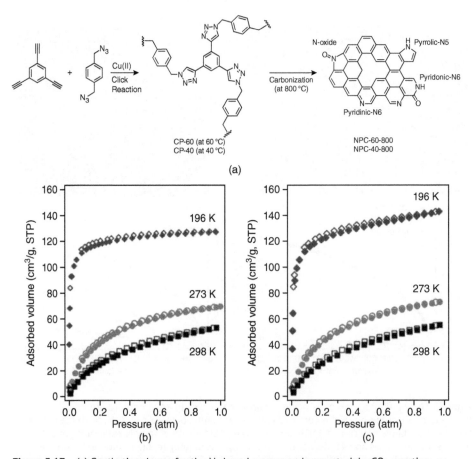

Figure 3.17 (a) Synthetic scheme for the N-doped porous carbon materials. CO_2 sorption isotherms of (b) NPC-60-800 and (b) NPC-40-800 measured at three different temperatures: 196, 273, and 298 K. Source: reproduced from Ref. [102] with permission from Elsevier.

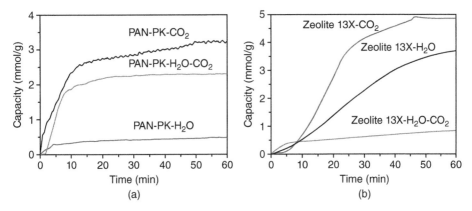

Figure 3.18 The effect of H_2O on the adsorption of CO_2 by (a) PAN-PK and (b) Zeolite 13X at 25 °C. Source: reproduced from Ref. [103] with permission from the Royal Society of Chemistry.

level of products by simply changing the degree of polymerization for the carbon precursors. The respective isosteric heats of adsorption (Q_{st}) values are 48.4 kJ mol⁻¹ for NPC-60-800 and 47.0 kJ mol⁻¹ for NPC-40-800, which supports the key role of the N-doping for stronger interactions with CO_2 molecules in the materials. As expected, the N-contents and surface area can be tuned to 4.30 wt% and 423 m² g⁻¹ after pyrolysis under 800 °C. Based on the developed pores and high N content, the sample can adsorb 126.8 cm³ g⁻¹ (5.66 mmol g⁻¹), 69.6 cm³ g⁻¹ (3.10 mmol g⁻¹) of CO_2 at 196, 273 and 298 K, respectively (Figure 3.17b and c).

Similarly, Shen et al. prepared a series of hierarchical porous carbon fibers with a BET surface area of 2231 m² g⁻¹ and a pore volume of 1.16 cm³ g⁻¹ [103]. In this synthesis, the polyacrylonitrile nanofibers (prepared by dry–wet spinning) were selected as precursors, and pre-oxidation and chemical activation were involved to get the developed porosities. This type of material contained a large amount of nitrogen-containing groups (N content >8.1 wt%) and consequently basic sites, resulting in a faster adsorption rate and higher adsorption capacity for CO_2 than the commercial zeolite 13X that is conventionally used to capture CO_2, in the presence of H_2O (Figure 3.18).

As discussed earlier, the construction of crosslinking bridges has been proven to play a decisive role in obtaining porous carbons using the template method. Under optimized conditions, the as-prepared carbons can have a high BET surface and large pore volume. Moreover, the template-free fabrication method avoids the use of any hard or soft templates, thereby endowing the as-prepared porous carbon with a very competitive price-to-performance ratio. Therefore, it is believed that this simple and effective template-free method will open new avenues for nanostructure design and fabrication of various types of carbon materials from low-cost polystyrenes and other polymers.

3.3 Porous Carbons Derived from Ionic Liquids for CO_2 Capture

The unique features of ILs, such as negligible vapor pressure and versatile solvation properties, have led to the wide use of ILs in many emerging areas: for example, as green solvents for synthesis, as media for advanced separation, and as precursors for porous carbons [104, 105]. Recently, porous carbons made by directly annealing of ILs or using appropriate porous templates have been an emerging field [106–111]. By choosing different ILs, materials with various heteroatoms doping and good pore properties can be produced [112]. The attractive features of IL-derived materials such as facile synthesis, high specific surface area, and nitrogen content make them promising candidates for CO_2 capture. Exceptional CO_2 separation performance can be achieved by these facilely made carbonaceous adsorbents. Thus, in this section, recent research progress on IL-derived carbonaceous materials and their potential CO_2 separation application is summarized.

Traditional carbonaceous adsorbents synthesis involves the carbonization of low-vapor pressure polymeric precursors derived from either synthetic (e.g. polyacrylonitrile [PAN], phenolic resins) or natural sources such as pitch and shell nuts. These polymeric species possess low vapor pressures so that cross-linking reactions can proceed with concomitant char formation and without vaporization of the corresponding precursor units. Nonpolymeric carbon sources are rarely used to form carbon because of their uncontrolled vaporization during high-temperature pyrolysis. Recently, ILs with cross-linkable functional groups, namely task-specific ionic liquids (TSILs), have been considered highly promising precursors for the synthesis of functional carbonaceous materials due to their negligible volatility and molecular tunability. The intrinsic nonvolatility suggests favorable conditions for an intriguing carbonization process based on well-behaved cross-linking reactions of monomeric TSIL precursor units with minimal loss of reactant.

The key structural prerequisite of TSIL precursors is the presence of certain functional groups that can undergo cross-linking reactions under pyrolysis conditions. Given the tunability of TSILs, either cations or anions can be functionalized with cross-linking groups. To date, nitrile groups, the key factors in determining the high carbon yields of PAN under charring conditions, have been mostly appended onto the structure of ILs because of their cyclotrimerization of triazine rings at high temperatures [113, 114]. In addition, TSILs further allow for the preparation of graphitizable carbons with heteroatom-doping (such as nitrogen and boron, with their ratios in the carbon materials controlled by their amounts initially present in the cross-linkable ions).

As shown in Figure 3.19, Dai and co-workers report a strategy for forming functional porous carbon and carbon-oxide composite materials from conventional ILs by confined carbonization [109]. This method does not require the ILs to have cross-linkable groups but instead utilizes the space confinement of ILs inside oxide networks (e.g. silica and titania)

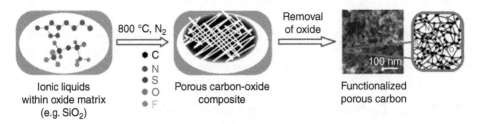

Figure 3.19 Schematic illustration of synthesis of ionic liquids (IL)-derived carbon materials. Source: reproduced from Ref. [109] with permission from John Wiley & Sons. (*See color plate section for color representation of this figure*).

to convert the ILs with no char residue into efficient carbon precursors with carbonization yields approaching the theoretical limits. Another interesting feature of this approach is that it allows a rational tuning of the pore structure of the corresponding carbon-oxide composites and the derived carbon materials, which ranges from microporous to mesoporous and macroporous architectures. Furthermore, other elements, such as nitrogen and boron, which are crucial to modify the physical and chemical properties of the carbon materials, can also be easily doped into carbon frameworks by employing heteroatom-substituted ILs.

Antonietti and Yuan reported a novel type IL-derived carbon with site-specific n-doping and biphasic heterojunction, which showed enhanced CO_2 capture performance. They used a poly(IL) as a "soft" activation agent to deposit nitrogen species exclusively on the surface of commercial microporous carbon fibers. In this configuration, this type of carbon-based biphasic heterojunction amplifies the interaction between carbon fiber and CO_2 molecule for unusually high CO_2 uptake (30 wt%). Their protocol allows for the simultaneous optimization of surface functionalities and the porous structure of carbon, and the proposed carbon/carbon heterojunction is believed to serve as a useful structural tool for solving material-based problems [115].

The formation of eutectic mixtures (described as either deep eutectic solvents [DESs] by Abbott and coworkers [116], or low transition-temperature mixtures [LTTMs] by Kroon and coworkers [117]) using some of the most typical synthetic precursors for carbon preparation has also opened interesting perspectives in this yield. The use of DESs is attractive because, when compared to nitrile-containing TSILs, they are less expensive and easy to prepare, owing to a wide range of compounds such as regular carbonaceous precursors (e.g. resorcinol). For instance, the use of DESs based on mixtures of resorcinol (R) and choline chloride (ChCl) has allowed, upon polycondensation with formaldehyde and without the use of further additives, the formation of monolithic carbons consisting of highly cross-linked clusters that aggregated and assembled into a stiff and interconnected hierarchical structure [118]. The application of resorcinol is by no means trivial, because it provides high carbonization yields (up to ca. 85 wt%). This feature, in addition to the capability of recovering the second component of DES that is not involved in carbon formation (e.g. choline chloride), makes synthetic processes based on DESs especially attractive in terms of efficiency and sustainability [119, 120]. These advantages may allow for efficient synthesis of CO_2 adsorbents with good CO_2 separation performance.

For example, del Monte and his co-workers have further explored the versatility of DES-assisted syntheses for the preparation of hierarchical N-MC molecular sieves [121]. DESs were composed of resorcinol and 3-hydroxypyridine (as hydrogen donors) and tetraethylammonium bromide (as a hydrogen acceptor) so that 3-hydroxypyridine acted as the nitrogen source of the resulting carbons while the presence of tetraethylammonium bromide determined the formation of a molecular sieve structure. The final carbon exhibited CO_2 adsorption capacities of up to 3.7 mmol g^{-1} and CO_2-N_2 selectivities of up to 14.4 from single component gas data, 95.9 in the Henry law regime, and 63.1 from ideal adsorption solution theory (IAST) simulations. It is worth noting that the main factor governing the CO_2 adsorption capacity was the micropore volume, while nitrogen content played a more critical role for the preferential adsorption of CO_2 versus other gases.

Further, using deep eutectic salts either as solvents or as carbonaceous precursors and structure-directing agents, Monte's group prepared carbon monoliths with high yield (80%) and tailored mesopore diameters [122, 123]. Sotiriou-Leventis, Leventis, and coworkers, in recent years, have developed several new polymerization systems such as isocyanate-cross-linked RF gels, polyurea gels, polyimide gels, etc., which offer a high degree of flexibility in producing monolithic carbons [124–126]. The carbon products show interconnected hierarchical pore networks and 3D bicontinuous morphology, high surface area, and large pore volume. For example, polyurea (PUA) gels, which eventually convert to highly porous (up to 98.6% v/v) aerogels over a very wide density range, can be prepared by carefully controlling the relative Desmodur RE (isocyanate)/water/triethylamine (catalyst) ratios in acetone. These materials are worth further exploration of their applications as CO_2 capture materials in future research.

N-MCs were obtained by polycondensation of resorcinol and 3-hydroxypyridine with formaldehyde and subsequent carbonization [119]. Moreover, by changing the composition of DESs, the pore architectures of the synthesized carbon monoliths can also be facilely modified. The morphology of the resulting carbons (e.g. $C_{RHC221\text{-}DES}$ and $C_{RHC111\text{-}DES}$, no matter the thermal treatment) consisted of a bicontinuous porous network built of highly cross-linked clusters that aggregated and assembled into a stiff, interconnected structure (Figure 3.20). The combination of N-MC and high microporous surface area made them especially suitable for CO_2 capture, with outstanding CO_2-adsorption capacities (up to 3.3 mmol g^{-1}) in addition to remarkable recyclability and stability.

Overall, features such as facile and low-cost synthesis of carbonaceous materials from ILs open interesting perspectives for the application of carbons in separation technologies for CO_2 low-pressure post-combustion processes and natural gas sweetening.

3.4 Porous Carbons Derived from Porous Organic Frameworks for CO_2 Capture

Porous organic frameworks (POFs), especially MOFs or porous coordination polymers (PCPs), have attracted much attention as a new type of carbon precursor for porous carbons due to their many merits [127–131]. Considering the large carbon content in the organic ligands of MOFs, they could be used as both templates and carbon precursors at the same time without the need for any additives [132, 133]. Moreover, the pore characteristics of POFs play a crucial role in determining the pore texture of the resultant porous carbon. Direct transformation of POF-related materials (PCPs, MOFs, ZIFs, PAFs, etc.) can generate porous carbon with nanopores of a precise and uniform size, which are very important in selective capture of CO_2 [134, 135].

Motivated by a unique system of "direct conversion of MOFs or PCPs," Yamauchi et al. have pioneered the synthesis of a novel NPC with highly developed porosity (5500 m^2 g^{-1} surface area and 4.3 cm^3 g^{-1} pore volume) by a direct carbonization of Al-PCPs (Al(OH)(1,4-NDC)·2H$_2$O) [132]. The obtained carbon materials had a similar fiber-like morphology. It is noteworthy that such a fiber-like morphology was retained even after leaching of metal species. In some parts, large cracks/voids were formed after the calcination followed by HF treatment, which were most likely caused by the large weight loss during thermal decomposition of organic components of the Al-PCPs (Figure 3.21a and b). By applying the appropriate carbonization temperature, both high surface area (5500 m^2 g^{-1})

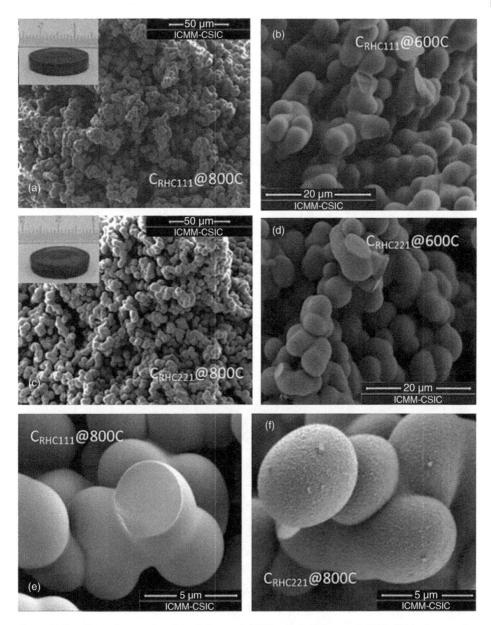

Figure 3.20 SEM micrographs of (a, e) C_{RHC111}@800C and (c, f) C_{RHC221}@800C. SEM micrographs of (b) C_{RHC111}@600C and (d) C_{RHC221}@600C. Insets in (a) and (c) show pictures of the $C_{RHC111-DES}$@800C and $C_{RHC221-DES}$@800C monoliths. Source: reproduced from Ref. [119] with permission from the Royal Society of Chemistry.

and large pore volume (4.3 cm³ g⁻¹) are realized for the first time, and the porosity is much higher than other carbon materials (such as ACs and mesoporous carbons).

By performing the direct carbonization of MOFs, Chaikittisilp [136], Yang [137], and Zou [134] et al. have synthesized NPC materials that exhibited promising gas storage

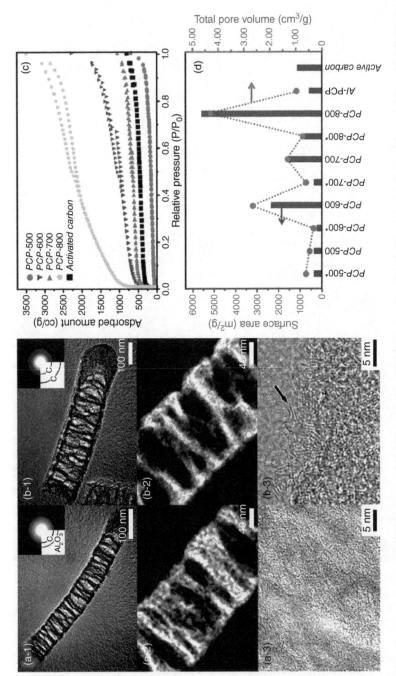

Figure 3.21 Bright- and dark-field TEM images of Al-PCP calcined at 800 °C (a-1, a-2, a-3) before and (b-1, b-2, b-3) after washing with HF. The insets in (a-1) and (b-1) are the corresponding ED pattern. (c) N_2 adsorption–desorption isotherms for the obtained samples. (d) Summary of surface areas and total pore volumes for the obtained samples. For comparison, the samples before HF treatment are also shown, as indicated by (*). Source: reproduced from Ref. [132] with permission from the American Chemical Society.

performance. Porous carbons with hierarchical pore structures could also be achieved by direct carbonization of the selected isoreticular metal-organic frameworks (IRMOF-1, -3, and -8), indicating the tunable pore characteristics of MOF-derived NPCs [137]. Ariga et al. reported a NPC with high surface area (up to $1110\,m^2\,g^{-1}$) and narrow pore-size distributions that are close to its parent ZIF-8 [136]. By using three types of ZIFs as a precursor without any additional carbon sources, ZIF-8, ZIF-68, and ZIF-69 with different topological structures and functional imidazolate-derived ligands have been directly carbonized to prepare porous carbon materials at 1000 °C [134]. The BET surface areas of the carbon materials that were activated with fused KOH are 2437 (CZIF8a), 1861 (CZIF68a), and $2264\,m^2\,g^{-1}$ (CZIF69a). CZIF69a has the highest CO_2 uptake of $4.76\,mmol\,g^{-1}$ at 1 atm and 273 K, owing to its local structure and pore chemical environment. Based on all of these reported results, MOFs are really strong candidates as templates or precursors for the preparation of porous carbon materials with high surface areas and large pore volumes.

A family of NPCs has also been prepared by thermal decomposition of guest-free MOFs (even non-porous MOFs) by Kim and colleagues [138]. They found that the porosity of the carbon materials depend linearly on the Zn/C ratio of MOFs precursors, which allows precise control of the porosity of the carbon materials in a predictable manner. Using this strategy, Gadipelli Srinivas has reported a new type of HPC structure with simultaneously high surface area and high pore volume, which was synthesized from carefully controlled carbonization of in-house optimized MOFs (Figure 3.22) [139]. The HPCs were studied for carbon dioxide capture at high pressure and temperature, up to 30 bar and 75 °C. This new class of HPCs (HPC5, HPC74, HPC53) are derived from the direct carbonization of three different MOF structures (MOF-5, MOF-74, MIL-53).

Changes in synthesis conditions led to millimeter-sized MOF-5 crystals in a high yield. Subsequent carbonization of the MOFs yielded HPCs with simultaneously high surface area, up to $2734\,m^2\,g^{-1}$, and exceptionally high total pore volume, up to $5.53\,cm^3\,g^{-1}$. In the HPCs, micropores are mostly retained, and meso- and macropores are generated from defects in the individual crystals, which is made possible by structural inheritance from the MOF precursor. The SEM images presented in Figure 3.23 reveal that the HPCs retain mostly similar crystallite shapes and surface structures to their MOF precursors. It is also noted that the parent MOF crystallites shrink after carbonization, as evident from the microcracks and the sponge-like surface morphology. The resulting HPCs show a significant amount of CO_2 adsorption, over $27\,mmol\,g^{-1}$ (119 wt%) at 30 bar and 27 °C, which is one of the highest values reported in the literature for porous carbons. The findings are comparatively analyzed with the literature.

Through a simple pyrolysis of crystalline polymer (PAF-1), Qiu et al. have prepared a series of NPCs having high surface area and narrow micropore size distributions [140]. The carbonized (at 450 °C) sample PAF-1-450 showed excellent adsorption capacities ($4.5\,mmol\,g^{-1}$) for CO_2 compared to the original PAF-1 at ambient conditions (Figure 3.24). These aforementioned works exemplify a striking indication that the facile, one-step pathway replicated from crystalline microporous materials is highly efficient toward highly nanoporous carbons.

Recently, Yan et al. reported a space-confined synthesis of ZIF-67 nanoparticles (NPs) in hollow carbon nanospheres (ZIF-67@HCSs) for CO_2 adsorption. In their method, direct

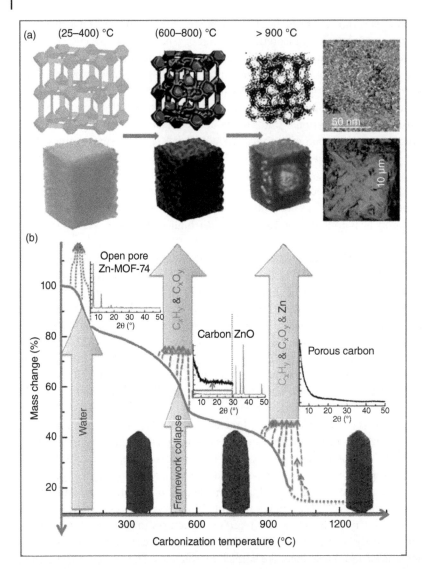

Figure 3.22 (a) Schematic illustration for steps in fabricating hierarchically porous carbons (HPCs) from a precursor metal organic framework (MOF)-5; (b) mechanism involved in the carbonization process of Zn-MOF-74. The plot represents the MOF mass change vs. carbonization temperature. Source: reproduced from Ref. [139] with permission from *Energy & Environmental Science*.

pyrolysis of polystyrene@polypyrrole composite nanospheres lead to the formation of hollow carbon spheres (HCSs). The introduction of ZIF-67 in the hollow voids of HCSs was through the melting-diffusion strategy by subsequent infiltration of 2-methylimidazole and $Co(NO_3)_2$ into the HCSs, which initiated the coordination between the ligand of 2-methylimidazole and metal ions of Co^{2+}. The resultant ZIF-67@HCSs show integrated pore structures and high pore volume, and thus exhibit outstanding adsorption capacity for CO_2 [141].

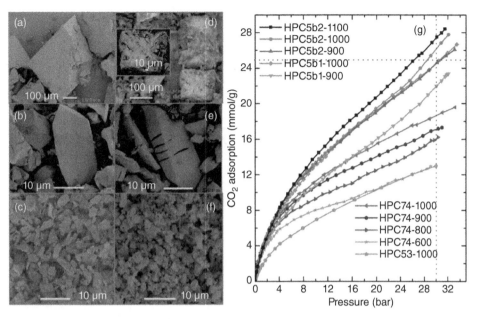

Figure 3.23 SEM images of (a) metal organic framework (MOF)-5b1, (b) MOF-74, (c) MIL-53 and their hierarchically porous carbons (HPCs): (d) HPC5b1–1000, (e) HPC74–1000, (f) HPC53–1000. (g) CO$_2$ adsorption isotherms of HPCs measured at 27 °C. The numbers between 600 and 1100 represent the carbonization temperature (unit: °C). HPC74-600, HPC74-800, and HPC53-1000 are after acid treatment. Source: reproduced from Ref. [139] with permission from *Energy & Environmental Science.*

Unambiguously, the controlled carbonization of porous organic polymers, a facile approach to replace high-temperature treatment and chemical activation, opens up an avenue to construct hierarchically porous frameworks of carbon from porous MOFs, which shows great potential in gas storage and electrical applications. We believe that this approach provides a simple way of preparing more hierarchically porous materials for application in the fields of CCS.

3.5 Porous Carbons Derived from Sustainable Resources for CO$_2$ Capture

Another new but rapidly expanding research area is to produce porous carbons from sustainable resources (e.g. collagen, cellulose, starch) [142–144]. The production of carbon materials from biomass (sugars, polysaccharides, etc.) is also a relatively new but rapidly expanding research area. A wide range of biomass precursors, such as coconut husk, bamboo, wood, peat, cellulose, and lignite can be used for fabrication of porous carbons. Interestingly, for this category, as new precursors are discovered, new types of ACs can be created through carbonization and activation. For example, the precursor ranges can be extended to microorganism, celtuce leaves, fungi, algae, bean dregs, and so on [145–147]. And thus, such a carbon "family" is enriched and will be further expanded due to the widely available carbon precursors, and their high effectiveness in CO$_2$ capture.

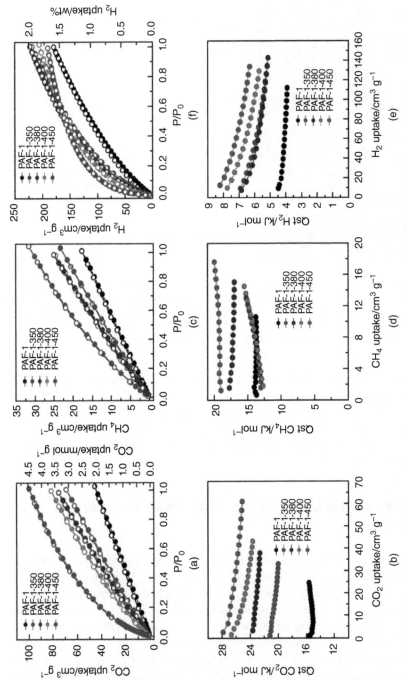

Figure 3.24 (a) CO_2 adsorption (solid symbols) and desorption (open symbols) isotherms of PAF-1 and carbonized samples at 273 K; (b) Q_{st} CO_2 of PAF-1 and carbonized samples as a function of the amount of CO_2 adsorbed. (c) CH_4 adsorption (solid symbols) and desorption (open symbols) isotherms of PAF-1 and carbonized samples at 273 K; (d) Q_{st} CH_4 of PAF-1 and carbonized samples as a function of the amount of CH_4 adsorbed. (e) H_2 adsorption (solid symbols) and desorption (open symbols) isotherms of PAF-1 and carbonized samples at 77 K; (f) Q_{st} H_2 of PAF-1 and carbonized samples as a function of the amount of H_2 adsorbed. Source: reproduced from Ref. [140] with permission from the Royal Society of Chemistry. *(See color plate section for color representation of this figure).*

Up to now, many kinds of AC have been well commercialized in gas sorption/separation including CO$_2$ capture [148–151]. For example, the BPL AC with a specific area of 1141 m^2 g^{-1} is able to adsorb 7 mmol g^{-1} CO$_2$ under conditions of 25 °C and 35 bar; while under the same conditions, MAXSORB AC with a specific area of 3250 m^2 g^{-1} can capture up to 25 mmol g^{-1} [152]. In this section, we summarize and compare two recently reported routes to the preparation of porous carbon materials derived from sustainable resources, which mainly are divided into the direct pyrolysis and/or activation method and the sol–gel process and hydrothermal carbonization method.

3.5.1 Direct Pyrolysis and/or Activation

The direct pyrolysis combined activation process is considered a promising approach for the fabrication of porous carbons from renewable sources, such as waste culture leaves, bamboo, etc. [153]. As shown in Table 3.1, a series of sustainable nitrogen-containing granular porous carbons with developed porosities and controlled surface chemical properties were prepared from sustainable resources [157]. For example, waste celtuce leaves were used to prepare porous carbons by air-drying, pyrolysis at 600 °C in argon, followed by KOH activation. The as-prepared porous carbons show a very high specific surface area of 3404 m^2 g^{-1} and a large pore volume of 1.88 cm^3 g^{-1}. They show an excellent CO$_2$ adsorption capacity at 1 bar, which is up to 6.04 and 4.36 mmol g^{-1} at 0 and 25 °C, respectively.

Wang et al. reported a series of porous carbons with adjustable surface areas and narrow micropore size distribution by KOH activation of fungi-based carbon sources [161]. The high CO$_2$ uptake of 5.5 mmol g^{-1} and CO$_2$/N$_2$ selectivity of 27.3 at 1 bar, 0 °C of such fungi-based carbons made them promising for CO$_2$ capture and separation. Similarly, Shen and co-workers found that yeast is a promising carbon precursor for the synthesis of hierarchical microporous carbons, which show a high CO$_2$ adsorption capacity (4.77 mmol g^{-1}) and a fast adsorption rate (equilibrium within 10 minutes) at 25 °C [162]. This may stem from their large surface area and hierarchical pore systems as well as the surface-rich basic sites.

To avoid several disadvantages of fabricating carbon materials reported in the literature, Fan et al. employed chitosan, which is the second-largest nitrogenous natural

Table 3.1 Comparison of structure textures and CO$_2$ adsorption performance of porous carbon adsorbents derived from sustainable resources.

Sustainable resource	S$_{BET}$ (m^2/g)	V$_{total}$ (cm^3/g)	CO$_2$ uptake^a (mmol/g)	Selectivity (CO$_2$/N$_2$)	Ref.
Chitosan	1381	0.57	3.86	21	[154]
Bamboo	930	0.41	4.0	11.1	[155]
Sawdust	1260	0.62	4.8	5.4	[156]
Poplar anthers	1123	0.47	4.12	NG^b	[157]
Eucalyptus wood	2079	1.292	3.22	NG	[158]
London plane leaves	2000	1.0	4.41	NG	[159]
Lignocellulosic fiber	355.4	0.232	2.43	18.7	[160]

Note: a. 25 °C, 1 atm. b. NG: not given.

organic matter only after protein, as a carbon precursor and K_2CO_3 as a mild activator to prepare CO_2 sorbents by a simple chemical activation method [154]. The textural and chemical properties of the porous carbons could be easily tuned by changing the ratio of K_2CO_3/chitosan and activation temperature. Due to their large pore volume, well-defined microporosity, and relatively high nitrogen content, these porous carbons were applied as adsorbents for CO_2 capture and demonstrated excellent CO_2 uptake performance. In particular, the sample prepared at 635 °C with K_2CO_3/chitosan ratio = 2 shows a CO_2 uptake as high as 3.86 mmol g^{-1} at 25 °C, 1 atm. Furthermore, the CO_2 uptake remains almost constant in five consecutive adsorption–desorption cycles, indicating this material has great stability and recyclability as a CO_2 sorbent. In addition, an extraordinary separation selectivity against N_2 (CO_2/N_2 selectivity of ca. 21) was observed.

3.5.2 Sol–Gel Process and Hydrothermal Carbonization Method

One successful example is polysaccharide-derived Starbon carbon, which exhibits outstanding mesoporous textural properties. More importantly, its pore volumes and sizes are comparable to materials prepared via the hard-template routes or soft-template methods based on the self-assembly and polymerization of aromatic precursors (e.g. phenols). In this technology, three main stages are involved. Selected precursors are first gelatinized by heating in water. Then, the water inside the gel is exchanged with the lower-surface-tension solvent (e.g. ethanol). After drying, the porous gel is doped with a catalytic amount of acid and pyrolyzed under vacuum, resulting in highly porous carbons.

The sol–gel method is indeed a simple and direct approach for the synthesis of bulky carbons, and is already widely used in both the laboratory and in industry. However, the major disadvantage is the long synthesis period and the rigorous drying process of the wet gel (i.e. solvent exchange or supercritical drying), in which slight variations may cause drastic variations in the structural features, and hence properties [163]. In addition, pore blocking and sometimes uncontrolled dispersion of active sites both on the surface and in carbon pore walls remain to be solved.

Concurrently with the Starbon technology, Antonietti et al. have been particularly active in the development and production of useful carbonaceous materials from sugar-based biomass via a hydrothermal carbonization approach (HTC, Figure 3.25) [165]. HTC is a spontaneous, exothermic process, producing materials where the majority of the original carbons are incorporated into the final structure. The initial products of the sugar dehydration (e.g. furfuryl derivatives) are thought to polymerize to form condensed spherical functional carbons after hydrothermal processing at 180 °C for 20–24 hours. Manipulation of particle size was possible via the utilization of different sugar-based carbon sources, whilst the surface and bulk chemical structure of the material may be directed by the utilization of hexose- or pentose-based biomass, as demonstrated by ^{13}CP/MAS NMR investigations [166]. HTC is relatively straightforward, affording small colloidal CSs, the surface texture and chemistry of which can be controlled via the introduction of co-monomers and selection of biomass precursor. However, HTC materials demonstrate low or negligible surface areas, very small particle size, and little developed or structured porosity.

The CO_2 adsorption behavior of HTC-based porous carbons has recently been investigated [164]. For example, Sevilla et al. reported a series of sustainable porous carbon

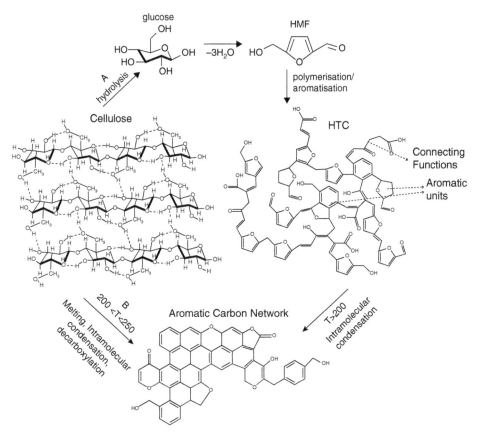

Figure 3.25 Conversion of cellulose into HTC: (a) via HMF, resulting in a furan-rich aromatic network; (b) via direct aromatization. Source: reproduced from Ref. [164] with permission from the Royal Society of Chemistry.

capture materials that are produced from the chemical activation of hydrothermally treated precursors (polysaccharides and biomass) using KOH as an activating agent [156]. The CO_2 adsorption properties, kinetics, and regeneration of these materials were investigated. Compared to raw HTC materials, the chemical-activated counterparts show a significant increase of micropores, delivering a high surface area of 1260 and 2850 $m^2\,g^{-1}$ depending on the activation conditions. The CO_2 capture properties at 0, 25, and 50 °C and 1 bar are shown in Table 3.2. As can be seen, these HTC-based porous carbons show a high capacity even up to 4.8 $mmol\,g^{-1}$ at 25 °C and 1 atm. The researchers found that the remarkable CO_2 capture capacity is due to the presence of rich, narrow micropores (<1 nm), and the surface area plays a less-important role. More interestingly, they found that the type of porous carbons showed very fast adsorption kinetics. Approximately 95% CO_2 uptake can be achieved in two minutes. Under the same conditions, the N_2 adsorption uptake is one-ninth that of CO_2, indicating a CO_2/N_2 selectivity of ca. 9.

Subsequently, Sevilla and co-workers prepared highly porous N-doped carbon through chemical activation of hydrothermal carbon derived from mixtures of algae and

Table 3.2 CO_2 capture capacities of the porous carbons at different adsorption temperatures and 1 atm [156].

HTC precursor	Chemical activation		CO_2 uptake, mmol g^{-1} (mg g^{-1})		
	T (°C)	KOH/HTC	0 °C	25 °C	50 °C
Starch	700	4	5.6 (247)	3.5 (152)	2.2 (196)
Cellulose	700	4	5.8 (256)	3.5 (155)	1.8 (79)
Eucalyptus sawdust	600	4	5.2 (230)	2.9 (128)	—
	700	4	5.5 (243)	2.9 (128)	1.8 (79)
	800	4	5.2 (227)	3.0 (130)	—
	600	2	6.1 (270)	4.8 (212)	3.6 (158)
	650	2	6.0 (262)	4.7 (206)	3.3 (145)
	700	2	6.6 (288)	4.3 (190)	2.6 (116)
	800	2	5.8 (255)	3.9 (170)	3.1 (136)

glucose [167]. They demonstrated that control of the activation conditions (temperature and amount of KOH) allows the synthesis of exclusively microporous biomass-based materials. These materials possess surface areas in the 1300–2400 m^2 g^{-1} range and pore volumes up to 1.2 cm^3 g^{-1}. They additionally exhibited N content in the 1.1–4.7 wt% range with the heteroatoms mainly present as pyridone-type structures. When tested as CO_2 sorbents at sub-atmospheric conditions, they show a large CO_2 capture capacity of up to 7.4 mmol g^{-1} at 0 °C and 1 bar, among the highest for porous materials. However, the results indicate that the large CO_2 capture capacity is exclusively due to their high volume of narrow micropores and not to the high surface areas or pore volumes, nor to the presence of heteroatoms.

Similarly, Sevilla et al. reported a chemically activated synthesis (KOH as activating agent) of highly porous N-doped carbons for CO_2 capture [168]. In their synthesis, polypyrrole (PPy) was selected as carbon precursor. The activation process was carried out under severe (KOH/PPy = 4) or mild (KOH/PPy = 2) activation conditions at different temperatures in the 600–800 °C range. Mildly ACs have two important characteristics: (i) they contain a large number of nitrogen functional groups (up to 10.1 wt% N) identified as pyridonic-N with a small proportion of pyridinic-N groups; and (ii) they exhibit, in relation to the carbons prepared with KOH/PPy = 4, narrower micropore size. These two properties ensure the mildly ACs have large CO_2 adsorption capacities. Furthermore, the capture of CO_2 in this type of carbon takes place at high adsorption rates, with more than 95% of the CO_2 being adsorbed in ca. two minutes. In contrast, N_2 adsorption occurs at slower rates; approximately 50 minutes are necessary to attain maximum adsorption uptake (0.77 mmol N_2 g^{-1}).

In summary, carbons and carbonaceous materials synthesized from carbohydrates, biomass, and other sustainable sources have been discussed. It is believed that such natural source–derived carbons will play an increasingly important role in the fabrication of nanostructured carbon materials in the future. Heteroatoms (especially nitrogen) can be successfully incorporated within the framework of these materials, and their chemical environment can also be controlled with post-annealing temperatures. Such dopants confer very special electronic as well as adsorption capability to the porous carbons.

3.6 Critical Design Principles of Porous Carbons for CO$_2$ Capture

For CO$_2$ capture, the adsorption capacity and selectivity of the adsorbent are the central issues. The critical parameters of adsorption capacity are pore size/volume and surface features of an adsorbent. Compared to other porous materials, porous carbons continue to attract much attention because of their outstanding thermal (in inert atmosphere) and chemical stability, as well as tunable pore surface functionalization/modification. In this section, several important aspects related to this topic are introduced based on current research progress, including: (i) pore structures, (ii) surface chemistry, (iii) crystalline degree, and (iv) functional integration and reinforcement.

3.6.1 Pore Structures

Appropriate pore structure is always the first consideration in the design and synthesis of porous material sorbents for selective gas separation. Considering the key points that influence the capture efficiency, i.e. adsorption capacity, kinetics, and selectivity, the following design principles should be considered. (i) The adsorbents should have abundant microporosity with suitable size matching the CO$_2$ molecules and well-dispersed surface functional groups that can polarize the CO$_2$ molecules. (ii) The pore size, quantity, and mutual matching degree and mutual interconnected modes of micro-, meso-, macropores should be considered for a highly interconnected hierarchical pore system that determines the adsorption diffusion dynamics of CO$_2$. Consequently, a primary focus and challenge has been to prepare porous carbons with multiscale, controllable, highly interconnected pores, tailored pore surface, and good mechanical properties suitable for actual application.

The porosity of porous carbon can be partially controlled by selecting particular precursor chemistry, activation method, and conditions. Activation method is the first choice of researchers. Current methods for the preparation of porous carbons are often classified into two categories: physical (or thermal) activation and chemical activation. Synthesis of ACs by physical activation commonly involves two steps: carbonization of a precursor (removal of noncarbon species by thermal decomposition in inert atmosphere) and gasification (development of porosity by partial etching of carbon during annealing with an oxidizing agent, such as CO$_2$, H$_2$O, or a mixture of both). In some cases, low-temperature oxidation in air (at temperatures of 250–350 °C) is performed on polymer precursors to increase the carbon yield.

To discuss the key parameters of materials for CO$_2$ capture, a type of model carbon, carbide-derived carbon (CDC), has been fabricated. The most attractive aspects of these materials lie in their precisely controlled micropore size (with a sub-angstrom accuracy) and tunable specific surface area (up to 3200 m^2 g^{-1}) [169–171]. These characteristics allow them to serve as carbon models for the fundamental investigation of the influence of micropore size on CO$_2$ adsorption. Gogotsi et al. systematically investigated CO$_2$ adsorption at atmospheric and sub-atmospheric pressures at near-ambient temperature (0 °C) on the basis of a series of CDCs with well controlled pore size distribution (PSD) and surface areas synthesized from TiC powders [170]. They found that the average pore size and the total pore volume are not adequate measures to predict the CO_2 uptake of microporous carbon sorbents, and the pore volume of micropores strongly governs the amount of adsorbed CO_2. Neither high-surface-area CDC after chemical activation (surface area: 3101 m^2 g^{-1}) nor high-pore-volume nano-TiC-CDC (V_{total}: 1.61 cm^3 g^{-1}) correspond with the highest CO_2 sorption capacity. At ambient pressure, the CO_2 uptake closely follows a linear correlation with the volume of pores smaller or equal to a diameter of 1.5 nm. Pores smaller than 0.5 nm contribute to the amount of adsorbed CO_2, but the best correlation is found for pore volume smaller than 0.8 nm (Figure 3.26). The correlation between the amount of adsorbed CO_2 at low partial pressures and the volume of smaller pores is the basis for the well-known application of CO_2 sorption as a method to calculate the pore characteristics of microporous materials. Sub-atmospheric pressures are of particular interest for industrial applications, where partial pressure of CO_2 is below 1 bar, and here the best prediction of the CO_2 uptake capacity at 0.1 bar would be based on the volume of pores smaller than or equal to a diameter of 0.5 nm (Figure 3.26). This correlation can be used to design better CO_2 sorbents and CCS devices.

The important role of micropores on CO_2 capture has been widely accepted. Therefore, in addition to the previously mentioned approaches, i.e. hard-template methods (e.g. zeolites as templates), CDC pathways, and so on, new strategies to fabricate such pore systems are highly anticipated. For instance, Lu et al. report a novel synthesis approach for the fabrication of hierarchical carbon materials (HCMs) by using discrete chelating-zinc species as dynamic molecular porogens to create extra micropores that enhance their CO_2 adsorption capacity and selectivity [172]. During the carbonization process, the evaporation of the in situ formed Zn species would create additional nanopaths that contribute to the additional micropore volume for CO_2 adsorption (Figure 3.27).

The resulting HCMs show an increased number of micropores with sizes in the range 0.7–1.0 nm, and a high CO_2 adsorption capacity of 5.4 mmol·g^{-1} (23.8 wt%) at 273 K and 3.8 mmol·g^{-1} (16.7 wt%) at 298 K and 1 atm, which are superior to most carbon-based adsorbents with N-doping or high specific surface area. As shown in Figure 3.28, the dynamic gas separation measurement, using 16% (v/v) CO_2 in N_2 as feedstock, demonstrates that CO_2 can be effectively separated from N_2 under ambient conditions and shows a high separation factor ($S_{CO2/N2}$ = 110) for CO_2 over N_2, reflecting a strongly competitive CO_2 adsorption capacity. When the feedstock contains water vapor, the dynamic capacity of CO_2 is almost identical to that measured under dried conditions, indicating the carbon material has an excellent tolerance to humidity. An easy CO_2 release can be realized by purging an argon flow through the fixed-bed adsorber at 298 K, indicating good regeneration ability.

Yuan et al. reported a new type of spherical nitrogen-containing polymer and microporous carbon material for CO_2 adsorption analysis [173]. The microporous CSs exhibit a high surface areas of 528–936 m^2 g^{-1} with a micropore size of 0.6–1.3 nm. The synthesized microporous carbons show a good CO_2 capture capacity, which is mainly due to the presence of nitrogen-containing groups and a large number of narrow micropores (<1.0 nm). At 1 atm, the equilibrium CO_2 capture capacities of the obtained microporous carbons are in the range of 3.9–5.6 mmol g^{-1} at 0 °C and 2.7–4.0 mmol g^{-1} at 25 °C. Further, the authors normalized the CO_2 capture capacities in accordance with narrow micropore volume and

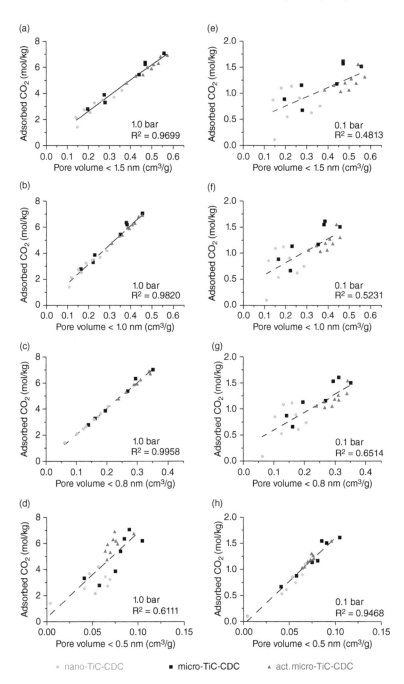

Figure 3.26 CO_2 uptake at 0 and 1.0 bar (a–d) and 0.1 bar (e–h) for the volume of pores smaller 1.5 nm (a, e), 1.0 nm (b, f), 0.8 nm (c, g), and 0.5 nm (d, h). Source: reproduced from Ref. [170] with permission from the Royal Society of Chemistry.

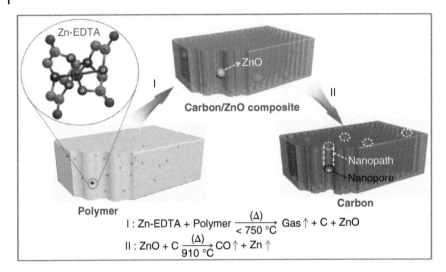

Figure 3.27 Synthesis principle of microporous carbon adsorbent using zinc species as dynamic molecular porogens for the creation of abundant microporosity. Source: reproduced from Ref. [172] with permission from John Wiley & Sons.

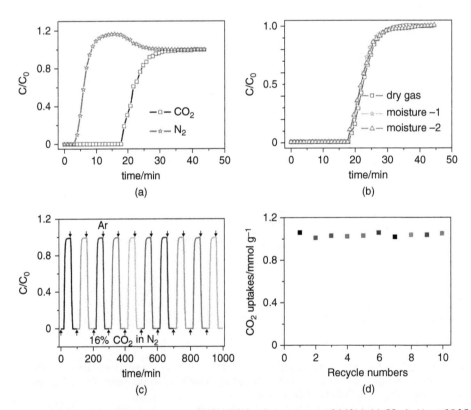

Figure 3.28 (a) Breakthrough curve of HCM-ZC-1 using a stream of 16% (v/v) CO_2 in N_2 at 25 °C; (b) breakthrough curve of CO_2 under a moisture condition; (c) recycle runs of CO_2 adsorption–desorption on HCM-ZC-1 at 25 °C, using a stream of 16% (v/v) CO_2 in N_2, followed by a regeneration under argon flow; (d) CO_2 uptake of each adsorption run. Source: reproduced from Ref. [172] with permission from John Wiley & Sons.

Table 3.3 CO_2 adsorption capacities of prepared adsorbents at 25 °C, normalized by pore volume, narrow micropore volume, and nitrogen content [173].

Sample	Normalized CO_2 capture capacities at 25 °C	
	W0 (mmol CO_2 cm^{-3})	N content (mmol CO_2 [mmol N] $-^1$)
HMT-80-600	15.0	2.88
HMT-80-700	14.2	2.07
HMT-80-800	13.8	1.62
HMT-80-900	11.8	2.80
HMT-80-1000	11.8	4.31

nitrogen content (Table 3.3), with the aim to determine the influence of both textural and surface chemistry properties on the capture performance. The normalization of the CO_2 capture capacities by the narrow micropore volume shows the effect of surface chemistry properties of the samples on the CO_2 uptake, while the normalization by nitrogen content exhibits the contribution of structure properties. From the results, one can assume that samples prepared from low treatment temperature, i.e. 600 °C, exhibit the greatest capacity per narrow micropore volume, while high-temperature pyrolyzed samples show increased contribution by the micropores.

In contrast to these approaches, some studies demonstrate that the structural modification of carbon composites by carbon nanotubes (CNTs) is also an effective way to create narrow micropores. For example, Su et al. reported a new type of CNT-modified carbon monolith that was prepared from a commercial phenolic resin mixed with just 1 wt% of CNTs followed by carbonization and physical activation with CO_2 [174]. The products possess a hierarchical macroporous–microporous structure and superior CO_2 adsorption properties. In particular, they show the top-ranked CO_2 capacity (52 mg CO_2 per g adsorbent at 25 °C and 114 mmHg) under low CO_2 partial pressures, which is of more relevance for flue gas applications. In contrast to established understanding, North group found the effective role of mesoporous pores in improving both selectivity and capacity [175]. They compared mesoporous bio-resource-derived carbon with Norit activated charcoal (AC) for CO_2 adsorption. Mesoporous carbons have ca. 50% lower microporosities but adsorb up to 65% more CO_2. The authors attribute the enhancement to the presence of interconnected micropores and mesopores.

Different from microstructure tuning, macrostructure (form, density, etc.) modification is also crucial for reducing pressure drop, mitigating adsorption heat, as well as enhancing volumetric capture capacity. Linares-Solano et al. systematically investigated this issue by using carefully selected carbon monoliths (A series, M3M, and K1M) [176]. From the systematic CO_2 adsorption investigation, the authors found that (i) the gravimetric storage capacities of the adsorbents depend on their textural properties, while the volumetric adsorption capacity is directly related to their textural properties and densities. It is worth noting that the density shows the most important impact on gas storage capacity. (ii) Due to their singular high density, a series of carbon monoliths, as well as CO_2 AC monoliths, present exceptionally high volumetric storage capacity for CO_2 at room temperature.

3.6.2 Surface Chemistry

In addition to the pore structures, the nature of the interaction between gas molecules and the pore surface is also important in determining the adsorption selectivity of an adsorbent. Compared with non-polar or weakly polar N_2, CH_4, and H_2, CO_2 is highly quadrupolar and weakly acidic. This means there are profound differences in the interaction between these gas molecules and the pore surface of a porous material. By introducing basic functional groups into the carbon framework, the heteroatom-doped carbon can polarize CO_2 molecules; meanwhile, narrow and interconnected micropores can lead to selective recognition of CO_2 molecules [177]. This can be taken into consideration when modifying the surface properties of carbon materials to enhance their adsorption and separation ability toward different gases. For this issue, the surface properties can be tuned not only by the predesign of precursors, but also by the post-modification of existing carbons.

Among the doping heteroatoms, the influence of N-doping has been reported most [178–180]. A wide range of N-doped carbons with diverse morphologies have been developed and tested for CO_2 capture. The methods of N-doping can be broadly classified into two categories: (i) using nitrogen-containing precursors and pyrolysis and (ii) high-temperature reaction and transformation based on premade carbons, i.e. NH_3 activation. For the two approaches, the introduction of nitrogen-containing functional groups has little effect on the pore structures, and these functional groups are highly dispersed either in the surface or in the carbon matrix. The role of heteroatom-involving sites, particularly N-containing groups in CO_2 capture, remains controversial. At present, there are three viewpoints regarding this issue: acid–base interactions, hydrogen-bonding interaction, and electrostatic interactions. The role of heteroatoms is discussed along with the review of each type of porous carbons.

3.6.2.1 Nitrogen-Containing Precursors

The rationally selected polymeric carbon precursors mainly involve *p*-diaminobenzene [181], polyacrylonitrile [182], melamine [183], and so on [184]. Han et al. have prepared a series of N-MC materials with a very high nitrogen doping concentration (ca. 13 wt%) and rich micropores (<1 nm), which lend them highly desired characteristics for selective CO_2 capture [185]. For example, they report a nitrogen-doped mesoporous carbon material that can be nanocasted from tri-continuous mesoporous silica (IBN-9) by using a mixed carbon precursor comprising nitrogen-containing p-diaminobenzene and nitrogen-free furfural alcohol. After a chemical activation process, a high content of nitrogen and a large proportion of micropores in one material are successfully integrated. The optimized material exhibited excellent adsorption properties in terms of both CO_2 uptake and CO_2/N_2 selectivity. Particularly, its CO_2 adsorption uptake in the typical condition of flue gas (e.g. CO_2 partial pressure of 0.2 bar, 25 °C) is as high as 1.75 mmol g^{-1}, which is far superior to most reported carbon materials. These new materials showed high CO_2 adsorption heats (ca. 40 kJ mol^{-1} at initial adsorption stages), suggesting an enhanced physical adsorption effect by nitrogen doping [181].

Later, they designed and prepared a series of porous carbons, including microporous carbon and mesoporous carbon used for selective CO_2 capture. The authors found that the combination of a high N-doping concentration (>10 wt %) and extra-framework cations endowed N-doped microporous carbons with exceptional CO_2 adsorption capabilities, especially at low pressures (CO$_2$ uptake of 1.62 mmol g^{-1} at 25° C and 0.1 bar) [185]. Single-component adsorption isotherms indicated that its CO$_2$/N$_2$ selectivity was 48, which also significantly surpasses the selectivity of conventional carbon materials. Furthermore, the dynamical breakthrough experiments using CO$_2$/N$_2$ (10 : 90 v/v) mixtures reveal that the CO$_2$/N$_2$ selectivity was as high as 44, comparable to that predicted from equilibrium adsorption data. More interesting, they conducted theoretical calculations that correlate the polarizing capabilities of various functional groups (K$^+$, Cl$^-$ ions as well as N-containing sites; see Figure 3.29) with their enhancement effects on CO$_2$ adsorption, and demonstrated that such effects are essentially based on electrostatic interactions. This represents a new perspective to explain the positive role of heteroatoms in contributing to a high CO$_2$ uptake.

Kowalewski et al. reported another type of nitrogen-enriched porous carbon nanostructure as CO$_2$ capture materials, prepared via the carbonization of polyacrylonitrile containing block copolymer [186]. The typical sample exhibited good selectivity for CO$_2$ manifested by a 7- to 10-fold larger amount of adsorbed CO$_2$ over N$_2$ (Figure 3.30a and b). The analysis of isosteric heats of CO$_2$ adsorption also leads to similar conclusions about the role of surface nitrogens. The characteristic initial sharp decrease to the plateau observed in these curves (Figure 3.30c) is likely indicative of initial adsorption driven by more active nitrogen surface sites. As shown in Figure 3.30b and d, for any given nitrogen content, the N-doped sample exhibited higher selectivity and Q$_{st}$ than the non-N doped counterpart. By these

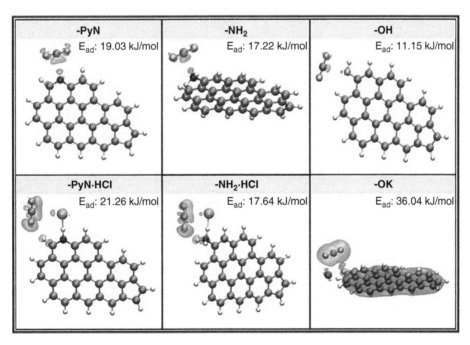

Figure 3.29 Optimized configurations of CO$_2$ adsorption on carbon clusters with different polar groups (cyan, C; white, H; blue, N; red, O; light green, K) and corresponding contour plots of the differential charge density. The contour value is ±0.001 au. The purple and lime regions represent the charge accumulation and charge depletion regions, respectively. Source: reproduced from Ref. [185] with permission from the American Chemical Society. (*See color plate section for color representation of this figure*).

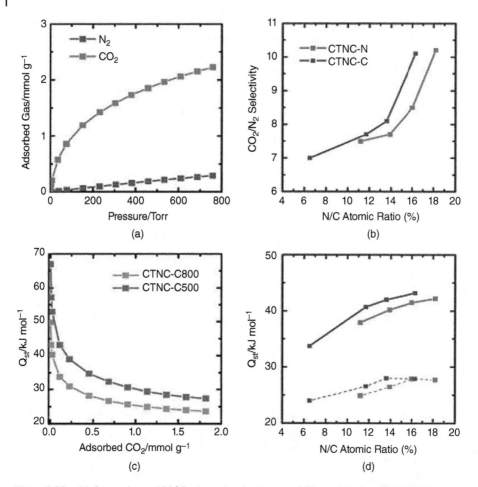

Figure 3.30 (a) Comparison of 25 °C adsorption isotherms of CO_2 and N_2 for CTNC-N700; (b) correlation between CO_2/N_2 selectivity and N/C atomic ratio; (c) isosteric heats of CO_2 adsorption (Q_{st}) by CTNC-C500 and CTNC-C800; (d) correlation between Q_{st} and N/C atomic ratio at different CO_2 coverages (red: CTNC-N; blue: CTNC-C; solid line: 0.1 mmol g^{-1} of CO_2 adsorbed; dash line: 1.8 mmol g^{-1} of CO_2 adsorbed). Source: reproduced from Ref. [186] with permission from the Royal Society of Chemistry. (*See color plate section for color representation of this figure*).

in-depth analysis, they believe that the adsorption capacity can be increased by enlarging the surface area under CO_2 treatment at lower temperatures (<700 °C), while CO_2 treatment at a higher temperature (800 °C) produced an even more significant increase of surface area and CO_2 capacity but reduced selectivity, due to the loss of nitrogen.

Nandi and coworkers have fabricated a series of highly porous N-doped carbon monoliths as CO_2 capture materials. This series of N-doped carbons was obtained from the mesoporous PAN monolith via thermal treatment in two steps (Figure 3.31) [182]. First, the monoliths were pretreated in air at 503 K for activation, which led to cyclization inside the polymer framework, generating a ladder polymer. Then the ladder polymer gradually underwent aromatization, generating an aromatic ladder. In the next heating step, the aromatized polymer was converted to carbon with a lamellar phase by carbonization

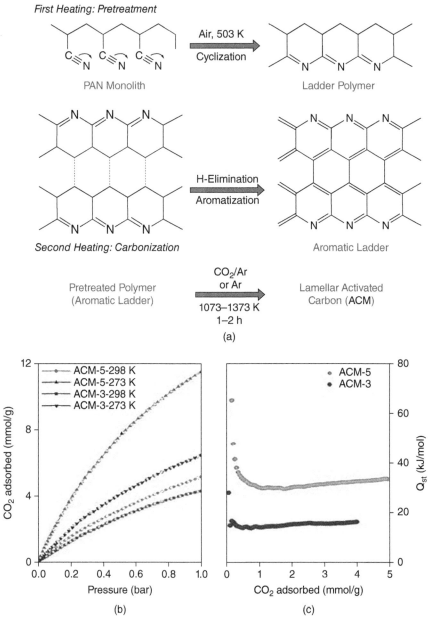

Figure 3.31 (a) Carbonization of PAN monolith; (b) CO$_2$ sorption up to 1 bar, at 273 and 25 °C: adsorption (filled symbols) and desorption (empty symbols); (c) isosteric heat of CO$_2$ adsorption (Q$_{st}$) as a function of CO$_2$ adsorbed. Source: reproduced from Ref. [182] with permission from the Royal Society of Chemistry. (*See color plate section for color representation of this figure*).

in Ar or an Ar–CO_2 mixture. CO_2 adsorption isotherms (Figure 3.31b) show reversible adsorption characteristics indicating weak interaction of CO_2 molecules with the pore walls. More impressive, these carbon monoliths show unprecedentedly high CO_2 uptake of 5.14 mmol g^{-1} at ambient pressure and temperature and 11.51 mmol g^{-1} at ambient pressure and 0 °C. As shown in Figure 3.31b, the typical sample showed a high initial Q_{st} value of up to 65.2 kJ mol^{-1}. High initial isosteric heats of adsorption (Q_{st}) values indicate strong adsorbent–adsorbate interaction between the N-containing carbon framework and CO_2 molecules.

Fan and co-workers have reported a simple, low-cost method for synthesizing a series of nitrogen-enriched porous carbons, in which petroleum coke was used as a carbon precursor and consequently modified by urea and activated by KOH under varying conditions [187]. Upon urea modification, a significant number of nitrogen groups were introduced into the carbon matrix. During KOH activation, some of the nitrogen groups reacted with KOH to facilitate KOH penetration into deeper layers of the carbon, leading to a greater development of the micropore structure, while other nitrogen groups were transformed into more thermally stable species and built into the carbon structure. The resulting porous carbons thus featured a high fraction of fine micropores (<1 nm) and some degree of basic nitrogen-containing groups. Consequently, these nitrogen-doped porous carbons showed a maximum CO_2 adsorption capacity of 4.40 mmol g^{-1} and 6.75 mmol g^{-1} at 25 °C and 0 °C under atmospheric pressure (1 bar), respectively, together with good stability and high CO_2/N_2 selectivity. This study further underscores the importance of nanoporous structures and nitrogen-containing groups to achieve high CO_2 adsorption capacity and CO_2/N_2 selectivity of carbon materials under ambient conditions.

Wang and coworkers fabricated N-doped pillaring layered carbon (NC) and N, S-co-doped honeycomb carbon (NSC) through a one-pot pyrolysis process of a mixture containing glucose, sodium bicarbonate, and urea or thiourea. The materials with N, S-codoping surface show better CO_2/N_2 selectivity compared with N-doped carbon [188].

3.6.2.2 High-Temperature Reaction and Transformation

Treating as-synthesized porous carbons with gaseous ammonia under high temperature (e.g. 900 °C) is another popular method used for the preparation of N-doped carbon. In principle, the reaction with ammonia is expected to take place at carboxylic acid sites formed by the oxidation of side groups and the ring system. At a high temperature, ammonia decomposes to form radicals such as NH_2, NH, and H [189, 190]. These radicals may react with the carbon surface to form functional groups, such as —NH_2—, —CN, pyridinic, pyrrolic, and quaternary nitrogen.

Many researchers believe that the introduction of nitrogen will increase the basicity of carbon thus facilitating the removal of trace amounts of acidic gases including CO_2. For example, Przepiórski et al. found that high-temperature ammonia treatment of activated carbon clearly enhanced CO_2 adsorption [191]. In their work, the ammonia treatment was performed for two hours at elevated temperatures ranging from 200–1000 °C. The CO_2 capture tests confirm that absorption of CO_2 was enhanced by the ammonia treatment. The enhancement was attributed to the presence of C—N and C=N groups. And further, the largest CO_2 uptake was found to be at 400 °C of the ammonia treatment temperature.

In their opinion, the higher-temperature treatment may cause the closure of micropores or changes in the size of pores.

Similarly, nitrogen-doped phenolic resin-based CSs were prepared by a slightly modified Stöber method using ammonia as the nitrogen source [192]. The as-synthesized phenolic resin spheres and the CSs obtained by carbonization of polymer spheres at 600 °C showed spherical morphology with an average diameter of 600 and 550 nm, respectively. A direct KOH activation of polymeric spheres produced carbons with small micropores (<0.8 nm) and large specific surface area (2400 m [2] g^{-1}), which are able to adsorb an unprecedented amount of CO_2 (up to 8.9 mmol g^{-1}) at 0 °C and ambient pressure.

Pevida and co-workers demonstrated that ammonia treatment at temperatures higher than 600 °C incorporated nitrogen mainly into the aromatic rings, while at lower temperatures nitrogen was introduced into more labile functionalities such as amide-like functionalities [193]. The CO_2 capture capacities at 25 °C of the treated carbons increased with respect to the parent carbons. In particular, after ammonia modification (at 800 °C), the CO_2 capture capacities of wood-derived carbons rose from 7 to 8.4 wt%. It is worth pointing out that it is the specific nitrogen-functionalities rather than the total nitrogen content that are responsible for increasing the CO_2-adsorbent affinity. The same group also investigated the ammonia treatment of pristine carbons in the presence of air (ammoxidation) [194]. They found that, during ammoxidation (amination in the presence of air), the formation of nitriles and amide-like functionalities is favored, and a greater amount of nitrogen is incorporated onto the carbon surface. Further, nitrogen uptake by ammoxidation at 300 °C was found to be proportional to the oxygen content of the starting carbon. Ammonia seems to react preferentially with the CO_2-evolving groups of the starting carbon, such as carboxyls, while the remaining oxygen mainly forms part of CO-evolving groups, such as amides or lactams. CO_2 capture results show that CO_2 capture capacity is related to the narrow micropore volume of the samples, and this relationship is approximately linear at room temperature. However, above room temperature, the trend deviates from linearity due to the possible influence of surface basicity.

3.6.2.3 Oxygen-Containing or Sulfur-Containing Functional Groups

In addition to the most commonly investigated N-containing functional groups, oxygen-containing and sulfur-containing functional groups have also been investigated to enhance CO_2 adsorption in carbon materials, particularly in the absence of water vapor, and hydrated graphite was found to hinder CO_2 adsorption [195].

For example, Liu and Wilcox [196] theoretically analyzed the role of oxygen-containing groups in CO_2 capture based on an assumption in which complex pore structures for natural organic materials (e.g. coal and gas shale) and carbon-based porous materials are modeled as a collection of independent, non-interconnected, functionalized graphitic slit pores with surface heterogeneities. Using simulation techniques (electronic structure calculations and grand canonical Monte Carlo simulations), the authors summarized the results as shown in Figure 3.32. The main points include: (i) the CO_2 molecules are more organized and aligned when they are adsorbed in the functionalized slit pores, and thus the adsorption capacity is enhanced by the more efficient side-by-side packing; and (ii) in the ultramicropores (less than 7 Å), due to the overlapping potentials from the strong pore wall–wall interactions and the strong CO_2 – wall interaction, the condensed adsorbed-CO_2 density is even higher than

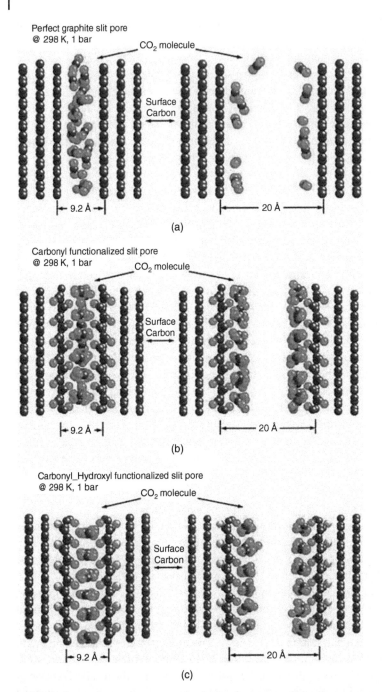

Figure 3.32 Comparisons of CO_2 adsorbed in functionalized micropores with that in the perfect graphite slit pore: left, side views of adsorbed CO_2 in various functionalized graphite slit pores with pore width of 9.2 Å; right, side views of adsorbed CO_2 in various functionalized graphite slit pores with pore width of 20 Å. Source: reproduced from Ref. [196] with permission from the American Chemical Society. (*See color plate section for color representation of this figure*).

that of the larger pores. In general, as the pore width decreases, the surface functionalities dictate the adsorption, and thus the surface functionalities play a more important role in increasing the CO_2 adsorption capacity. The surface heterogeneity changes the adsorbates' accumulation configuration by changing the geometry of the pore surface and the charge distribution of the surface, which is consistent with the Bader charge results of density functional theory (DFT) study.

Research on the adsorption behavior of pure CO_2 gas and mixtures (e.g. CO_2–CH_4, CO_2–N_2, etc.) in porous carbons with surface functional groups is conducted continuously [197–199]. Tenney et al. [200] performed structurally and chemically heterogeneous modifications on the graphite surface and observed that CO_2 adsorption generally increased with the increase of the surface oxygen content because of the enhanced adsorbate–adsorbent interactions. Wilcox et al. [201] indicated that the introduction of O-containing functional groups on the graphite surface could enhance the adsorption capacity of CO_2 and the selectivity of CO_2 over CH_4 and N_2. More interestingly, the embedded positions of functional groups have a significant effect on their gas adsorption behaviors.

Very recently, Lu et al. systematically investigated the effect of edge-functionalization on the competitive adsorption of a binary CO_2–CH_4 mixture in NPCs by combining DFT and grand canonical Monte Carlo (GCMC) simulation [202]. In their work, four functional groups were considered to improve the gas adsorption capacity and selectivity performance of NPCs: i.e. hydrogen (H–), hydroxyl (OH–), amine (NH₂–), and carboxyl (COOH–), as shown in Figure 3.33. A coronene-shaped graphitic basis unit was chosen

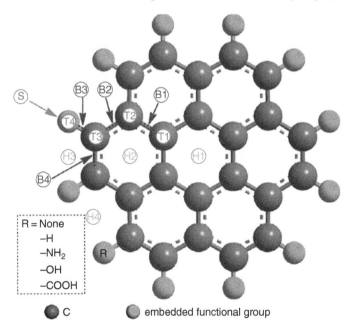

Figure 3.33 Initial configurations of CO_2–CH_4 adsorption on the edge-functionalized basis unit. Nomenclature: H, position above the center of a benzene ring in the basis unit; T, position at the top of the C atom or the atom connected to the functional group; B, position above the bond center; and S, side position in the plane of the functional group. Source: reproduced from Ref. [202] with permission from the Royal Society of Chemistry. (*See color plate section for color representation of this figure*).

as the electron structure model for quantum-chemistry calculations. Results show that the introduction of functional groups has a significant influence, through the inductive effect, on gas adsorption on the basis unit surface. For the basis unit, the adsorption energy of CO_2 is 34.77 meV, which is slightly lower than 38.64 meV for CH_4 (Figure 3.34). This difference indicates that the surface site is energetically favorable for CH_4 with respect to CO_2 [203]. However, edge-functionalization significantly enhances CO_2 adsorption but has less influence on CH_4 adsorption for single-component CO_2–CH_4 adsorption, therefore significantly improving the selectivity of CO_2 over CH_4, in the order of NH_2–NPC > COOH–NPC > OH–NPC > H–NPC > NPC at low pressure.

Taking the NH_2–group as an example, the introduction of the NH_2-group improves the adsorption energy of CO_2 up to 73.57 meV, but has only a slight influence on the adsorption energy of CH_4 to 41.69 meV (Figure 3.34). This contribution clearly arises from the cooperative effect of electronegative N atoms and electropositive H atoms, which serve as basic adsorption sites on the basal plane, and function as Lewis bases by donating their

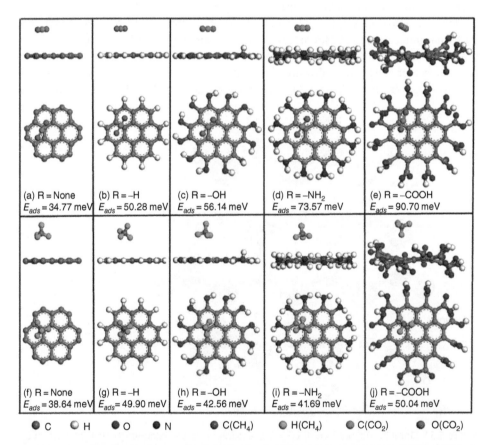

Figure 3.34 Stable adsorption configurations (side view [up] and top view [down]) of (a–e) CO_2 and (f–j) CH_4 on the edge-functionalized basis unit at the B1 site. Source: reproduced from Ref. [202] with permission from the Royal Society of Chemistry. (*See color plate section for color representation of this figure*).

electron to the acidic C atom of the CO_2 molecules. Therefore, embedding the NH_2-group could effectively increase the discrimination between CO_2 and CH_4 adsorption. This work not only highlights the potential of edge-functionalized NPCs as excellent candidates for the competitive adsorption, capture, and separation of a binary CO_2–CH_4 mixture, but also provides an effective and superior alternative strategy in the design and screening of adsorbent materials for CCS applications. Similarly, Shi et al. found the importance of pyridinic N with neighboring OH or NH_2 species in enhancing the CO_2 adsorption capacity and selectivity [204].

3.6.3 Crystalline Degree of the Porous Carbon Framework

In addition to amorphous carbons, graphitic porous carbons are also widely investigated as adsorbents for understanding CO_2 capture behavior due to their ordering at the atomic scale. Electronic structure calculations coupled with van der Waals-inclusive corrections have been performed to investigate the electronic properties of functionalized graphitic surfaces [196]. With Bader charge analysis, electronic structure calculations can provide the initial framework comprising both the geometry and corresponding charge information required to carry out statistical modeling. Also, other molecular-level simulations of CO_2 sorption behavior in the micropores of porous carbons show that heteroatom doping greatly enhances CO_2 uptakes and selectivity at low coverage, while CO_2 capture performance at high pressure is largely dispensed by the textural parameters [205].

The emergence of graphene nanosheets has opened up an exciting new field in the science and technology of two-dimensional nanomaterials [206, 207]. Graphene oxide (GO) is a derivative of graphene and consists of oxygen functional groups on their basal planes and edges, so surface modification of GO with amines or amine-containing molecules takes place easily through the corresponding nucleophilic substitution reactions. If polyamines covalently attach to its layers, the residual unreacted amine groups can react with CO_2 and have potential for the removal of CO_2. Consequently, Zhao et al. prepared GO-amine composites based on the intercalation reaction of GO with amines, including EDA, diethylenetriamine (DETA), and triethylene tetramine (TETA). Dynamic CO_2 breakthrough tests revealed that the aminated GO was an efficient adsorbent for CO_2 capture. For example, a typical sample of GO/EDA showed an adsorption capacity of $53.62\,mg\,g^{-1}$ [208]. Interestingly, Koenig et al. pioneered one type of graphene membrane-based molecular sieve [209]. In their synthesis, ultraviolet-induced oxidative etching can create pores in micrometer-sized graphene sheets. A pressurized blister test and mechanical resonance are used to measure the transport of a range of gases (H_2, CO_2, Ar, N_2, CH_4, and SF_6) through the pores. The experimental results reveal the realization of graphene gas separation membranes by molecular sieving, and represent an important step toward the realization of size-selective porous graphene membranes.

Another familiar crystalline carbon nanostructure is CNTs. Since the first synthesis of CNTs via arcing between graphite-like electrodes by Iijima in 1991 [210], CNTs and their composite materials have been extremely researched, due to their outstanding properties such as excellent chemical and thermal stability, electronic properties, etc. One of their potential applications is as sorbents for CO_2 capture due to their shortened size, easy functionalization, and/or integration with foreign active species for selective

CO$_2$ recognition [211, 212]. For example, Barron and coworkers reported the covalent attachment of branched polyethyleneimine (PEI) to the sidewalls of SWNTs through the use of fluorinated single-wall carbon nanotubes as precursors [213]. The structural integrity of the original purified SWNT is maintained upon covalent functionalization with PEI. Solid-state ^{13}C NMR shows the presence of carboxylate substituents due to carbamate formation as a consequence of the reversible CO$_2$ absorption to the primary amine substituents of the PEI. Desorption of CO$_2$ is accomplished by heating under argon at 75 °C, while the dependence of the quantity of CO$_2$ absorbed on temperature and the molecular weight of the PEI is also observed.

In addition to these experiment investigations, theoretical research on CO$_2$ capture over CNT has also been reported. For example, Liu et al. have shown, from molecular dynamics simulations, that windowed carbon nanotubes are able to separate CO$_2$ from the CO$_2$/CH$_4$ mixture with a CO$_2$ permeance several orders of magnitude higher than conventional analogues (Figure 3.35) [214].

Asai et al. reported a new CO$_2$ sorption behavior over graphitic nanoribbons, which is distinctly different from the behavior of nanoporous carbon and carbon black (Figure 3.36) [215]. They found a remarkable irreversibility in the adsorption of CO$_2$ and H$_2$O on such graphitic nanoribbons at ambient temperature. The irreversible adsorptions of both CO$_2$ and H$_2$O are due to the large number of sp^3-hybridized carbon atoms located at the edges. The authors believe that the observed irreversible adsorptivity of the edge surfaces of graphitic nanoribbons for CO$_2$ and H$_2$O indicates a high potential in the fabrication of novel types of catalysts and highly selective gas sensors.

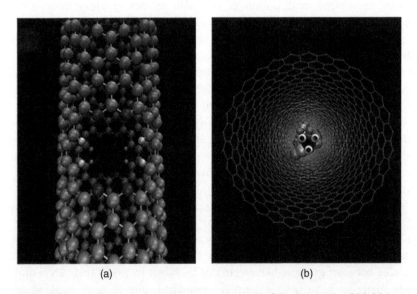

(a)   (b)

Figure 3.35 (a) 4N4H windows or pores on the wall of the inner tube. (b) Initial setup of the simulation where CO$_2$/CH$_4$ gas mixture is inside the windowed inner tube; on the outside is a pristine tube. Source: reproduced from Ref. [214] with permission from the American Chemical Society.

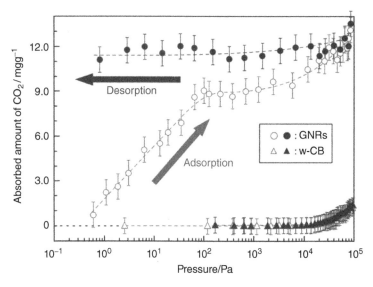

Figure 3.36 Adsorption isotherms of CO_2 on GNRs and well-crystalline CBs at 303 K. Source: reproduced from Ref. [215] with permission from the American Chemical Society.

Through molecular dynamics simulations, Mantzalis and coworkers investigated the layering behavior of carbon dioxide transported through carbon nanoscrolls [216]. The layering arrangements are investigated for carbon nanoscrolls with intralayer distances spanning from 4.2 to 8.3 Å at temperature of 300 K and pressures ranging from 5 to 20 bars. It is shown that the number of layers, their relative strength, and the starting point of bifurcation phenomena vary as a function of the nanoscrolls' interlayer distance, core radius, CO_2 density, and gas structure interactions. It is also shown that the number of carbon dioxide molecules adsorbed per scroll's carbon particles is a function of the scroll's surface-to-volume ratio and is maximized under certain structural configurations.

The aforementioned examples open the door for the design and preparation of highly effective carbonaceous CO_2 adsorbents with controlled pore features and tailored surface chemistry. These porous carbons would combine the merits of designed synthesis (controlled pore structure and task-specific surface chemistry) and intrinsic properties (excellent chemical and thermal stability, developed porosity) of carbon materials, and meet the complex requirements of efficient adsorbents for CO_2 capture.

3.6.4 Functional Integration and Reinforcement of Porous Carbon

To strengthen the performance of porous carbon, new components such as amines and metal oxides, which have strong interactions with CO_2, have often been introduced to provide efficient and active CO_2 binding sites. There are some successful illustrations of metallic and non-metallic modifications through in situ polymerization or a post-treatment method. Post-modification is a versatile method for the preparation of advanced carbons with powerful functions through processes such as chemical vapor deposition [217, 218], impregnation [219–221], and metal transfer reactions [222].

Inspired by silica-based hybrid sorbents (molecular basket) with grafted or impregnated amine groups on porous silica substrates [223, 224], Zhao et al. reported an aminated adsorbent generated from sustainable biomass (glucose) [225]. Two steps are involved in this synthesis: (i) hydrothermal carbonization of glucose, and (ii) transformation into a porous carbon-amine composite by a post-synthetic modification with a branched tetramine. The authors first prepared the substrate carbons with a novel raspberry morphology, which may be beneficial for the loading of liquid amines. Interestingly, the raspberry morphology is maintained after modification of the carbons with grafted polyamines. And FT-IR spectra prove the presence and the external accessibility of amino groups on the surface of our materials. CO_2 capture results show a very high CO_2 uptake (up to 4.3 mmol g^{-1}) at $-20\,°C$. More importantly, this type of composite delivered a very high CO_2 selectivity at low ($-20\,°C$, CO_2/N_2 of 65–85) and high ($70\,°C$, CO_2/N_2 of 90–110) temperatures. These high capacities and selectivities are consistent with high amine loadings. The high capacity is remarkable given that the prepared absorbents have only a moderately large specific surface area. Clearly, some of the CO_2 absorbs within the amine-rich, liquid-grafted surface layer. Thus, a more ideal CO_2 capture material can be imagined with a high active amine content combined with a more optimized channel.

It should be noted that the introduction of liquid amines through impregnation might result in other negative effects such as the blockage of pores and an instability of the basic sites on the surface in long time cycling. To address this issue, Tour's group developed a route to synthesize polymer-mesocarbon composites that would lead to higher degrees of CO_2 sorption by the in situ polymerization of amine species to produce PEI and PVA inside the mesocarbon CMK-3 (Figure 3.37) [226]. This structured composite exhibits high stability due to the formation of interpenetrating composite frameworks between the entrapped polymers and the mesocarbon CMK-3. CO_2 uptake measurements showed that the 39% PEI-CMK-3 composite had ca. 12 wt% CO_2 uptake capacity, and the 37% PVA-CMK-3 composite had ca. 13 wt% CO_2 uptake capacity at $30\,°C$ and 1 atm. More importantly, the composite can easily be regenerated at $75\,°C$ and cycles stably (even up to 500 minutes; see Figure 3.37). In this way, a high concentration of functional groups can be obtained; however, a large part of the pores will be blocked by filling of liquid amines.

To improve CO_2 uptake using carbon materials, a desirable approach is to incorporate basic sites, ensuring improved adsorption capacity for acidic CO_2 gas by consolidating nitrogen functionalities into the carbon framework. The effectiveness of alkali metal-based solid sorbents for CO_2 capture from flue gas has been investigated by many researchers [227, 228]. Highly porous sodium-impregnated and N-doped carbon sorbents were prepared by KOH activation of polyacrylonitrile (PAN), followed by NaOH impregnation of the activated N-doped carbons [229]. Characterization indicated that Na_2CO_3 in the porous carbon structures plays an important role as effective basic sites in acidic CO_2 gas capture. The AC exhibited an adsorption capacity of 6.84 and 4.48 mmol g^{-1} at 0 and $25\,°C$ under ambient pressure. Among the carbon sorbents reported to date, it showed the highest CO_2 uptake of 3.03 and 1.90 mmol g^{-1} at 0 and $25\,°C$ under a typical pressure condition of post-combustion flue gas (0.15 bar CO_2). The enhanced CO_2 uptake is due to high porosity caused by KOH activation, enriched pyridonic/pyrrolic nitrogen content, and strong basic sites generated by NaOH impregnation.

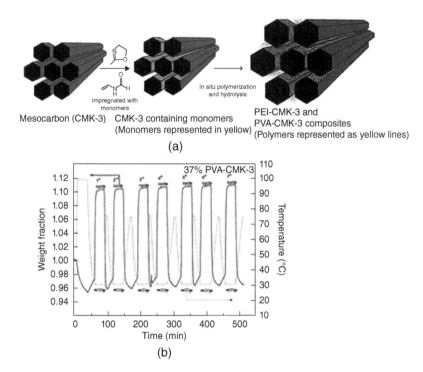

(a)

(b)

Figure 3.37 (a) Synthesis processes to produce mesoporous polymer-carbon composites PEI-CMK-3 and PVA-CMK-3. (b) Sorption cycles of CO$_2$ studied by TGA at 30 °C on the 37% PVA-CMK-3 sorbent. Source: reproduced from Ref. [226] with permission from the American Chemical Society.

In addition to sodium, MgO and/or Mg(OH)$_2$-incorporated mesoporous carbon composite [230, 231] has been recognized as a promising candidate for CO$_2$ capture from the diluted flue gas stream due to its high CO$_2$ adsorption capacity and relatively low regeneration temperature. Jang et al. reported a one-pot synthesis of Mg-OMC for CO$_2$ adsorption [38]. The maximum CO$_2$ adsorption capacity of Mg-OMC-1 was found to be 92 mg g^{-1} at 25 °C. This complete desorption of adsorbed CO$_2$ at 200 °C supports the potential of Mg-OMC as a low-temperature swing sorbent. Mg-OMC is reusable, selective, and thermally stable. Zhang et al. reported a one-pot solvent-free method for the first time to prepare a series of nitrogen and magnesium co-doped mesoporous carbon materials (NMgCs) (Figure 3.38) [232]. The strategy involves facile mixing, grinding, and thermal treatment of structure-directing agent and precursors. The obtained composites show CO$_2$ capture capacity with encouraging uptakes of 2.45 mmol g^{-1} at 25 °C. Interestingly, CO$_2$ adsorption at 75 °C still showed a capacity as high as 1.17 mmol g^{-1}. The enhanced CO$_2$ adsorption in comparison with the parent material can be attributed to the doping of highly dispersed MgO sites and nitrogen atoms in the composite that are beneficial for adsorbing acidic CO$_2$ molecules.

CNTs are unique, one-dimensional macromolecules that have thermal and chemical stability. These nanomaterials have been proven to possess good potential as superior

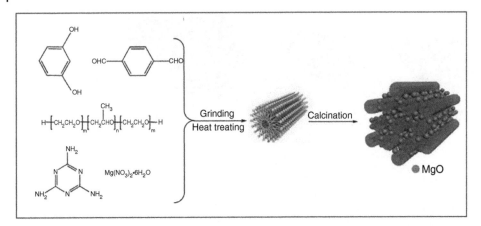

Figure 3.38 Representative schematic diagram of the one-pot solvent-free synthesis of nitrogen and magnesium co-doped mesoporous carbons. Source: reproduced from Ref. [232] with permission from the American Chemical Society.

adsorbents for removing many kinds of organic and inorganic pollutants in air streams or from aqueous environments. The large adsorption capacity of pollutants by CNTs is mainly attributed to their pore structure and the existence of a wide spectrum of surface functional groups that can be achieved by chemical modification or thermal treatment to make CNTs that possess optimum performance for particular purposes. Therefore, CNTs are considered another component for functional reinforcement of porous carbon in capturing CO_2.

In addition to post-modification, direct copolymerization not only is suitable for introduction of molecular functional groups to the carbon products, but also enables the dispersion of nanoparticles throughout the carbon framework. From an applications perspective, volumetric capacities are more important than gravimetric capacities due to the limited volume of the gas storage tank. Under this consideration, Lu et al. optimized the structural features of hierarchical porous carbon monoliths by incorporating the advantages of MOFs $(Cu_3[BTC]_2)$) to maximize the volumetric-based CO_2 capture capability (CO_2 capacity in cm^3 cm^{-3} adsorbent) [233]. The mesoscopic structure of the HCM-$Cu_3[BTC]_2$ composites and the parent materials (HCM and $Cu_3(BTC)_2$) were characterized by SEM. The SEM micrograph (Figure 3.39) clearly displays that $Cu_3(BTC)_2$ crystallites are generated within the macropores of the HCM matrix. The sponge-like skeleton of HCM before and after MOF growth remains unchanged. The octahedral $Cu_3(BTC)_2$ crystallites are well dispersed within the HCM matrix. The equilibrium CO_2 adsorption measurements were carried out at 25 °C, and the results reveal that HCM-$Cu_3(BTC)_2$ composites exhibit an obvious increment in CO_2 adsorption capacity on a volumetric basis compared with the original HCM. The HCM-$Cu_3(BTC)_2$-3 composite with the highest $Cu_3(BTC)_2$ loading can achieve maximum CO_2 uptake of 22.7 cm^3 STP per cm^3 at ~1 bar, which is almost twice the uptake of HCM (12.9 cm^3 STP per cm^3) under the same conditions. This result motivates a new principle for the rational design of CO_2 capture material by maximizing the capacity on a volumetric basis.

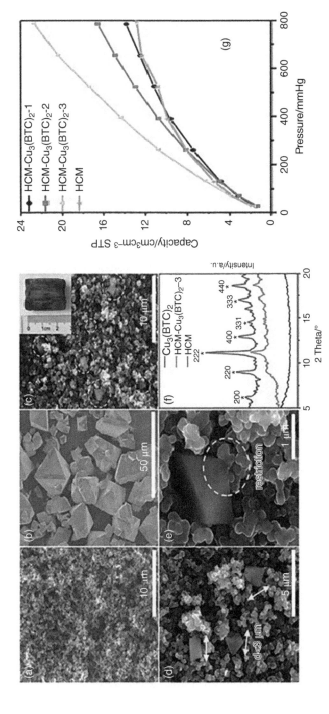

Figure 3.39 SEM micrographs of (a) HCM, (b) Cu₃(BTC)₂, and (c–e) HCM-Cu₃(BTC)₂-3. (f) XRD patterns of HCM, Cu₃(BTC)₂, and HCM-Cu₃(BTC)₂-3. (g) CO₂ adsorption isotherms on a volumetric basis. Source: reproduced from Ref. [233] with permission from the American Chemical Society. (*See color plate section for color representation of this figure*).

3.7 Summary and Perspective

In summary, porous carbons for CO_2 capture have undergone rapid development during the last several decades and will continue to blossom. To date, various kinds of new carbon materials with defined structural properties as well as tailored surface chemistry have been synthesized for specific CO_2 capture processes. This is because porous carbons always have high specific surface area, high amenability to pore structure modification and surface functionalization, and relative ease of regeneration. In this chapter, the first four sections sought to highlight advanced synthesis strategies and CO_2 capture performance of porous carbons, according to the classification of precursors. Then, the critical design principles of carbon adsorbents were summarized and provided, which may be beneficial for the interested reader.

Although much progress has been made, there are many challenges from both the scientific and technological viewpoints. It is worth noting that the requirements of CO_2 capture vary significantly depending on specific processes. In general, carbonaceous adsorbents are attractive for pre-combustion CO_2 capture. However, the selectivity and capacity of carbonaceous adsorbents is too low for post-combustion applications. Moreover, selectivity may be critical in some applications but less so in others, tolerance to other components in the gas stream such as water and H_2S may or may not be required, and long-term chemical and mechanical stability may be more or less important. In addition, CO_2 adsorption on carbon materials is "physical" and weak, which makes these adsorbents sensitive to temperature and relatively poor in selectivity.

Clearly, no unique material or solution exists currently to solve the problem of CO_2 capture. Multiscale modulation and function integration of porous carbons with other components are needed in the future, such as MOF-related materials, inorganic nanomaterials (MgO, CaO, etc.), functional macromolecules (e.g. PEI), etc. The use of materials that complement each other with advantages between different materials may facilitate porous carbons with improved adsorption capacity, selectivity, diffusion rates, and mechanical strength for efficient CO_2 capture.

Furthermore, in order to satisfy the requirements for practical applications, researchers should simultaneously consider the adsorption capacity per unit mass or volume, easy regeneration, and good mechanical stability. (i) Porous carbons should have abundant microporosity with a suitable size that matches CO_2 molecules, especially when the capture is performed at high CO_2 pressure. (ii) Pore sizes, pore population, and pore interconnection modes should be considered for a highly interconnected hierarchical pore system that determines the adsorption diffusion dynamics of CO_2. Bio-inspired pore systems, like the structure of the lung, are beneficial for mass diffusion during adsorption. Moreover, adsorbents should have good mechanical strength to meet the demand of practical applications. With high mechanical strength, porous carbons can be good CO_2 adsorbents because of their long-term structural stability under dynamically consecutive cycles and tolerance to moisture at elevated temperature. For instance, Guo et al report a novel porous carbon-based composite prepared from an interfacial assembling strategy using nanoclay LAPONITE®, resorcinol, and formaldehyde as the precursors. This kind of composite displayed a record-high CO2/N2 selectivity (114.3) at 70 °C due to the preferable adsorbent–CO_2 interactions [234].

With respect to new carbon-based materials, the key scientific challenges are the development of a level of molecular control, and the development of modern characterization and computational methods that will support, guide, and provide further refinement to the most promising structures. Therefore, along with developed characterizations, computational prediction based on an ideal structure may also be needed. Such techniques will enable large-scale screening of new materials. Ultimately, a clear understanding of the structure–function relationships would direct experimental efforts toward a new generation of porous carbons with improved CO_2 capture abilities [235]. Beyond these considerations, the engineering economics of the new materials must be evaluated upon scaling up the materials for industrial applications, and economic models must be established to cover lifecycle CO_2 separation, capture, and sequestration costs for various technologies. In spite of numerous challenges surrounding CO_2 capture using carbon materials as adsorbents, and the various political, regulatory, and economic drivers that will ultimately dictate the time to deployment for new CCS schemes, the time is ripe for us as a scientific community to play a central role in developing new carbons for solving the CO_2 capture problem.

References

1 Patel, H.A., Byun, J., and Yavuz, C.T. (2017). *ChemSusChem* 10: 1303.

2 Oschatz, M. and Antonietti, M. (2018). *Energy Environ. Sci.* 11: 57.

3 Sreenivasulu, B. and Sreedhar, I. (2015). *Environ. Sci. Technol.* 49: 12641.

4 Zhang, X.-Q., Li, W.-C., and Lu, A.-H. (2015). *New Carbon Mater.* 30: 481.

5 Al Mesfer, M.K. and Danish, M. (2018). *J. Environ. Chem. Eng.* 6: 4514.

6 Zhao, H., Shi, L., Zhang, Z. et al. (2018). *ACS Appl. Mater. Interfaces* 10: 3495.

7 He, L., Li, W.-C., Xu, S., and Lu, A.-H. (2019). *Chem. A Eur. J.* 25: 3209.

8 Nakanishi, K. and Tanaka, N. (2007). *Acc. Chem. Res.* 40: 863.

9 Brun, N., Prabaharan, S.R.S., Morcrette, M. et al. (2009). *Adv. Funct. Mater.* 19: 3136.

10 Alvarez, S., Esquena, J., Solans, C., and Fuertes, A.B. (2004). *Adv. Eng. Mater.* 6: 897.

11 Yang, H., Shi, Q., Liu, X. et al. (2002). *Chem. Commun.* 2842.

12 Wang, X., Bozhilov, K.N., and Feng, P. (2006). *Chem. Mater.* 18: 6373.

13 Xia, Y. and Mokaya, R. (2007). *J. Phys. Chem. C* 111: 10035.

14 Ma, T.-Y., Liu, L., and Yuan, Z.-Y. (2013). *Chem. Soc. Rev.* 42: 3977–4003.

15 Taguchi, A., Smått, J.-H., and Lindén, M. (2003). *Adv. Mater.* 15: 1209.

16 Lu, A.-H., Smått, J.-H., and Lindén, M. (2005). *Adv. Funct. Mater.* 15: 865.

17 Lu, A.-H., Smått, J.-H., Backlund, S., and Lindén, M. (2004). *Microporous Mesoporous Mater.* 72: 59.

18 Hu, Y.-S., Adelhelm, P., Smarsly, B.M. et al. (2007). *Adv. Funct. Mater.* 17: 1873.

19 Liu, N., Yin, L., Wang, C. et al. (2010). *Carbon* 48: 3579.

20 Sánchez-Sánchez, Á., Suárez-García, F., Martínez-Alonso, A., and Tascón, J.M.D. (2014). *ACS Appl. Mater. Interfaces* 6: 21237–21247.

21 Vinu, A. (2008). *Adv. Funct. Mater.* 18: 816.

22 Li, Q., Yang, J., Feng, D. et al. (2010). *Nano Res.* 3: 632.

23 Han, B.-H., Zhou, W., and Sayari, A. (2003). *J. Am. Chem. Soc.* 125: 3444.

24 Nishihara, H. and Kyotani, T. (2012). *Adv. Mater.* 24: 4473.

25 Pachfule, P., Biswal, B.P., and Banerjee, R. (2012). *Chem. A Eur. J.* 18: 11399.

26 Deng, H., Jin, S., Zhan, L. et al. (2012). *New Carbon Mater.* 27: 194.

27 Almasoudi, A. and Mokaya, R. (2012). *J. Mater. Chem.* 22: 146.

28 Lee, K.T., Lytle, J.C., Ergang, N.S. et al. (2005). *Adv. Funct. Mater.* 15: 547.

29 Adelhelm, P., Hu, Y.-S., Chuenchom, L. et al. (2007). *Adv. Mater.* 19: 4012.

30 Gierszal, K.P. and Jaroniec, M. (2006). *J. Am. Chem. Soc.* 128: 10026.

31 Zhang, S., Chen, L., Zhou, S. et al. (2010). *Chem. Mater.* 22: 3433.

32 Fang, B., Kim, M.-S., Kim, J.H. et al. (2010). *J. Mater. Chem.* 20: 10253.

33 Liang, Y., Liang, F., Wu, D. et al. (2011). *Phys. Chem. Chem. Phys.* 13: 8852.

34 Morishita, T., Soneda, Y., Tsumura, T., and Inagaki, M. (2006). *Carbon* 44: 2360.

35 Morishita, T., Ishihara, K., Kato, M., and Inagaki, M. (2007). *Carbon* 45: 209.

36 Konnoa, H., Onishia, H., Yoshizawab, N., and Azumia, K. (2010). *J. Power Sources* 195: 667.

37 Meng, L.-Y. and Park, S.-J. (2012). *J. Colloid Interface Sci.* 366: 125–129.

38 Bhagiyalakshmi, M., Hemalatha, P., Ganesh, M. et al. (2011). *Fuel* 90: 1662.

39 Czyèwski, A., Kapica, J., Moszyǹski, D. et al. (2013). *Chem. Eng. J.* 226: 348–356.

40 Meng, L.-Y. and Park, S.-J. (2012). *Mater. Chem. Phys.* 137: 91.

41 Gaweł, B., Gaweł, K., and Øye, G. (2010). *Materials* 3: 2815.

42 Kadib, A.E., Chimenton, R., Sachse, A. et al. (2009). *Angew. Chem. Int. Ed.* 48: 4969.

43 Davis, M.E. (2002). *Nature* 417: 813.

44 Chowdhury, S. and Balasubramanian, R. (2016). *Ind. Eng. Chem. Res.* 55: 7906.

45 Lu, A.-H. and Schüth, F. (2006). *Adv. Mater.* 18: 1793.

46 Lee, J., Kim, J., and Hyeon, T. (2006). *Adv. Mater.* 18: 2073.

47 Hoheisel, T.N., Schrettl, S., Szilluweit, R., and Frauenrath, H. (2010). *Angew. Chem. Int. Ed.* 49: 6496.

48 Tao, Y., Endo, M., and Kaneko, K. (2009). *J. Am. Chem. Soc.* 131: 904.

49 Silva, A.M.T., Machado, B.F., Figueiredo, J.L., and Faria, J.L. (2009). *Carbon* 47: 1670.

50 Stein, A., Wang, Z., and Fierke, M.A. (2009). *Adv. Mater.* 21: 265.

51 Yu, J., Guo, M., Muhammad, F. et al. (2014). *Carbon* 69: 502.

52 Liang, C.D., Hong, K.L., Guiochon, G.A. et al. (2004). *Angew. Chem. Int. Ed.* 43: 5785.

53 Valkama, S., Nykänen, A., Kosonen, H. et al. (2007). *Adv. Funct. Mater.* 17: 183.

54 Liang, C.D. and Dai, S. (2006). *J. Am. Chem. Soc.* 128: 5316.

55 Saha, D. and Deng, S. (2010). *J. Colloid Interface Sci.* 345: 402.

56 Wang, X., Liang, C., and Dai, S. (2008). *Langmuir* 24: 7500.

57 Liang, C. and Dai, S. (2009). *Chem. Mater.* 21: 2115.

58 Huang, Y., Cai, H., Feng, D. et al. (2008). *Chem. Commun.* 2641.

59 Wei, J., Zhou, D., Sun, Z. et al. (2013). *Adv. Funct. Mater.* 23: 2322–2328.

60 Liu, L., Wang, F.-Y., Shao, G.-S., and Yuan, Z.-Y. (2010). *Carbon* 48: 2089.

61 Gutiérrez, M.C., Picó, F., Rubio, F. et al. (2009). *J. Mater. Chem.* 19: 1236.

62 Zhao, X., Wang, A., Yan, J. et al. (2010). *Chem. Mater.* 22: 5463.

63 Hao, G.-P., Li, W.-C., Wang, S. et al. (2011). *Carbon* 49: 3762.

64 Hao, G.-P., Li, W.-C., Qian, D. et al. (2011). *J. Am. Chem. Soc.* 133: 11378.

65 To, J.W.F., He, J., Mei, J. et al. (2016). *J. Am. Chem. Soc.* 138: 1001.

66 Wang, Z., Li, F., Ergang, N.S., and Stein, A. (2006). *Chem. Mater.* 18: 5543.

67 Deng, Y., Liu, C., Yu, T. et al. (2007). *Chem. Mater.* 19: 3271.

68 Xue, C., Tu, B., and Zhao, D. (2008). *Adv. Funct. Mater.* 18: 3914.

69 Wei, H., Lv, Y., Han, L. et al. (2011). *Chem. Mater.* 23: 2353.

70 Liu, C.Y., Li, L.X., Song, H.H., and Chen, X.H. (2007). *Chem. Commun.* 757.

71 Feng, D., Lv, Y.Y., Wu, Z.X. et al. (2011). *J. Am. Chem. Soc.* 133: 15148.

72 Rodriguez, A.T., Li, X.F., Wang, J. et al. (2007). *Adv. Funct. Mater.* 17: 2710.

73 Liang, C.D. and Dai, S. (2006). *J. Am. Chem. Soc.* 128: 5316.

74 Meng, Y., Gu, D., Zhang, F.Q. et al. (2005). *Angew. Chem. Int. Ed.* 44: 7053.

75 Yoshimune, M., Yamamoto, T., Nakaiwa, M., and Haraya, K. (2008). *Carbon* 46: 1031.

76 Hao, G.-P., Jin, Z.-Y., Sun, Q. et al. (2013). *Energy Environ. Sci.* 6: 3740.

77 Jin, Z.Y., Xu, Y.Y., Sun, Q., and Lu, A.H. (2015). *Small* 11: 5151.

78 Zhang, L.H., He, B., Li, W.C., and Lu, A.H. (2017). *Adv. Energy Mater.* 7: 1701518.

79 Zhang, L.H., Li, W.C., Tang, L. et al. (2018). *J. Mater. Chem. A* 6: 24285.

80 Zhang, L.H., Li, W.C., Liu, H. et al. (2018). *Angew. Chem. Int. Ed.* 57: 1632.

81 Fang, Y., Gu, D., Zou, Y. et al. (2010). *Angew. Chem. Int. Ed.* 49: 7987.

82 Liu, J., Qiao, S.Z., Liu, H. et al. (2011). *Angew. Chem. Int. Ed.* 50: 5947.

83 Tien, B.M., Xu, M.W., and Liu, J.F. (2010). *Mater. Lett.* 64: 1465.

84 Horikawa, T., Hayashi, J., and Muroyama, K. (2004). *Carbon* 42: 169.

85 Fujikawa, D., Uota, M., Sakai, G., and Kijima, T. (2007). *Carbon* 45: 1289.

86 Sun, X. and Li, Y. (2005). *J. Colloid Interface Sci.* 291: 7.

87 Li, Y., Chen, J., Xu, Q. et al. (2009). *J. Phys. Chem. C* 113: 10085.

88 Wickramaratne, N.P. and Jaroniec, M. (2013). *ACS Appl. Mater. Interfaces* (5): 1849.

89 Jiang, P., Bertone, J.F., and Colvin, V.L. (2001). *Science* 291: 453.

90 Lu, A.-H., Hao, G.-P., and Sun, Q. (2011). *Angew. Chem. Int. Ed.* 50: 9023.

91 Wang, S., Li, W.-C., Hao, G.-P. et al. (2011). *J. Am. Chem. Soc.* 133: 15304.

92 Wang, S., Li, W.-C., Zhang, L. et al. (2014). *J. Mater. Chem. A* 2: 4406.

93 Zou, C., Wu, D., Li, M. et al. (2010). *J. Mater. Chem.* 20: 731.

94 Zeng, Q., Wu, D., Zou, C. et al. (2010). *Chem. Commun.* 46: 5927.

95 Han, F.-D., Bai, Y.-J., Liu, R. et al. (2011). *Adv. Energy Mater.* 1: 798.

96 Biener, J., Stadermann, M., Suss, M. et al. (2011). *Energy Environ. Sci.* 4: 656.

97 Pekala, R.W. (1989). *J. Mater. Sci.* 24: 3221.

98 Fairén-Jiménez, D., Carrasco-Marín, F., and Moreno-Castilla, C. (2008). *Langmuir* 24: 2820.

99 Wan, Y., Qian, X., Jia, N. et al. (2008). *Chem. Mater.* 20: 1012.

100 Sepehri, S., García, B.B., Zhang, Q., and Cao, G. (2009). *Carbon* 47: 1436.

101 Hao, G.-P., Li, W.-C., Qian, D., and Lu, A.-H. (2010). *Adv. Mater.* 22: 853.

102 Gu, J.-M., Kim, W.-S., Hwang, Y.-K., and Huh, S. (2013). *Carbon* 56: 208.

103 Shen, W., Zhang, S., He, Y. et al. (2011). *J. Mater. Chem.* 21: 14036.

104 Zhang, S., Miran, M.S., Ikoma, A. et al. (2014). *J. Am. Chem. Soc.* 136: 1690.

105 Jin, Z.-Y., Lu, A.-H., Xu, Y.-Y. et al. (2014). *Adv. Mater.* 26: 3700.

106 Ma, Z., Yu, J., and Dai, S. (2010). *Adv. Mater.* 22: 261.

107 Paraknowitsch, J.P. and Thomas, A. (2012). *Macromol. Chem. Phys.* 213: 1132.

108 Fechler, N., Fellinger, T.-P., and Antonietti, M. (2013). *Adv. Mater.* 25: 75.

109 Wang, X. and Dai, S. (2010). *Angew. Chem. Int. Ed.* 49: 6664.

110 Zhai, Y., Dou, Y., Zhao, D. et al. (2011). *Adv. Mater.* 23: 4828.

111 Zhao, L., Hu, Y.-S., Li, H. et al. (2011). *Adv. Mater.* 23: 1385.

112 Watanabe, M., Thomas, M.L., Zhang, S. et al. (2017). *Chem. Rev.* 117: 7190.

113 Lee, J.S., Wang, X., Luo, H. et al. (2009). *J. Am. Chem. Soc.* 131: 4596.

114 Paraknowitsch, J.P., Zhang, J., Su, D. et al. (2010). *Adv. Mater.* 22: 87.

115 Gong, J., Antonietti, M., and Yuan, J. (2017). *Angew. Chem. Int. Ed.* 56: 7557.

116 Abbott, A.P., Capper, G., Davies, D.L. et al. (2003). *Chem. Commun.* 70.

117 Francisco, M., van den Bruinhorst, A., and Kroon, M.C. (2013). *Angew. Chem. Int. Ed.* 52: 3074.

118 Carriazo, D., Gutiérrez, M.C., Ferrer, M.L., and del Monte, F. (2010). *Chem. Mater.* 22: 6146.

119 Gutierrez, M.C., Carriazo, D., Ania, C.O. et al. (2011). *Energy Environ. Sci.* 4: 3535.

120 Patino, J., Gutierrez, M.C., Carriazo, D. et al. (2012). *Energy. Environ. Sci.* 5: 8699.

121 Patino, J., Gutierrez, M.C., Carriazo, D. et al. (2014). *J. Mater. Chem. A* 2: 8719.

122 Gutiérrez, M.C., Rubio, F., and del Monte, F. (2010). *Chem. Mater.* 22: 2711.

123 Carriazo, D., Gutiérrez, M.C., Ferrer, M.L., and del Monte, F. (2010). *Chem. Mater.* 22: 6146.

124 Mulik, S., Sotiriou-Leventis, C., and Leventis, N. (2008). *Chem. Mater.* 20: 6985.

125 Leventis, N., Sotiriou-Leventis, C., Chandrasekaran, N. et al. (2010). *Chem. Mater.* 22: 6692.

126 Chidambareswarapattar, C., Larimore, Z., Sotiriou-Leventis, C. et al. (2010). *J. Mater. Chem.* 20: 9666.

127 Gao, X., Zou, X., Ma, H. et al. (2014). *Adv. Mater.* 26: 3644.

128 Ma, T.Y., Dai, S., Jaroniec, M., and Qiao, S.Z. (2014). *J. Am. Chem. Soc.* 136: 13925.

129 Amali, A.J., Sun, J.K., and Xu, Q. (2014). *Chem. Commun.* 50: 1519.

130 Zhang, W., Wu, Z.-Y., Jiang, H.-L., and Yu, S.-H. (2014). *J. Am. Chem. Soc.* 136: 14385.

131 Fracaroli, A.M., Furukawa, H., Suzuki, M. et al. (2014). *J. Am. Chem. Soc.* 136: 8863.

132 Hu, M., Reboul, J., Furukawa, S. et al. (2012). *J. Am. Chem. Soc.* 134: 2864.

133 Wang, T., Kim, H.-K., Liu, Y. et al. (2018). *J. Am. Chem. Soc.* 140: 6130.

134 Wang, Q., Xia, W., Guo, W. et al. (2013). *Chem. Asian J.* (8): 1879.

135 Yang, W., Li, X., Li, Y. et al. (2018). *Adv. Mater.* 1804740.

136 Chaikittisilp, W., Hu, M., Wang, H. et al. (2012). *Chem. Commun.* 48: 7259.

137 Yang, S.J., Kim, T., Im, J.H. et al. (2012). *Chem. Mater.* 24: 464.

138 Lim, S., Suh, K., Kim, Y. et al. (2012). *Chem. Commun.* 48: 7447.

139 Srinivas, G., Krungleviciute, V., Guo, Z.-X., and Yildirim, T. (2014). *Energ. Environ. Sci.* 7: 335.

140 Ben, T., Li, Y., Zhu, L. et al. (2012). *Energ. Environ. Sci.* 5: 8370.

141 Li, Q., Guo, J., Zhu, H., and Yan, F. (2019). *Small* 1804874.

142 White, R.J., Budarin, V., Luque, R. et al. (2009). *Chem. Soc. Rev.* 38: 3401.

143 Gong, Y., Wei, Z., Wang, J. et al. (2014). *Sci. Rep.* 4: 6349.

144 Grzyb, B., Hildenbrand, C., Berthon-Fabry, S. et al. (2010). *Carbon* 48: 2297.

145 Olivares-Marín, M. and Maroto-Valer, M. (2012). *Greenhouse Gas Sci. Technol.* 2: 20.

146 Wang, R., Wang, P., Yan, X. et al. (2012). *ACS Appl. Mater. Interfaces* 4: 5800.

147 Xing, W., Liu, C., Zhou, Z. et al. (2012). *Energ. Environ. Sci.* 5: 7323.

148 Li, Y., Li, D., Rao, Y. et al. (2016). *Carbon* 105: 454.

149 Guo, L., Yang, J., Hu, G. et al. (2016). *ACS Sustainable Chem. Eng.* 4: 2806.

150 Lee, M.-S., Park, M., Kim, H.Y., and Park, S.-J. (2016). *Sci. Rep.* 23224.

151 Blankenship, T.S. and Mokaya, R. (2017). *Energy Environ. Sci.* 10: 2552.

152 Himeno, S., Komatsu, T., and Fujita, S. (2005). *J. Chem. Eng. Data* 50: 369.

153 Guo, L.P., Zhang, Y., and Li, W.C. (2017). *J. Colloid Interface Sci.* 493: 257.

154 Fan, X., Zhang, L., Zhang, G. et al. (2013). *Carbon* 61: 423.

155 Wei, H.R., Deng, S.B., Hu, B.Y. et al. (2012). *ChemSusChem* 5: 2354.

156 Sevilla, M. and Fuertes, A.B. (2011). *Energ. Environ. Sci.* 4: 1765.

157 Song, J., Shen, W., Wang, J., and Fan, W. (2014). *Carbon* 69: 255–263.

158 Heidari, A., Younesi, H., Rashidi, A., and Ghoreyshi, A.A. (2014). *Chem. Eng. J.* 254: 503.

159 Zhu, B., Qiu, K., Shang, C., and Guo, Z. (2015). *J. Mater. Chem. A* 3: 5212.

160 Parshetti, G.K., Chowdhury, S., and Balasubramanian, R. (2014). *RSC Adv.* 4: 44634.

161 Wang, J., Heerwig, A., Lohe, M.R. et al. (2012). *J. Mater. Chem.* 22: 13911.

162 Shen, W., He, Y., Zhang, S. et al. (2012). *ChemSusChem* 5: 1274.

163 ElKhatat, A.M. and Al-Muhtaseb, S.A. (2011). *Adv. Mater.* 23: 2887.

164 Titirici, M.-M., White, R.J., Falco, C., and Sevilla, M. (2012). *Energ. Environ. Sci.* 5: 6796.

165 Titirici, M.-M., Thomas, A., and Antonietti, M. (2007). *New J. Chem.* 31: 787–789.

166 Titirici, M.-M., Antonietti, M., and Baccile, N. (2008). *Green Chem.* 10: 1204.

167 Sevilla, M., Falco, C., Titirici, M.-M., and Fuertes, A.B. (2012). *RSC Adv.* 2: 12792.

168 Sevilla, M., Valle-Vigón, P., and Fuerte, A.B. (2011). *Adv. Funct. Mater.* 21: 2781.

169 Rose, M., Korenblit, Y., Kockrick, E. et al. (2011). *Small* 7: 1108.

170 Presser, V., McDonough, J., Yeon, S.-H., and Gogotsi, Y. (2011). *Energy Environ. Sci.* 4: 3059.

171 Chmiola, J., Largeot, C., Taberna, P.L. et al. (2010). *Science* 328: 480.

172 Qian, D., Lei, C., Wang, E.-M. et al. (2014). *ChemSusChem* 7: 291.

173 Liu, L., Deng, Q.-F., Hou, X.-X., and Yuan, Z.-Y. (2012). *J. Mater. Chem.* 22: 15540.

174 Jin, Y., Hawkins, S.C., Huynh, C.P., and Su, S. (2013). *Energy Environ. Sci.* 6: 2591.

175 Durá, G., Budarin, V.L., Castro-Osma, J.A. et al. (2016). *Angew. Chem. Int. Ed.* 55: 9173.

176 Marco-Lozar, J.P., Kunowsky, M., Suárez-García, F. et al. (2012). *Energy Environ. Sci.* 5: 9833.

177 Hao, G.-P., Mondin, G., Zheng, Z. et al. (2015). *Angew. Chem. Int. Ed.* 54: 1941.

178 Thote, J.A., Iyer, K.S., Chatti, R. et al. (2010). *Carbon* 48: 396.

179 Zhang, X., Lin, D., and Chen, W. (2015). *RSC Adv.* 5: 45136.

180 Talapaneni, S.N., Lee, J.H., Je, S.H. et al. (2017). *Adv. Funct. Mater.* 27: 1604658.

181 Zhao, Y., Zhao, L., Yao, K.X. et al. (2012). *J. Mater. Chem.* 22: 19726.

182 Nandi, M., Okada, K., Dutta, A. et al. (2012). *Chem. Commun.* 48: 10283.

183 Chen, C., Kim, J., and Ahn, W.-S. (2012). *Fuel* 95: 360.

184 Guo, L.P., Hu, Q.T., Zhang, P. et al. (2018). *Chem. A Eur. J.* 24: 8369–8374.

185 Zhao, Y., Liu, X., Yao, K.X. et al. (2012). *Chem. Mater.* 24: 4725.

186 Zhong, M., Natesakhawat, S., Baltrus, J.P. et al. (2012). *Chem. Commun.* 48: 11516.

187 Bai, R., Yang, M., Hu, G. et al. (2015). *Carbon* 81: 465.

188 Tian, W., Zhang, H., Sun, H. et al. (2016). *Adv. Funct. Mater.* 26: 8651.

189 Boehm, H.P., Mair, G., Stoehr, T. et al. (1984). *Fuel* 63: 1061.

190 Stohr, B., Boehm, H.P., and Schlogl, R. (1991). *Carbon* 29: 707.

191 Przepiórski, J., Skrodzewicz, M., and Morawski, A.W. (2004). *Appl. Surf. Sci.* 225: 235.

192 Wickramaratne, N.P. and Jaroniec, M. (2013). *J. Mater. Chem. A* (1): 112.

193 Pevida, C., Plaza, M.G., Arias, B. et al. (2008). *Appl. Surf. Sci.* 254: 7165.

194 Plaza, M.G., Rubiera, F., Pis, J.J., and Pevida, C. (2010). *Appl. Surf. Sci.* 256: 6843.

195 Xia, Y., Zhu, Y., and Tang, Y. (2012). *Carbon* 50: 5543.

196 Liu, Y. and Wilcox, J. (2012). *Environ. Sci. Technol.* 46: 1940.

197 Kumar, K.V., Müller, E.A., and Rodríguez-Reinoso, F. (2012). *J. Phys. Chem. C* 116: 11820.

198 Jain, S.K., Pikunic, J.P., Pellenq, R.J.-M., and Gubbins, K.E. (2005). *Adsorption* 11: 355.

199 Palmer, J.C., Moore, J.D., Roussel, T.J. et al. (2011). *Phys. Chem. Chem. Phys.* 13: 3985.

200 Tenney, C.M. and Lastoskie, C.M. (2006). *Environ. Prog.* 25: 343.

201 Liu, Y.Y. and Wilcox, J. (2012). *Int. J. Coal Geol.* 104: 83.

202 Lu, X., Jin, D., Wei, S. et al. (2015). *Nanoscale* 7: 1002.

203 Zhao, J.J., Buldum, A., Han, J., and Lu, J.P. (2002). *Nanotechnology* 13: 195.

204 Wang, M., Fan, X., Zhang, L. et al. (2017). *Nanoscale* 9: 17593.

205 Babarao, R., Dai, S., and Jiang, D. (2012). *J. Phys. Chem. C* 116: 7106.

206 Srinivas, G., Burress, J., and Yildirim, T. (2012). *Energy Environ. Sci.* 5: 6453.

207 Ghosh, A., Subrahmanyam, K.S., Krishna, K.S. et al. (2008). *J. Phys. Chem. C* 112: 15704.

208 Zhao, Y., Ding, H., and Zhong, Q. (2012). *Appl. Surf. Sci.* 258: 4301.

209 Koenig, S.P., Wang, L., Pellegrino, J., and Bunch, J.S. (2012). *Nat. Nanotechnol.* 7: 728.

210 Iijima, S. (1991). *Nature* 354: 56.

211 Kowalczyk, P., Furmaniak, S., Gauden, P.A., and Terzyk, A.P. (2010). *J. Phys. Chem. C* 114: 21465.

212 Su, F., Lu, C., and Chen, H.-S. (2011). *Langmuir* 27: 8090.

213 Dillon, E.P., Crouse, C.A., and Barron, A.R. (2008). *ACS Nano* 2: 156.

214 Liu, H., Cooper, V.R., Dai, S., and Jiang, D. (2012). *J. Phys. Chem. Lett.* 3: 3343.

215 Asai, M., Ohba, T., Iwanaga, T. et al. (2011). *J. Am. Chem. Soc.* 133: 14880.

216 Mantzalis, D. and Asproulis, N. (2011). *Phys. Rev. E84*: 066304.

217 Su, F., Zhao, X.S., Wang, Y., and Lee, J.Y. (2007). *Microporous Mesoporous Mater.* 98: 323.

218 Wang, X., Bozhilov, K.N., and Feng, P. (2006). *Chem. Mater.* 18: 6373.

219 Huwe, H. and Froeba, M. (2007). *Carbon* 45: 304.

220 Wikander, K., Hungria, A.B., Midgley, P.A. et al. (2007). *J. Colloid Interface Sci.* 305: 204.

221 Jang, J.H., Han, S., Hyeon, T., and Oh, S.M. (2003). *J. Power Sources* 123: 79.

222 Kim, H., Kim, P., Joo, J.B. et al. (2006). *J. Power Sources* 157: 196.

223 Angeletti, E., Canepa, C., Martinetti, G., and Venturello, P. (1989). *J. Chem. Soc.* 105.

224 Yue, M.B., Chun, Y., Cao, Y. et al. (2006). *Adv. Funct. Mater.* 16: 1717.

225 Zhao, L., Bacsik, Z., Hedin, N. et al. (2010). *ChemSusChem* 3: 840.

226 Hwang, C.-C., Jin, Z., Lu, W. et al. (2011). *ACS Appl. Mater. Interfaces* 3: 4782.

227 Min, Y., Hong, S.-M., Kim, S. et al. (2014). *Korean J. Chem. Eng.* 31: 1668.

228 Sitthikhankaew, R., Chadwick, D., Assabumrungrat, S., and Laosiripojana, N. (2013). *Chem. Eng. Commun.* 201: 257.

229 Kim, Y.K., Kim, G.M., and Lee, J.W. (2015). *J. Mater. Chem. A* 3: 10919.

230 Bhagiyalakshmi, M., Lee, J.Y., and Jang, H.T. (2010). *Int. J. Greenhouse Gas Control* 4: 51.

231 Siriwardane, R.V. and Stevens, R.W. Jr., (2009). *Ind. Eng. Chem. Res.* 48: 2135.

232 Zhang, Z., Zhu, C., Sun, N. et al. (2015). *J. Phys. Chem. C* 119: 9302.

233 Qian, D., Lei, C., Hao, G.-P. et al. (2012). *ACS Appl. Mater. Interfaces* 4: 6125.

234 Guo, L.P., Li, W.C., Qiu, B. et al. (2019). *J. Mater. Chem. A* 7: 5402.

235 Zhou, J., Su, W., Sun, Y. et al. (2016). *J. Chem. Eng. Data* 61: 1348.

4

Porous Aromatic Frameworks for Carbon Dioxide Capture

Teng Ben¹ and Shilun Qiu²

¹*Department of Chemistry, Jilin University, Changchun, China*
²*State Key Laboratory of Inorganic Synthesis and Preparative Chemistry, Jilin University, Changchun, China*

CHAPTER MENU

4.1 Introduction

The world climate and environmental problems such as global warming, sea level rise, and an irreversible increase in the acidity levels of the oceans caused by excessive greenhouse gases emissions have attracted widespread public concern in recent years. Carbon dioxide is the major greenhouse gas that is responsible for 63% of the warming attributable to all greenhouse gases. As measured by Scripps Institute of Oceanography, the CO_2 concentration increased from c. 315 ppm in March 1958 to 391 ppm in January 2011, and close to 398 ppm in January 2014 [1]. These emissions are mainly generated from the combustion of coal, oil, and natural gas, the main energy resources for our daily life. For example, a 500 MW coal power plant will generate three million tons of CO_2 per year [2]. Though there are many efforts to develop new techniques for alternative energy, such as wind and solar, it is predicted that this trend of increasing atmospheric CO_2 concentration will not be altered within the next several decades because fossil fuels will remain the dominant energy source. Thus, it is critical to decrease the CO_2 concentration in the atmosphere.

Carbon capture and sequestration (CCS) has been explored as a means to decrease the CO_2 concentration, and a variety of materials have been developed for this purpose [3]. Especially in recent decades, a number of porous organic frameworks such as metal-organic frameworks (MOFs) [4–9], hyper-crosslinked polymers (HCPs) [10–12],

Materials for Carbon Capture, First Edition. Edited by De-en Jiang, Shannon M. Mahurin and Sheng Dai.
© 2020 John Wiley & Sons Ltd. Published 2020 by John Wiley & Sons Ltd.

polymers of intrinsic microporosity (PIMs) [13–16], covalent organic frameworks (COFs) [17–24], conjugated microporous polymers (CMPs) [25–29], and porous aromatic frameworks (PAFs) [30–36] have been reported for CCS. Among these porous materials, MOFs are composed of coordination bonds, and all the others are composed of covalent bands. HCPs represent a broad class of porous materials – one of the first kinds of pure microporous organic materials –, which are mainly prepared by Friedel-Crafts alkylation chemistry. PIMs have good solubility and can easily be used to form membranes. COFs are a type of crystalline organic porous material, and their high surface area and structural diversity make them very important. CMPs are conjugated porous materials and show great potential in photoelectric devices. PAFs are an important milestone in porous polymeric materials that have demonstrated for the first time that disordered materials can possess high surface area while also exhibiting excellent physicochemical stability, which is another obvious advantage. In this chapter, we will discuss the progress of PAFs along with their potential application in CO_2 capture and separation.

4.2 Carbon Dioxide Capture of Porous Aromatic Frameworks

There are primarily three different CO_2 capture systems: (i) pre-combustion removal of CO_2 from H_2 or CH_4 of syngas; (ii) post-combustion separation of CO_2 from flue gases after the combustion of primary fuel in air; and (iii) oxy-fuel combustion, which isolates CO_2 from water, the other combustion product [3]. The components of flue gas in the post-combustion process are primarily nitrogen (N_2, >70%) and CO_2 (10–15%), which accounts for roughly 33–40% of global CO_2 emissions. In the post-combustion model, adsorption of CO_2 occurs at 1 bar and desorption occurs at 0.1 bar. In comparison with the post-combustion carbon capture, the pre-combustion process is carried out at a higher pressure and also has a higher concentration of CO_2 (15–60%). The exhaust from oxygen combustion has a very high concentration of CO_2, requiring sorbents that are stable in high humidity.

The ideal CO_2 capture system should have sufficient space for CO_2 storage and an appropriate binding capacity between the CO_2 molecules and the sorbent material. Generally, a large pore volume and a high surface area contribute to the high storage uptake, especially in the high-pressure range. Matching the pore size with the kinetic diameter of CO_2 and a suitable binding energy leads to optimal interactions between the gas molecules and the material. In this regard, PAFs show potential for carbon dioxide capture when applied in pressure swing adsorption (PSA), temperature swing adsorption (TSA), or vacuum swing adsorption (VSA) systems. To improve the CO_2 sorption capacity, different strategies such as increasing the surface area, heteroatom doping, tailoring the pore size, and post-modification have been developed and will be discussed in detail.

4.3 Strategies for Improving CO_2 Uptake in Porous Aromatic Frameworks

4.3.1 Improving the Surface Area

For porous materials used for gas storage, the surface area of the material is always one of the most important factors. One obvious advantage of PAFs is their extremely high surface area. The first PAF, PAF-1, exhibited an ultrahigh surface area of 5600 $m^2 g^{-1}$ using BET calculation theory, making it an excellent candidate for gas storage. High-pressure H_2 storage of PAF-1 shows 7.0 wt% excess hydrogen uptake at 77 K, 48 bar, which equals 10.7 wt% absolute uptake. High-pressure CO_2 adsorption results also indicate a high uptake of 1300 mg g^{-1} CO_2 at 40 bar and room temperature, suggesting its potential application for CO_2 capture [30]. Using an improved Yamamoto type Ullmann cross-coupling method, another porous polymer network named PPN-4 was reported by Zhou et al.; this structure is also known as PAF-3 [37]. PPN-4 has a higher surface area than PAF-1, with a value of 6461 $m^2 g^{-1}$, and its high-pressure gas storage is even greater. Experimental results show that the total CO_2 storage capacity could reach 2121 mg g^{-1}at 295 K and 50 bar. Though this value is not the highest among all the porous materials and many MOFs show higher surface area and greater CO_2 storage capacity, the weak coordination in MOFs makes them rather unstable in harsh conditions and not comparable with PAFs for long-term storage and cycling.

4.3.2 Heteroatom Doping

Though PAF-1 shows very high surface area and could potentially store a large amount of CO_2 at high pressure, the weak interaction with CO_2 results in low gas storage at low pressure, which is unsuitable for CO_2 capture that generally occurs at low CO_2 concentration. Heteroatom doping seems to be a good method to improve its ability to store gases at low pressure.

Recently, PAFs with the same topology as PAF-1 but different heteroatoms containing PAF-3 and PAF-4 (PAF-3 containing Si and PAF-4 containing Ge) were reported and investigated for gas storage [38]. Low-pressure CO_2 storage was performed, and the result indicated that high CO_2 sorption capacities of 9.1 wt%(46 $cm^3 g^{-1}$) for PAF-1, 15.3 wt% (78 $cm^3 g^{-1}$) for PAF-3, and 10.7 wt% (54 $cm^3 g^{-1}$) for PAF-4 were achieved at 273 K and 1 atm, respectively (see Figure 4.1). When the CO_2 sorption behavior was measured at 298 K and 1 atm, PAF-1, PAF-3, and PAF-4 adsorbed 4.8, 8.0, and 5.1 wt%, respectively. Depending on the degree of CO_2 loading at 273 K and 298 K, the zero-coverage isosteric enthalpy of CO_2 adsorption (Q_{stCO2} for short) is 15.6, 19.2, and 16.2 kJ mol^{-1} for PAF-1, PAF-3, and PAF-4 respectively, which clearly suggests that the interaction between the adsorbate–adsorbent is maximized by both aromatic rings in PAFs and overlap of the potential energy fields from the micropore walls. Thus we conclude that at high pressure, the surface area of a material plays a more important role in gas sorption because the total volume determines the number of gas molecules that can go into the pores; at low pressure, the interaction between adsorbate and adsorbent (reflected as heat of adsorption) is more important. Especially at ambient conditions, the effect of the isosteric heat of adsorption (Qst) often surpasses the effect of surface area. So it must be noted that for post-combustion capture of CO_2 during CCS, designing materials with higher Qst is more efficient than designing materials with higher surface area, because the CO_2 concentration in this condition is typically less than 15%.

PAF-3 and PAF-4 did show better selective adsorption CO_2 ability than PAF-1. At 1 atm and 273 K, PAF-3 exhibits extraordinarily promising selectivity of 87/1 for the adsorption of CO_2 over N_2, and PAF-4 shows a selectivity of 44/1, while this value is only 38/1 for PAF-1.

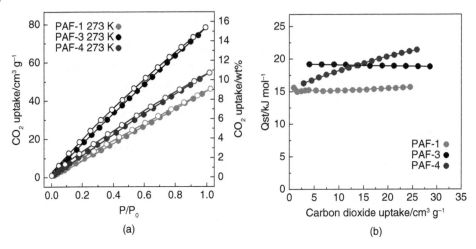

Figure 4.1 (a) CO_2 adsorption (solid symbols) and desorption (open symbols) isotherms of PAF-1, PAF-3, and PAF-4 at 273 K; (b) $QstCO_2$ of PAF-1, PAF-3, and PAF-4 as a function of the amount of CO_2 adsorbed. Source: Reproduced from Ref. [38] with permission from the Royal Society of Chemistry. (*See color plate section for color representation of this figure*).

Introducing basic nitrogen groups into porous material has been used to increase the interaction with acid CO_2 gas. Recently, a nitrogen-rich PAF has been synthesized by using 1,4-bis(diphenylamino) benzene as monomer through Yamamoto coupling [39] (see Figure 4.2). The obtained nitrogen-rich PAF (NPAF) shows a CO_2 uptake of 3.64 mmol g^{-1} at 273 K and 1 atm and CO_2/N_2 selectivity of 48 based on ideal adsorption solution theory (IAST) using a 0.15/0.85 mol ratio of CO_2/N_2. The relatively high CO_2 uptake and selectivity make the NPAF a promising material for CCS applications. In addition, hydrogen uptake of the NPAF at 77 K and 1 atm (1.87 wt%) is among the highest in porous polymers.

JUC-Z2 is also a N-containing PAF, and its low-pressure adsorption and desorption properties were determined using volumetric measurements at 273 and 298 K [39]. The absence of hysteresis in the adsorption and desorption isotherms demonstrates the reversible process of CO_2 storage in JUC-Z2. The electron-rich JUC-Z2 has an exceptionally high density of binding sites with a strong affinity for CO_2, and it is expected that 0.76 CO_2 occupied one structure building unit (SBU) of JUC-Z2 at 273 K and 760 mmHg, resulting in 71 cm^3 g^{-1} CO_2 uptake in that condition. Additionally, the quantity of CO_2 adsorbed in JUC-Z2 is higher than that in PAF-1 as a result of the Lewis acid–base interactions that exist between the electron-rich aromatic constituents and the electron-poorer CO_2 molecules, and also between lone pairs at the nitrogen atoms and CO_2. Based on the two CO_2 sorption isotherms at 273 K and 298 K, JUC-Z2 has a heat of adsorption of 23.1 kJ mol^{-1}, which is stronger than physical sorption interactions (about 5–20 kJ mol^{-1}) for storage of considerable greenhouse gases, but weaker than strong chemisorptive interactions (>50 kJ mol^{-1}) for regeneration of sorbents. Additionally, the heat of sorption curve does not change significantly with the increase of CO_2 loaded. These impressive uptake and adsorption enthalpies of JUC-Z2 exhibit potential application in CCS.

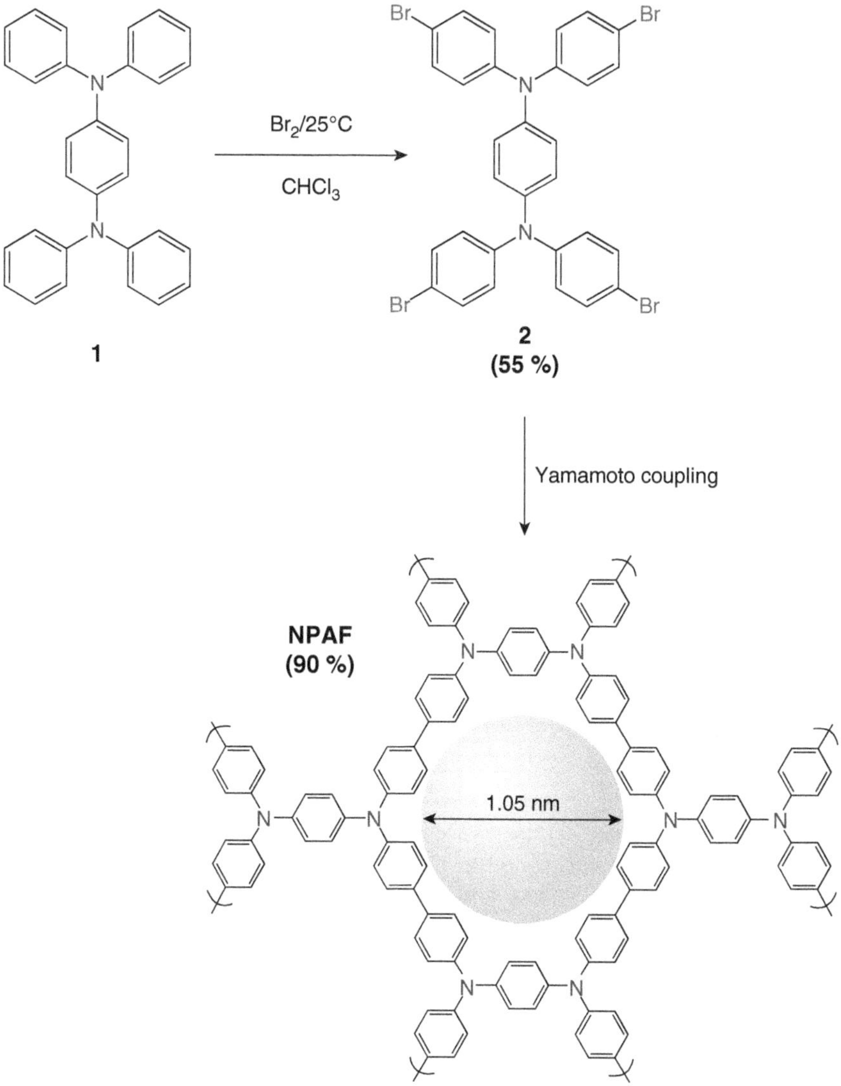

Figure 4.2 Synthetic route for the NPAF. Source: Reproduced from Ref. [39] with permission from the Royal Society of Chemistry.

Another nitrogen-rich porous aromatic framework, PAF-30, was also synthesized by Zhao et al. [40]. It is very interesting that PAF-30 shows very small pore size of 3.30–3.54 Å, which is beneficial for increasing the heat of adsorption. In the narrow-pore structures, CO_2 molecules can interact with multiple pore surfaces simultaneously. Further, the high N content in PAF-30 will also favor strong interaction with acidic CO_2. Thus, PAF-30 shows a high CO_2 heat of adsorption with a value of 36.9 kJ mol^{-1}. This results in the highly selective adsorption of CO_2 from mixed gases. IAST predictions show that the CO_2/CH_4 selectivity for PAF-30 is 24.3–63.2 for 50% CO_2 and 50% CH_4, and in the range of 4–30 for 5% CO_2 and 95% CH_4.

4.3.3 Tailoring the Pore Size

From the previous results, we can see that high surface area is very beneficial for gas storage at high pressure, while the interaction between the framework and gas molecules is more important for gas storage at low pressure. While heteroatom doping is one method to increase the interaction between gas molecules and the framework, tailoring the pore size is another potential approach that has been demonstrated by both theoretical and experimental methods. For example, Cao et al. simulated CO_2 storage in PAF-30X (X = 1–4) [41] and showed that CO_2 uptake at low pressure decreases in the order of PAF-301 > PAF-302 > PAF-303 > PAF-304; this is opposite the pore size but follows the isosteric heats, indicating that the CO_2 isosteric heats in PAFs follow the same trend as the CO_2 uptake (see Figure 4.3). This suggests that the uptake at low pressure is determined by the isosteric heats.

PAF-301 shows much higher CO_2 uptake than the other three PAFs at p < 1 bar due to its small pore size of 5.2 Å. Recent research has revealed that a suitable pore size comparable with the kinetic diameter of a CO_2 molecule is desirable for improving the CO_2 capture at room temperature. The pore size of PAF-301 is 5.2 Å, which is comparable with the kinetic diameter of CO_2, meaning PAF-301 might be a promising material for CO_2 separation at low pressure. The CO_2 uptake of PAF-301 is 275 mg g^{-1} (equivalent to 6.2 mmol g^{-1}) at 1 bar.

The CO_2 uptakes of PAFs at high pressure are shown in Figure 4.3b, where different PAFs possess the highest uptake in four pressure ranges, i.e. PAF-301 at p = 0–10 bar, PAF-302 at p = 10–32 bar, PAF-303 at 32–51 bar, and PAF-304 at p > 51 bar. At p > 55 bar, CO_2 uptake follows the order PAF-304 > PAF-303 > PAF-302 > PAF-301, which is entirely the opposite of the uptake order at low pressure. This observation reflects the fact that the uptake correlates with isosteric heats at low pressure, accessible surface areas at intermediate pressure, and free volumes at high pressure (Figure 4.4). With increasing pore sizes of PAFs, the saturated pressures for CO_2 uptake increase and are 9, 30, 52, and 60 bar, respectively. Because of the large free volume of PAF-303 and PAF-304, the uptakes of CO_2 in PAF-303 and PAF-304 reach 3432 and 3124 mg g^{-1} at 298 K and 50 bar, respectively.

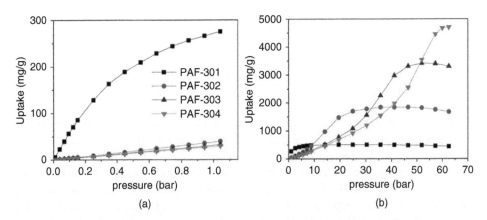

Figure 4.3 CO_2 adsorption isotherms at 298 K in porous aromatic frameworks (PAFs): (a) at low pressure; (b) at high pressure. Source: Reproduced from Ref. [41] with permission from the American Chemical Society.

Figure 4.4 Isoteric heats of CO_2 in porous aromatic frameworks (PAFs) at 298 K (1–55 bar). Source: Reproduced from Ref. [41] with permission from the American Chemical Society.

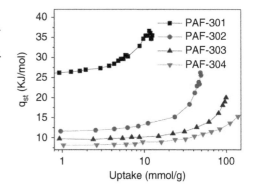

The influence of pore size on CO_2 adsorption capacity has also been demonstrated in carbonized PAF-1; this will be introduced in the next section [42]. In addition to increasing CO_2 uptake, tailoring the pore diameter of PAFs enhances their potential in gas separation. Using a condensation reaction, a pyridinium-type PAF denoted by Cl-PAF-50 has been synthesized by Yuan, while AgCl-PAF-50 can be readily obtained due to the size of Cl⁻ ions (5 Å) are large enough to accommodate Ag⁺ ions (2.52 Å) [43] (Figure 4.5). Through a feasible ion-exchange method, a family of positively charged quaternary pyridinium-type PAFs (X-PAF-50, X = F, Cl, Br, I) have been prepared and their pores systematically tuned on the angstrom scale from 3.4–7 Å. The designed pore sizes may bring benefits to capturing or sieving gas molecules with varied diameters to separate them efficiently by size-exclusion effects. By combining their specific separation properties, a five-component (hydrogen, nitrogen, oxygen, carbon dioxide, and methane) gas mixture can be separated completely.

4.3.4 Post Modification

Presumably, increasing the isosteric heat of CO_2 adsorption through the introduction of CO_2-philic moieties should have a great influence on CO_2-uptake capacity and CO_2/N_2 selectivity under ambient conditions. Indeed, significant increases of isosteric heats and CO_2/N_2 adsorption selectivity have been observed upon pre- or post-synthetic introduction of polar functionalities; however, this approach usually has a negative impact on the surface area that can lead to very low CO_2-uptake capacity if the surface area is severely compromised. One possible strategy to tackle this issue is to judiciously select porous polymers with ultrahigh surface areas and physicochemical stability as starting materials. Therefore, sufficient surface area can be retained after the introduction of CO_2-philic moieties.

PAF-1 was an appropriate material that met both requirements of high surface area and excellent physicochemical stability. By reaction with chlorosulfonic acid, PPN-6 was modified to give PPN-6-SO₃H, which was further neutralized to produce PPN-6-SO₃Li [44] (Figure 4.6). Nitrogen gas adsorption/desorption isotherms were collected at 77 K. Notably, the large desorption hysteresis in PPN-6 disappeared, and almost-ideal type-I isotherms were obtained for sulfonate-grafted PPN-6. The BET surface areas obtained from the experimental data were 4023, 1254, and 1186 m² g⁻¹ for PPN-6, PPN-6-SO₃H, and PPN-6-SO₃Li, respectively. Along with the decrease in surface area, the pore size became

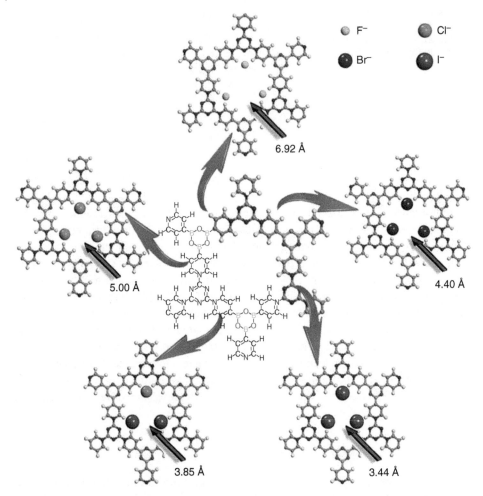

Figure 4.5 Scheme of preparation of F-PAF-50, Br-PAF-50, 2I-PAF-50, and 3I-PAF-50 from Cl-PAF-50. Source: Reproduced from Ref. [43] with permission from Springer Nature. (*See color plate section for color representation of this figure*).

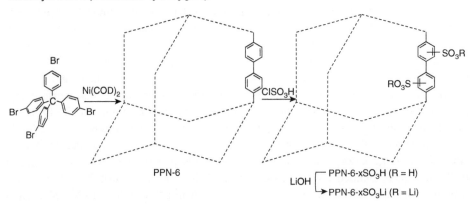

Figure 4.6 Synthesis and grafting of PPN-6. Source: Reproduced from Ref. [44] with permission from the American Chemical Society.

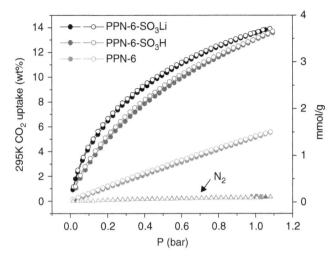

Figure 4.7 Gravimetric CO_2 and N_2 adsorption/desorption isotherms at 295 K. Source: Reproduced from Ref. [44] with permission from the American Chemical Society. (*See color plate section for color representation of this figure*).

progressively smaller with aromatic sulfonation and ensuing lithiation. The relatively small pore sizes of both sulfonate-grafted PPN-6 networks fall into the 5.0–10.0 Å range, which is believed to be suitable for CO_2 uptake and thus for CO_2 separation from other gases with relatively larger kinetic diameters, such as N_2 and CH_4. Non-grafted PPN-6 has a gravimetric CO_2 uptake of 5.1 wt % at 295 K and 1 bar, whereas sulfonate-grafted PPN-6 showed remarkable increases in gravimetric CO_2 uptake, with values of 13.1 and 13.5 wt % (equivalent to 3.6 and 3.7 mmol g^{-1}) for PPN-6-SO$_3$H and PPN-6-SO$_3$Li, respectively (Figure 4.7). Sulfonate-grafted PPN-6 exhibited exceptionally high adsorption selectivity for CO_2 over N_2 at 295 K and 1 bar (Sads = 150 for PPN-6-SO$_3$H and 414 for PPN-6-SO$_3$Li). (i) Functionalization of all-carbon-scaffold frameworks was expected to create electric fields on the surface that impart to the networks a strong affinity toward CO_2 molecules through its high quadrupole moment. (ii) Small pore size and polar functionalities will increase the heat of adsorption. At zero-loading, PPN-6-SO$_3$H and PPN-6-SO$_3$Li showed heats of adsorption reaching 30.4 and 35.7 kJ mol^{-1}, respectively, which are substantially higher than that of nongrafted PPN-6 (17 kJ mol^{-1}).

The metal cation in the framework acts as an open coordination site after full activation, which leads to stronger electrostatic interactions between CO_2 and the metal cation. The light element Li is a commonly used dopant. Lithiation was conducted by using PAF-1 as host material [45]. After lithiation, the BET surface area dropped from 3639 m^2 g^{-1} for PAF-1 to 1358 m^2 g^{-1} and 479 m^2 g^{-1} for 1%_Li@PAF-1 and 5%_Li@PAF-1, respectively. Lithiation leads to shrinkage of the pores and a decrease in the surface area. The pore size decreased from 14 Å in PAF-1 to 11 Å in 5%_Li@PAF-1. Low-pressure CO_2 sorption measurements at 273 and 298 K were performed on Li@PAF-1, 1%_Li@PAF-1, 2%_Li@PAF-1, 5%_Li@PAF-1, and 10%_Li@PAF-1 to obtain the CO_2 storage properties (Figure 4.8). The CO_2 uptakes at 273 K reveal a remarkable enhancement in CO_2 capture capacities upon reduction by lithiation. 5%_Li@PAF-1 (8.99 mmol g^{-1}) shows the highest CO_2 uptake at 273 K and 1.22 bar.

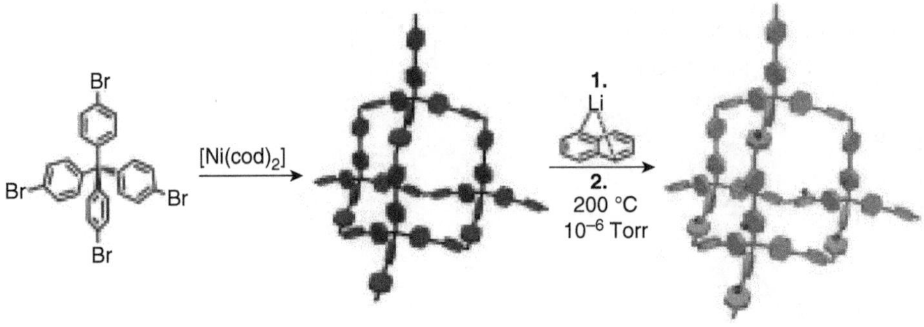

Figure 4.8 Synthetic route to Li-PAF-1. Source: Reproduced from Ref. [45] with permission from John Wiley & Sons.

CO$_2$ removal from flue gas by sorption and stripping with aqueous amines (commonly monoethanolamine [MEA], diethanolamine [DEA], and methyldiethanolamine [MDEA]) has been established since 1930 and is still believed to be a feasible technology. However, this process suffers from a series of inherent problems, including the corrosive nature of the amines, fouling of the process equipment, and high regeneration energy. To avoid these problems, solid adsorbents with organic amines loaded on certain support materials are intensively investigated.

PPN-6, which has an exceptionally high surface area and an extremely robust all-carbon scaffold based on biphenyl rings, is ideal for the introduction of CO$_2$-philic groups under harsh reaction conditions. Figure 4.9 shows the synthetic route to polyamine-tethered

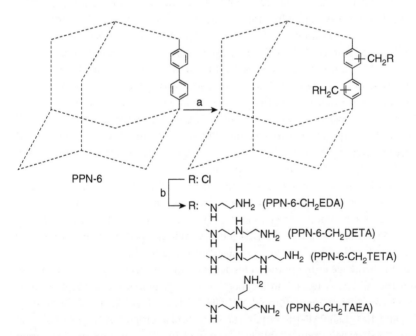

Figure 4.9 Synthetic route to polyamine-tethered PPNs. (a) CH$_3$COOH/HCl/H$_3$PO$_4$/HCHO, 90 °C, 3 days; (b) amine, 90 °C, 3 days. Source: Reproduced from Ref. [46] with permission from John Wiley & Sons.

PPNs [46]. Nitrogen gas adsorption/desorption isotherms were collected at 77 K. The calculated BET surface areas were 1740, 1014, 663, 634, and 555 $m^2 g^{-1}$ for PPN-6-CH$_2$Cl, PPN-6-CH$_2$EDA, PPN-6-CH$_2$TAEA, PPN-6-CH$_2$TETA, and PPN-6-CH$_2$DETA, respectively. Notably, the huge adsorption/desorption hysteresis in PPN-6-CH$_2$Cl disappeared into a nearly type I isotherm for the polyamine-tethered PPNs. Along with the decrease in surface area, the pores are smaller after tethering, which supports the reaction occurring within the cavities.

The CO_2 storage capacities were measured at 295 K in the low-pressure range (Figure 4.10). Interestingly, the highest CO_2 uptake at 295 K and 1 bar belonged to PPN-6-CH$_2$DETA at 4.3 mmol g^{-1} (15.8 wt%), despite it having the smallest surface area. The recorded CO_2 uptake value demonstrated that the strong interaction between CO_2 molecules and the polar species-modified PAFs played a more significant role than the surface area. As expected, the isosteric heats of adsorption (Qst) calculated from dual-site

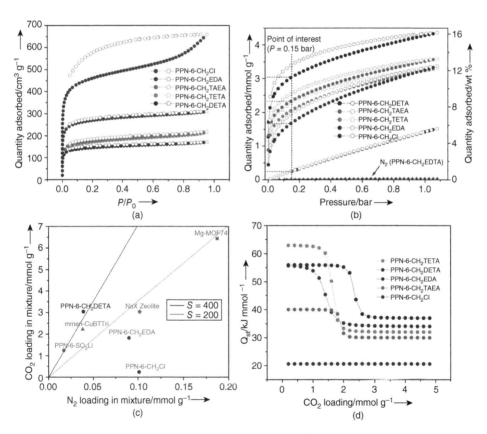

Figure 4.10 (a) N$_2$ adsorption and desorption isotherms at 77 K; (b) CO$_2$ adsorption and desorption isotherms, as well as PPN-6-CH$_2$DETA N$_2$ adsorption, at 295 K; (c) the component loadings of N$_2$ and CO$_2$ calculated by IAST with bulk gas-phase partial pressures of 85 kPa and 15 kPa for N$_2$ and CO$_2$, respectively, with PPN-6-CH$_2$DETA, PPN-6-CH$_2$EDA, PPN-6-CH$_2$Cl, PPN-6-SO$_3$Li, NaX zeolite, MgMOF-74, mmen-CuBTTri; loadings for calculated selectivities of 200 and 400 are shown as a guide; (d) isosteric heats of adsorption Qst for the adsorption of CO$_2$, calculated using the dual-site Langmuir isotherm fits. Source: Reproduced from Ref. [46] with permission from John Wiley & Sons. (*See color plate section for color representation of this figure*).

Langmuir fits of the polyamine-tethered PAFs are higher than that of PPN-CH$_2$Cl and PPN-6. Since the CO$_2$ constituent in flue gas is 15%, CO$_2$ uptake at 0.15 bar attracts much attention. At 295 K and 0.15 bar, PPN-6-CH$_2$Cl can store 0.25 mmol g^{-1} (1.1 wt%), while PPN-6-CH$_2$DETA takes up 3.0 mmol g^{-1} (11.8 wt%). Furthermore, the CO$_2$ uptake value dropped only slightly to 10.0 wt% when the temperature increased to 313 K. The high CO$_2$ uptake capacities that were retained at the elevated temperatures indicate that the materials are more useful under realistic flue-gas condition. Ideal adsorbed solution theory (IAST) modeling was used to evaluate the CO$_2$/N$_2$ selectivities. The poor N$_2$ uptake due to the low porosity of PPN-6-CH$_2$DETA leads to high CO$_2$/N$_2$ selectivity. The value is 442 and 115 for PPN-6-CH$_2$DETA and PPN-6-CH$_2$EDA, respectively, which is higher than that of PPN-6-CH$_2$Cl (S = 13).

Carbonization is also a promising method to improve the CO$_2$ capture ability of PAFs. By heat treatment of PAF-1 to a proper temperature, a series of carbonized PAF-1 s were obtained with enhanced gas storage capacities and isosteric heats of adsorption [42] (Figure 4.11). The carbonization of PAF-1 leads to a continuous shrinking of the pore size and surface area. The apparent surface areas calculated from BET models for relative pressure between 0.01 and 0.1 were 4033, 2881, 2292, and 1191 m^2 g^{-1} for PAF-1-350, PAF-1-380, PAF-1-400, and PAF-1-450, respectively. The total pore volumes of those samples calculated by using the density functional theory (DFT) method dropped from 2.43 cm^3 g^{-1} for PAF-1 to 0.53 cm^3 g^{-1} for PAF-1-450, while the pore size distribution shrank from 1.44 nm for PAF-1 to 1.00 nm for PAF-1-450 (Figure 4.12). With the increase in carbonization temperature, the pore size decreases continuously to 1 nm, which may be suitable for CO$_2$ capture. Furthermore, stronger interactions between the carbonized PAFs are desirable because the all-carbon-scaffold networks are expected to create an electric field onto the frameworks surface that imparts a strong affinity of the networks toward CO$_2$ due to its high quadrupole moment. Indeed, the carbonized PAF-1s exhibit unexpected enhanced CO$_2$ adsorption capacities, which show complete reversibility at ambient condition of 273 K and 1 bar. PAF-1-450 shows a remarkable increase in CO$_2$ uptake, with a value of 100 cm^3 g^{-1} (equivalent to 16.5 wt%, 4.5 mmol g^{-1}).

To further investigate the factors that affect the CO$_2$ uptakes of these carbonized PAF-1s, the heat of adsorption of CO$_2$ was measured. As expected, the Q$_{stCO2}$ of carbonized PAF-1 increased evidently from the original value of 15.6 kJ mol^{-1} for PAF-1 to 27.8 kJ mol^{-1} for PAF-1-450. That is to say, the carbonization causes a 78% increase in the heat of adsorption, which provides a potential mechanism for the fact that the CO$_2$ adsorption capacity of PAF-1-450 is larger than that of PAF-1, although the BET surface area of PAF-1-450 after carbonization decreases compared with the PAF-1.

Besides the improved gas uptake, the carbonized samples also show very high CO$_2$ selectivity toward other gases. On the basis of single-component isotherm data, the dual-site Langmuir–Freundlich adsorption model-based IAST prediction indicates that the CO$_2$/N$_2$ adsorption selectivity is as high as 209 at a 15/85 CO$_2$/N$_2$ ratio. Also, the CO$_2$/CH$_4$ adsorption selectivity is in the range of 7.8–9.8 at a 15/85 CO$_2$/CH$_4$ ratio at 0 < p < 40 bar, which is highly desirable for landfill gas separation. The calculated CO$_2$/H$_2$ adsorption selectivity is approximately 392 at 273 K and 1 bar for a 20/80 CO$_2$/H$_2$ mixture.

As a further work on carbonization of PAF-1, Qiu et al. used a high-temperature KOH activation method to produce special, bimodal microporous carbons (denoted

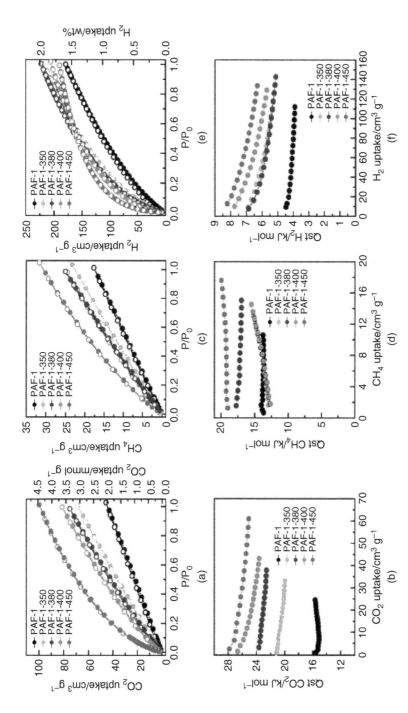

Figure 4.11 (a) CO_2 adsorption (solid symbols) and desorption (open symbols) isotherms of PAF-1 and carbonized samples at 273 K; (b) $QstCO_2$ of PAF-1 and carbonized samples as a function of the amount of CO_2 adsorbed. (c) CH_4 adsorption (solid symbols) and desorption (open symbols) isotherms of PAF-1 and carbonized samples at 273 K; (d) $QstCH_4$ of PAF-1 and carbonized samples as a function of the amount of CH_4 adsorbed. (e) H_2 adsorption (solid symbols) and desorption (open symbols) isotherms of PAF-1 and carbonized samples at 77 K; (f) $QstH_2$ of PAF-1 and carbonized samples as a function of the amount of H_2 adsorbed. Source: Reproduced from Ref. [42] with permission from the Royal Society of Chemistry. (See color plate section for color representation of this figure).

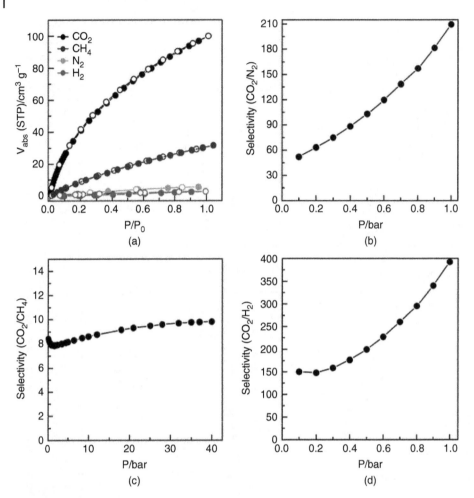

Figure 4.12 (a) CO_2, CH_4, N_2 and H_2 sorption of PAF-1-450 at 273 K; (b) IAST-predicted adsorption selectivity of CO_2/N_2 of PAF-1-450 using a 15/85 CO_2/N_2 ratio; (c) IAST-predicted adsorption selectivity of CO_2/CH_4 of PAF-1-450 using 15/85 CO_2/CH_4 ratio; (d) IAST-predicted adsorption selectivity of CO_2/H_2 of PAF-1-450 using 20/80 CO_2/H_2 ratio. Source: Reproduced from Ref. [42] with permission from the Royal Society of Chemistry.

by K-PAF-1-X, where X is the thermolysis temperature) with high surface areas [47] (Figure 4.13). Compared with the low-temperature carbonized PAF-1-X series, these high-temperature KOH activated carbons exhibited enhanced gas uptakes. Among them, K-PAF-1-600 showed the highest CO_2 uptake capacity of 161 cm^3 g^{-1} at 273 K and 1 bar (equivalent to 24.0 wt%, 7.2 mmol g^{-1}). K-PAF-1-750 exhibited the highest hydrogen uptake with a value of 342.2 cm^3 g^{-1} (15.3 mmol g^{-1}, 3.06 wt%) at 77 K and 1 bar. Furthermore, high-pressure gas adsorption measurements revealed that K-PAF-1-750 was able to store 1320 mg g^{-1} of carbon dioxide (56.9 wt%, 40 bar, RT), 207 mg g^{-1} of methane (17.1 wt%, 35 bar, RT) and 71.6 mg g^{-1} of hydrogen (6.68 wt%, 48 bar, 77 K). In addition, it also exhibited excellent methane storage ability (0.207 g g^{-1}) at room temperature. The authors

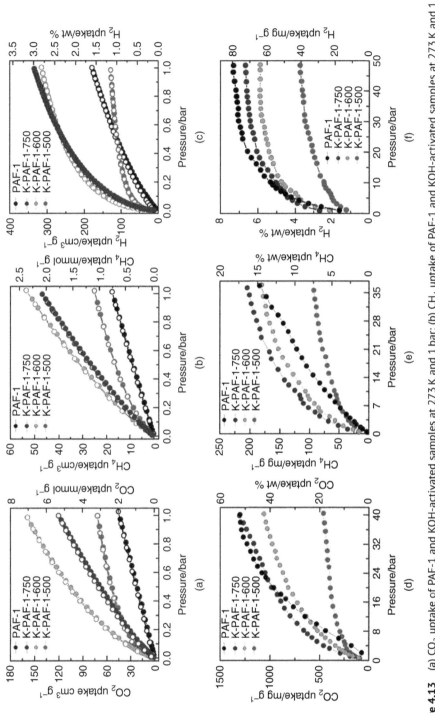

Figure 4.13 (a) CO_2 uptake of PAF-1 and KOH-activated samples at 273 K and 1 bar; (b) CH_4 uptake of PAF-1 and KOH-activated samples at 273 K and 1 bar; (c) H_2 uptake of PAF-1 and KOH-activated samples at 77 K and 1 bar; (d) high-pressure CO_2 uptake of KOH-activated samples at 298 K; (e) high-pressure CH_4 uptake of KOH-activated samples at 298 K; (f) high-pressure H_2 uptake of KOH-activated samples at 77 K. Source: Reproduced from Ref. [47] with permission from Springer Nature. (*See color plate section for color representation of this figure*).

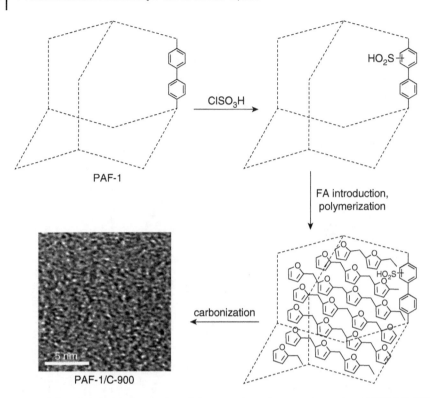

Figure 4.14 Schematic illustration of the procedures for the preparation of PAF-1/C-900. Source: Reproduced from Ref. [48] with permission from the Royal Society of Chemistry.

proposed that the carbon networks could be expected to create strong polarity on the surface of the frameworks, which imparts the networks with a strong affinity toward gas molecules such as hydrogen, methane, and carbon dioxide through their high quadrupole moments.

Ma et al. used a modified PAF-1 as the precursor to prepare microporous carbon materials [48]. As shown in Figure 4.14, PAF-1 was first post-grafted with sulfonic acid in order to introduce an extra carbon source into it because it has been shown that sulfonic acid can serve as the catalytic site for the polymerization of furfuryl alcohol. The resulting PAF-1-SO$_3$H was then stirred in FA for two days, during which the adsorbed FA was catalyzed by the grafted sulfonic acid to polymerize within the pores of PAF-1. After careful filtration and washing with ethanol to remove FA physically adsorbed on the exterior surface, the obtained FA-PAF-1-SO$_3$H composite was heated at 900 °C under an inert gas atmosphere for eight hours to synthesize the carbonized material PAF-1/C-900. Pore-size distribution analysis based on the Horvath–Kawazoe (HK) model, which is widely employed for micropore size analysis, indicates that the pore size of PAF-1/C-900 is predominantly distributed around 5.4 Å in comparison with 14.5 Å for PAF-1. As a result of the substantially reduced pore size, PAF-1/C-900 can adsorb a large amount of CO$_2$ with an uptake capacity of 93 cm^3 g^{-1} (equivalent to 4.1 mmol g^{-1} or 18.2 wt%) under 1 atm and 295 K, meaning enhancement by a factor of 2.4 compared to parent PAF-1. In addition, a dramatic increase in Qst of CO$_2$ by 11.6 kJ mol^{-1} at zero-loading compared to the parent

PAF-1 was observed. The strategy of pre-introducing extra carbon sources into the PAF materials followed by thermolysis represents a promising approach to create microporous carbon materials with very narrow pores (8 Å), which could hold promise for CO_2 capture applications.

Recently, another porous aromatic framework (PAF-6) containing 41% nitrogen in its scaffold was employed as a self-sacrificing template to prepare the porous nitrogen-rich carbon with furfuryl alcohol (FA) as a carbon precursor [49]. The carbon material obtained at 1000 °C (POF-C-1000) with high N content (about 6 wt%) displayed the highest surface area of 785 m^2 g^{-1} with a H$_2$ uptake of 1.6 wt% (77 K, 1 bar) and a CO_2 uptake of 3.5 mmol g^{-1} (273 K, 1 bar).

These results have shown that post modification of PAF-1 with functional groups could greatly increase its low-pressure CO_2 capture capacity and selectivity. However, the post modification usually must be performed under severe conditions that can cause a dramatic decrease in the surface area of the framework. In addition to post modification, the introduction of functional groups during synthesis is also a promising method to increase the gas storage ability. By choosing tetraphenyl methane and its amino and hydroxyl derivatives as building blocks, and formaldehyde dimethyl acetal (FDA) as external cross-linkers, PAF-32 was obtained in the presence of FeCl$_3$ as catalyst, and three PAF materials (PAF-32s) based on tetrahedral building blocks were successfully synthesized [50] (Figure 4.15). Amino and hydroxyl functional groups were successfully incorporated in PAF-32 networks as confirmed by various spectroscopic methods. The functionalized PAF-32 exhibited enhanced CO_2 capture capability because of the presence of amino and hydroxyl groups in the frameworks. This was also proven by the Qst of PAF-32-NH$_2$ and PAF-32-OH, which are both higher than PAF-32 due to the stronger interaction of CO_2 with the electron-rich functional groups.

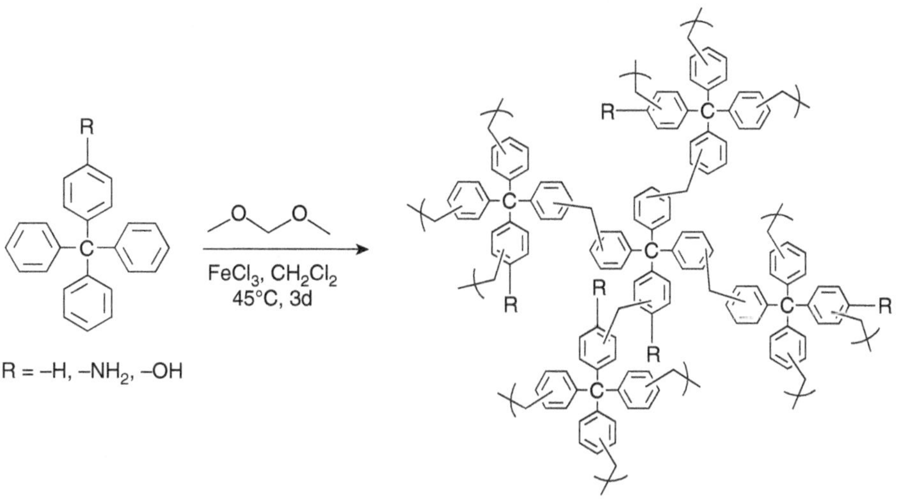

Figure 4.15 Schematic depiction of synthesis of PAF-32 and its functionalized analogues PAF-32-NH$_2$ and PAF-32-OH. Source: Reproduced from Ref. [50] with permission from the Royal Society of Chemistry.

4.4 Conclusion and Perspectives

In conclusion, as a new generation of porous materials, PAFs show great potential for CO_2 capture, and major progress has been made in the past several years in PAF synthesis. PAFs combine extremely high surface area and excellent physicochemical stability, which provide obvious advantages compared with other porous materials.

Though great progress has been made in the synthesis and application of PAFs, there are still some aspects that require further investigation before the application of PAFs becomes a reality. First, the Yamamoto-type Ullmann cross-coupling reaction seems to be the most effective method for the synthesis of PAFs. However, the high cost and harsh synthetic conditions of this reaction limit the ability to scale up the process to make the materials industrially valuable. Therefore, new low-cost, effective methods are critically needed for PAF synthesis. Second, post-modification of PAFs has been demonstrated to be an effective method to increase the gas sorption capacity; however, post-modification is usually conducted under harsh conditions, which also limits the ability to scale it up. Finally, all the PAFs are currently synthesized in powder form, which is a challenge for use in gas-separation applications. Their potential for gas separation would be improved if PAF membranes could be synthesized easily, but this is very challenging.

References

1 Scripps Institute of Oceanography. (2019). Atmospheric CO_2 data. http://scrippsco2.ucsd .edu/data/atmospheric_co2/.

2 Liu, W., King, D., Liu, J. et al. (2009). *JOM* 61: 36–41.

3 Metz, B., Davidson, O., de Coninck, H.C. et al. (eds.) (2005). *IPCC Special Report on Carbon Dioxide Capture and Storage*. Cambridge: Cambridge University Press.

4 Furukawa, H., Ko, N., Go, Y.B. et al. (2010). *Science* 329: 424–428.

5 Farha, O.K., Eryazici, I., Jeong, N.C. et al. (2012). *J. Am. Chem. Soc.* 134: 15016–15021.

6 Llewellyn, P.L., Bourrelly, S., Serre, C. et al. (2008). *Langmuir* 24: 7245–7250.

7 Saha, D., Bao, Z., Jia, F., and Deng, S. (2010). *Environ. Sci. Technol.* 44: 1820–1826.

8 Kim, H., Yang, S., Rao, S.R. et al. (2017). *Science* 356: 430–434.

9 Shen, K., Zhang, L., Chen, X. et al. (2018). *Science* 359: 206–210.

10 Davankov, V.A. and Tsyurupa, M.P. (1990). *React. Polym.* 13: 27–42.

11 Wood, C.D., Tan, B., Trewin, A. et al. (2007). *Chem. Mater.* 19: 2034–2048.

12 Li, B., Gong, R., Wang, W. et al. (2011). *Macromolecules* 44: 2410–2414.

13 McKeown, N.B. and Budd, P.M. (2006). *Chem. Soc. Rev.* 35: 675–683.

14 McKeown, N.B., Budd, P.M., Msayib, K.J. et al. (2005). *Chem. Eur. J.* 11: 2610–2620.

15 McKeown, N.B. and Budd, P.M. (2010). *Macromolecules* 43: 5163–5176.

16 Ghanem, B.S., Hashem, M., Harris, K.D.M. et al. (2010). *Macromolecules* 43: 5287–5294.

17 Chen, L., Furukawa, K., Gao, J. et al. (2014). *J. Am. Chem. Soc.* 136: 9806–9809.

18 Ding, S.-Y. and Wang, W. (2013). *Chem. Soc. Rev.* 42: 548–568.

19 Kuhn, P., Antonietti, M., and Thomas, A. (2008). *Angew. Chem. Int. Edit.* 47: 3450–3453.

20 El-Kaderi, H.M., Hunt, J.R., Mendoza-Cortes, J.L. et al. (2007). *Science* 316: 268–272.

21 Feng, X., Ding, X., and Jiang, D. (2012). *Chem. Soc. Rev.* 41: 6010–6022.

22 Jiang, J.-X., Su, F., Trewin, A. et al. (2008). *J. Am. Chem. Soc.* 130: 7710–7720.

23 Ma, T., Kapustin, E.A., Yin, S.X. et al. (2018). *Science* 361: 48–52.

24 Xu, H., Tao, S., and Jiang, D. (2016). *Nat. Mater.* 15: 722–726.

25 Schmidt, J., Werner, M., and Thomas, A. (2009). *Macromolecules* 42: 4426–4429.

26 Jiang, J.X., Su, F., Trewin, A. et al. (2007). *Angew. Chem. Int. Ed.* 46: 8574–8578.

27 Lee, J., Buyukcakir, O., Kwon, T.W., and Coskun, A. (2018). *J. Am. Chem. Soc.* 140: 10937–10940.

28 Marco, A.B., Cortizo-Lacalle, D., Perez-Miqueo, I. et al. (2017). *Angew. Chem. Int. Ed. Ed.* 56: 6946–6951.

29 Wang, L., Wan, Y., Ding, Y. et al. (2017). *Adv. Mater.*: 29.

30 Ben, T., Ren, H., Ma, S. et al. (2009). *Angew. Chem. Int. Ed.* 48: 9457–9460.

31 Ben, T. and Qiu, S. (2013). *CrystEngComm* 15: 17–26.

32 Pei, C., Ben, T., and Qiu, S. (2015). *Mater. Horiz.* 2: 11–21.

33 Jiang, L., Tian, Y., Sun, T. et al. (2018). *J. Am. Chem. Soc.* 140: 15724–15730.

34 Li, M., Ren, H., Sun, F. et al. (2018). *Adv. Mater.* 30: 1804169.

35 Yuan, Y., Yang, Y., Faheem, M. et al. (2018). *Adv. Mater.* 30: 1800069.

36 Das, S., Heasman, P., Ben, T., and Qiu, S. (2017). *Chem. Rev.* 117: 1515–1563.

37 Yuan, D., Lu, W., Zhao, D., and Zhou, H. (2011). *Adv. Mater.* 23: 3723–3725.

38 Pei, C., Ben, T., Cui, Y., and Qiu, S. (2012). *Adsorption* 18: 375–380.

39 Demirocak, D.E., Ram, M.K., Srinivasan, S.S. et al. (2013). *J. Mater. Chem. A* (1): 13800–13806.

40 Zhao, H., Jin, Z., Su, H. et al. (2013). *Chem. Commun.* 49: 2780–2782.

41 Yang, Z., Peng, X., and Cao, D. (2013). *J. Phys. Chem. C* 117: 8353–8364.

42 Ben, T., Li, Y., Zhu, L. et al. (2012). *Energy Environ. Sci.* 5: 8370–8376.

43 Yuan, Y., Sun, F., Li, L. et al. (2014). *Nat. Commun.* 5: 4260–4267.

44 Lu, W., Yuan, D., Sculley, J. et al. (2011). *J. Am. Chem. Soc.* 133: 18126–18129.

45 Konstas, K., Taylor, J.W., Thornton, A.W. et al. (2012). Lithiated porous aromatic frameworks with exceptional gas storage capacity. *Angew. Chem. Int. Ed.* 51: 6639–6642.

46 Lu, W., Sculley, J.P., Yuan, D. et al. (2012). Polyamine-tethered porous polymer networks for carbon dioxide capture from flue gas. *Angew. Chem. Int. Ed.* 51: 7480–7484.

47 Li, Y., Ben, T., Zhang, B. et al. (2013). *Sci. Rep.* 3: 2420–2426.

48 Zhang, Y., Li, B., Williams, K. et al. (2013). *Chem. Commun.* 49: 10269–10271.

49 Pachfule, P., Dhavale, V.M., Kdambeth, S. et al. (2013). *Chem. Eur. J.* 19: 974–980.

50 Jing, X., Zou, D., Cui, P. et al. (2013). *J. Mater. Chem. A* 1: 13926–13931.

5

Virtual Screening of Materials for Carbon Capture

Aman Jain[1,2], Ravichandar Babarao[1,3] and Aaron W. Thornton[1]

[1]*Manufacturing Flagship, Commonwealth Scientific and Industrial Research Organisation, Clayton, Victoria, Australia*
[2]*Indian Institute of Technology, Kanpur, Uttar Pradesh, India*
[3]*RMIT University, Melbourne, Victoria, Australia*

Materials for Carbon Capture, First Edition. Edited by De-en Jiang, Shannon M. Mahurin and Sheng Dai.
© 2020 John Wiley & Sons Ltd. Published 2020 by John Wiley & Sons Ltd.

5.1 Introduction

A material that efficiently captures carbon dioxide must harness many properties including selectivity, reversibility, porosity, stability, and many others. Most properties depend on the chemical structure of the material that must be accessible to CO_2 and capable of adsorbing CO_2 either chemically or physically. There has been much research on the synthesis of new structures, but the number of possible structures is endless. It is not practical to synthesize and test every single material in the laboratory. Therefore, efforts to estimate the CO_2 capture performance of candidate materials using molecular simulations and theoretical modeling play a very important role in identifying the most promising materials prior to experimental investigation. Here we review *in-silico* (or virtual) methods for predicting CO_2 capture efficiency that accelerate the discovery of new carbon capture materials.

The *in-silico* methods range from modeling electrons to the full operation of industrial plants. In this chapter, we cover predictions for two separations methods: adsorbent-based and membrane-based. Many of the predictions are useful for both separation methods, including:

- Monte-Carlo simulations for predicting adsorption, heat of adsorption and selectivity.
- Molecular dynamics (MD) for predicting the kinetics and diffusion of gas.
- Density functional theory (DFT) for predicting the electron configurations responsible for gas-material interactions, electronic dynamics, and structural equilibrium and stability.

The format of this chapter consists of a short introduction of the *in-silico* methods available and their application in CO_2 capture platforms including raw natural gas and pre-/post-combustion of fossil fuels. A summary of candidate materials is provided along with a list of criteria for selecting the most promising candidates. The chapter concludes with a snapshot of important insights discovered using *in-silico* virtual screening.

5.2 Computational Methods

5.2.1 Monte Carlo-Based Simulations

In principle, Monte Carlo simulation is a computerized mathematical technique that randomly samples a given parameter space with the goal of collecting enough statistics to determine the properties of a system. This can be used to account for risk in quantitative analysis and decision-making where the decision-maker is equipped with a range of possible outcomes and the probabilities that they will occur for any choice of action. For carbon capture, Monte Carlo-based simulations can be used to predict adsorption isotherms, heat of adsorption, selectivity, and accessible surface area for any potential material. For more details of Monte Carlo simulations, we recommend reading Frenkel and Smit, *Understanding Molecular Simulation – From Algorithms to Applications* [1].

To predict gas adsorption within a porous material where the structure is known, this Monte Carlo method can be used to determine the adsorption equilibria at a given composition, pressure, temperature, and other conditions. The method will trial individual adsorption sites within the material, and the adsorption energy is calculated at each attempt that is used to either accept or reject the trialed position. This type of Monte Carlo simulation for predicting adsorption equilibria is called the grand canonical Monte Carlo (GCMC) simulation because the acceptance criteria are based on the grand canonical ensemble unifying the key variables temperature, pressure, amount adsorbed, and chemical potential. See Figure 5.1 for a snapshot during a simulation of gas adsorption within a metal organic framework (MOF) at four different pressures, performed by Düren et al. [2].

GCMC simulations give detailed information about the molecular scale, as the positions and potential energies of all adsorbate molecules can be predicted over the duration of the simulation. This information can help the researcher understand and tune the energetic interactions, preferential adsorption sites, and reversibility, and exploit the adsorption mechanism.

For example, Karra and Walton performed GCMC simulations to reveal the role of open metal sites in MOFs for the separation of carbon monoxide from binary mixtures containing CH_4, N_2, or H_2. The adsorption mechanism was dominated by the electrostatic interactions between the CO dipole and the partial charges on the MOF atoms. The simulations also predicted that CuBTC is quite selective for carbon monoxide over hydrogen and nitrogen for three different mixture compositions at 298 K [3].

Yang and Zhong also performed a computational study of binary and ternary mixture adsorption of CO_2, CH_4, and C_2H_6 in CuBTC. They explored three scenarios: (i) all electrostatic interactions are switched off; (ii) only the electrostatic interactions between CO_2 molecules and the CuBTC framework are switched off, thus taking into account the CO_2-CO_2 electrostatic interactions; and (iii) all electrostatic interactions are switched on [4]. Their findings showed that electrostatic interactions resulting from the atomic partial charges of the MOF framework can greatly enhance the separation of mixtures of components with largely differing polarities. This effect is probably impossible to measure experimentally and therefore is a good example of the powerful insights that *in-silico* screening can reveal.

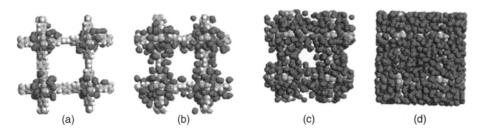

    (a)   (b)   (c)   (d)

Figure 5.1 Snapshots of gas adsorption in IRMOF-16 at four different pressures. Source: reprinted from Ref. [2] with the permission of the American Chemical Society.

Comparing simulation and experimental data requires a careful conversion because simulation produces the absolute number of molecules present in the framework while experimental results are often reported as the excess amount adsorbed. The excess number of molecules, N_{ex}, is the number of molecules that is present in "excess" of the molecules that would be present in the same (pore) volume at bulk conditions. Therefore, it is related to the absolute number of molecules, N_{abs}, by

$$N_{ex} = N_{abs} - V_g \rho_g$$

where ρ_g is the density of the bulk gas phase calculated with an equation of state and V_g is the pore volume of the framework. Figure 5.2 shows the comparison made by Walton et al. [5] between experimental and simulated adsorption isotherms for CO_2 in isoreticular metal organic framework (IRMOF)-1, where there is excellent agreement.

Accessible surface area is another critical property of a material that can indicate its adsorption performance. Surface area of some of the MOFs is reported up to $8000\,m^2\,g^{-1}$, which can equate to a mol of CO_2 per gram [6]. Accessible surface area can be defined as the area traced out by the center of a probe molecule as the probe is rolled across the surface of the framework atoms. Using Monte Carlo integration, the probe sphere is randomly inserted around the surface of each framework atom to determine the overlap between adjacent atoms (see Figure 5.3a) [8–10]. The fraction of probes that do not overlap with other framework atoms is used to calculate the accessible surface area, and the dimension of the probe corresponds with the size of the adsorbate. There is surprisingly good agreement between the accessible surface areas calculated geometrically and the BET surface areas obtained from the simulated isotherms based on the consistency criteria recommended by Rouquerol et al. [11] (see Figure 5.3b). The good agreement only holds true in the range between 2500 and $5000\,m^2\,g^{-1}$ where all the consistency criteria were fulfilled.

Heat of adsorption gives a measure of the CO_2 binding strength within a material. Heat of adsorption, ΔH, is related to the differential adsorption enthalpy, ΔU, as follows: $\Delta H = \Delta U - RT$. Note that the heat of adsorption is represented as a positive value, while the differential adsorption enthalpy is negative. Heat of adsorption can be directly calculated from the

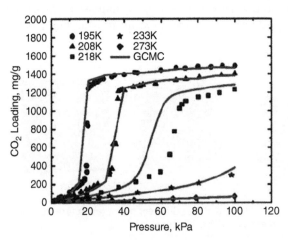

Figure 5.2 Comparison between experimental and simulated adsorption isotherms in IRMOF-1. Source: reprinted from Ref. [5] with the permission of the American Chemical Society.

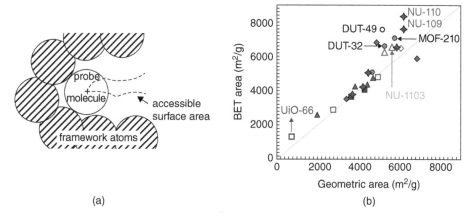

(a)   (b)

Figure 5.3 Geometrical surface area calculated using a N_2-sized probe. (a) Schematic illustration of geometry; (b) BET surface areas versus accessible surface areas of various types of MOFs that differ in topology: pcu (diamond), fcu (squares), ftw (triangles), rht (stars), and other topologies (circles). Filled symbols represent the achieved consistency criteria, while empty symbols relate to the unfilled consistency criteria [7]. Sources: (a) reproduced with the permission of the Royal Society of Chemistry; (b) reproduced with the permission of the American Chemical Society.

GCMC simulation using the following formula [12]:

$$-\Delta H_{ads} = -\Delta H_{res,bulk} - RT(1 - Z_{bulk}) + \frac{\langle V_{ads}N_{ads}\rangle - \langle N_{ads}\rangle\langle V_{ads}\rangle}{\langle N_{ads}^2\rangle - \langle N_{ads}\rangle\langle N_{ads}\rangle}$$

Here, ΔH_{ads} is the residual enthalpy of the bulk phase and Z_{bulk} is the compressibility factor of the bulk phase, both of which can be calculated from an equation of state; N_{ads} is the number of adsorbed molecules, and V_{ads} is their potential energy. If the gas phase can be treated as ideal, the residual enthalpy is equal to zero and the compressibility factor is equal to one, so that the equation reduces to the last term.

Monte-Carlo simulations can also disclose the density distribution of adsorbed CO_2 within the material. This can be useful to understand the adsorption mechanism and the nature of nanospace utilized for adsorption. For example, a density plot is shown in Figure 5.4, where color represents the probability of finding CO_2 and Na^+ ions at a

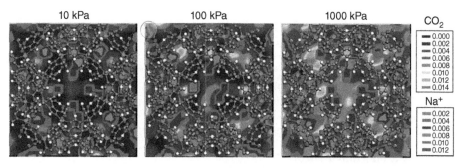

Figure 5.4 Density distribution contours of CO_2 molecules and Na^+ ions for a $15:85$ CO_2/H_2 mixture at (left) 10, (center) 100, and (right) 1000 kPa using GCMC simulations. Source: reprinted from Ref. [13] with the permission of the American Chemical Society. (*See color plate section for color representation of this figure*).

certain location. It can be seen that the concentration of molecules changes as a function of pressure within a CO_2/H_2 mixture in *rho*-ZMOF.

5.2.2 MD Simulation

The physical movements of atoms and molecules are important for understanding the kinetics and diffusion of carbon capture using materials that can be simulated using an approach referred to as MD. For a system of interacting particles, the trajectories of atoms and molecules are determined by numerically solving Newton's equations of motion. The forces between the particles and energy are defined by interatomic potentials or molecular mechanics force fields.

The advantage of MD over MC is that it gives an understanding of the dynamical properties of the system, including transport coefficients, time-dependent responses to perturbations, rheological properties, and spectra. For example, the trajectories of the particle positions and velocities with time in the system can be used to predict the diffusion coefficients. For practical uses, this information can be used to determine how long the material will take to adsorb CO_2 as well as the time for CO_2 to pass through the material over other components within the mixture.

The diffusion coefficients can be calculated from the mean squared displacement (MSD) over an average of multiple simulations, providing a reliable estimate according to the following Einstein relationship,

$$D_{self} = \frac{1}{6N} \lim_{t \to \infty} \frac{d}{dt} \sum_{i=1}^{N} \langle [r_i(t) - r_i(0)]^2 \rangle$$

where N is the number of gas molecules and $r_i(t)$ is the vector position at time t for gas molecule i.

Gas permeability and selectivity can also be predicted for mixed matrix membranes using a combination of MD combined with continuum models [14–16]. Keskin and Sholl [16] verified a modeling framework by incorporating the intrinsic permeability properties of MOFs from molecular simulation within either Maxwell's model [17] (accurate up to 20% volume fraction) or Bruggeman's model [18] (accurate up to 40% volume fraction). Bruggeman theory is described as follows,

$$\left(\frac{P_{MMM}}{P_p} \right)^{-1/3} \left[\frac{(P_{MMM}/P_p) - (P_{ZIF}/P_p)}{1 - (P_{ZIF}/P_p)} \right] = (1 - \phi_{ZIF})$$

where P_{MMM} is the total permeability within the mixed matrix membrane, P_{ZIF} is the simulated permeability within the ZIF crystals according to the equation, ϕ_{ZIF} is the volumetric fraction of ZIF crystals dispersed within the continuous polymer phase, and P_p is the experimental permeability of the polymer phase.

5.2.3 Density Functional Theory

DFT is a computational quantum mechanical modeling method used to investigate the electronic structure (principally the ground state) of many-body systems. With this theory, the properties of a many-electron system can be determined by using functionals, which

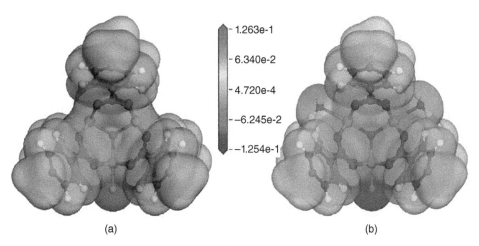

(a)   (b)

Figure 5.5 Electrostatic potential maps by density functional theory (DFT) calculations around the Cr3O trimer in (a) dehydrated and (b) hydrated MIL-101 [19]. These maps were drawn to investigate the adsorption of CO_2 and CH_4 in dehydrated and hydrated MIL-101 and the effect of terminal water molecules on adsorption Reprinted with the permission of the American Chemical Society. *(See color plate section for color representation of this figure).*

are functions of another function. DFT typically solves the time-independent Schrödinger equation to find minimum energy structures, binding energies, and details of the electronic structure. There are many different functions that allow researchers to calculate the charge of an overall cluster or a unit cell.

DFT addresses the challenging prediction of intermolecular interactions, especially van der Waals forces (dispersion), charge transfer excitations, transition states, global potential energy surfaces, dopant interactions, and some other strongly correlated systems. For instance, electrostatic potential map around a metal-oxide cluster is calculated based on DFT as shown in Figure 5.5. From Figure 5.5, we can see that the fluorine atom possesses large electronegativity in both dehydrated and hydrated chromium oxide clusters. The significant difference is the electric fields generated by the terminal H_2O molecules in the hydrated cluster. These maps illuminate regions within the material that will attract CO_2 at different strengths as a result of moisture.

5.2.4 Empirical, Phenomenological, and Fundamental Models

In principle, any model is useful if it provides insight or prediction. For CO_2 capture, there is a collection of models ranging from empirical to fundamental. Empirical models are based only on observations with no knowledge of underlying relations between the parameters observed. Phenomenological models are based on known underlying relationships from either empirical observations or purely logical conclusions or a combination of both, while fundamental models are established laws usually referred to as irrefutable facts with an overwhelming amount of evidence from both empirical observations and logical proofs. Here a short summary of models relevant for CO_2 capture is given.

5.2.4.1 Langmuir and Others

The Langmuir model mathematically describes adsorption of gas upon a surface that can be derived from both kinetic and thermodynamic principles. For CO_2 capture, CO_2 binding is treated as a chemical reaction between the adsorbate molecule, A_g, and an empty site, S, resulting in an adsorbed state A_{ad}. This leads to the following reaction whose equilibrium constant is K_{eq}:

$$A_g + S \rightleftharpoons A_{ad}$$

The local adsorption isotherm is given by the Langmuir model:

$$\theta_A = \frac{V}{Vm} = \frac{K_{eq}P}{1 + K_{eq}P}$$

where θ_A is the fractional occupancy of the adsorption sites, V_m is the volume of the monolayer, and P is pressure. A monolayer of gas molecules surrounding a solid is the conceptual basis for this adsorption model.

In principle, one can use the Langmuir in a predictive manner when the equilibrium constant is measured or derived. Langmuir can also be used to fit experimental results and reveal insights into the adsorption mechanism related to binding strength, site distribution and surface area. Langmuir-Freundlich, Brunauer–Emmett–Teller (BET), Toth, and SIP are other such models used for the same purpose.

5.2.4.2 Ideal Adsorbed Solution Theory (IAST)

IAST predicts the loading of the gas mixture within the material, which is difficult to predict with GCMC. IAST is based only on the knowledge of the pure adsorption isotherms of the individual components. This means the researchers only have to run the calculations or experiments for pure adsorption isotherms and use IAST to predict any composition ratios of mixed gas. For adsorption of a gas mixture, the equilibrium condition is the equality of fugacity in the gas and adsorbed phases

$$P * y_i * \theta_i = x_i * \gamma_i * f_i^0$$

where P is bulk pressure, and y and x are the molar fractions in the gas and adsorbed phases, respectively. The fugacity coefficient of component i in the gas phase is θ_i, f_i^0 is the fugacity of pure component i in a standard state, and γ_i is the activity coefficient of component i in the adsorbed phase. If a perfect mixing is assumed in the adsorbed phase, $\gamma_i = 1$. The standard state is specified by the surface potential $\emptyset_i$ given by the Gibbs adsorption approach,

$$\emptyset_i = -RT \int_0^{f_i^0} N_i^0(f_i)\mathrm{d}\ln(f_i)$$

where $N_i^0(f)$ is the adsorption isotherm of pure component i. The mixing process is carried out at a constant surface potential $\emptyset_1 = \emptyset_2 = \emptyset$.

Figure 5.6 shows the adsorption isotherms for an equimolar mixture of CH_4/CO_2 in MFI, C168 schwarzite, and IRMOF-1 as a function of total bulk pressure. The graphs were drawn in order to compare pure CO_2 and CH_4 and separation of their binary mixture in three different classes of nanostructured adsorbents: silicalite, C168 schwarzite, and IRMOF-1. CH_4 is represented as a spherical Lennard-Jones molecule, and CO_2 is represented as a rigid linear

Figure 5.6 Adsorption of an equimolar mixture of CH₄ and CO₂ in MFI, C168, and IRMOF-1 as a function of bulk pressure. The filled symbols are simulation results, and the lines are IAST predictions. Source: reprinted from Ref. [20] with the permission of the American Chemical Society.

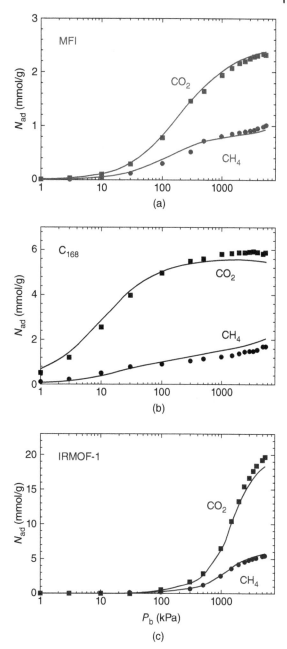

molecule with a quadrupole moment. For adsorption from an equimolar binary mixture, CO_2 is preferentially adsorbed in all three adsorbents. Predictions of mixture adsorption with the ideal-adsorbed solution theory on the basis of only pure component adsorption agree well with the simulation results, though IRMOF-1 has a significantly higher adsorption capacity than silicalite and C168 schwarzite.

5.2.5 Materials Genome Initiative

Materials discovery today still involves significant trial and error. It can require decades of research to identify a suitable material for a technological application, and longer still to optimize that material for commercialization. A principal reason for this long discovery process is that materials design is a complex, multidimensional optimization problem, and the data needed to make informed choices about which materials to focus on and what experiments to perform usually does not exist or is too complex to make sense of.

The Materials Genome Initiative (MGI), launched in 2011 in the United States, is a large-scale collaboration between material scientists (both experimentalists and theorists) to unite and deploy proven computational methodologies to predict, screen, and optimize materials at an unparalleled scale and rate. The time is right for this ambitious approach: it is now well established that many important materials properties can be predicted by solving equations or running simulations based on the fundamental laws. This virtual approach to developing and testing materials can be employed to design and optimize CO_2 capture materials *in-silico*.

The Nanoporous Materials Genome Center (NMGC) is a subgroup of the MGI that explores microporous and mesoporous materials including MOFs, zeolites, and porous polymers. The center collaboratively publishes many articles and develops new computational tools to support with genome initiative to accelerate materials discovery. These tools include:

- *CP2K*: Quantum chemistry package designed to perform MD and Monte-Carlo simulations of clusters and periodic systems.
- *PySCF*: Quantum chemistry and solid-state physics package to perform electronic structure calculations in molecular and periodic systems.
- *QMMM*: Program for performing single-point calculations (energies, gradients, and Hessians), geometry optimizations, and MD using combined quantum mechanics (QM) and molecular mechanics (MM) methods.
- *MCCCS-MN*: Monte Carlo software tailored for simulations of phase and adsorption equilibria in the Gibbs ensemble using the TraPPE force field.
- *RASPA*: Package for simulation of adsorption and diffusion of molecules in flexible nanoporous materials.
- *chemical_VAE*: Chemical Variational Autoencoder software for machine learning of molecular properties.
- *CoRE MOF*: Computation-Ready Experimental (CoRE) MOF is a database that enables high-throughput computational screening by using GitHub's versioning system to manager and curate the data.
- *Nanoporous Materials Explorer*: An app providing access to a database containing information about thousands of materials' computational properties.
- *Polymatic*: A set of codes for the structure generation of amorphous polymers via a stimulated polymerization algorithm.
- *pyIAST*: A user-friendly Python script that can fit data into analytical isotherm models or use interpolation to characterize the pure-component adsorption isotherms.
- *Zeo++*: Open source software for performing high-throughput geometry-based analysis of porous materials and their voids.

5.2.6 High-Throughput Screening

In the past, computational tools were applied to single structures with fine detail of atomic positioning and interactions with guest media. Now those tools have become fast and powerful enough to process thousands and even millions of materials, allowing high-throughput screening for various applications such as solar energy, water splitting, carbon capture, nuclear detection, topological insulators, piezoelectrics, thermoelectrics, catalysis, and energy storage [21]. Carbon capture is one of those applications that has benefited from high-throughput screening where there are millions of candidate structures worth considering but impossible to test experimentally. High-throughput screening relies on a database of candidate structures with diverse structural and chemical characteristics (or textural properties). Some of the databases available include:

- *Hypothetical MOFs*: Around 140 000 structures constructed from a library of 102 building blocks [22].
- *CoRE MOF*: CoRE MOFs derived from experimental data submitted to the Cambridge Structural Database. The structures were "cleaned" and assessed, resulting in around 5000 porous structures ready for computational screening [23].
- *Hypothetical zeolites*: Around 330 000 structures chosen from a set of energetically feasible structures from the Predicted Crystallography Open Database (PCOD) [24].
- *Ideal silica zeolites*: Around 200 experimental structures from the International Zeolite Association [25].
- *Hypothetical zeolitic imidazolate frameworks (ZIFs)*: Identical to the 330 000 hypothetical zeolite structures but with the Si—O bond replaced with Zn-imidazolate [26].
- *Hypothetical porous polymer networks (PPNs)*: 180 000 structures assembled from commercially available chemical building blocks and two experimentally known synthetic routes [27].

Here is a summary of high-throughput screening for carbon capture using these databases and a range of screening methods. One of the first screening studies was performed by Yazaydin et al. in 2009: they screened a diverse selection of 14 MOFs using a combined experiment and modeling approach [28]. This study brought confidence in computational screening because of the rigorous validation with experimental isotherms. They also were able to provide insights that were not possible when studying a single structure. Specifically, the study identified heat of adsorption as a key parameter to enhance CO_2 uptake at low pressures. The results were useful for CO_2 capture of flue gas.

Lin et al. launched a mega-screening study of hundreds of thousands of structures including zeolites and ZIFs [26]. With a focus on processing flue gas, the study covered a broad range of operating conditions including temperature and pressure to ensure that each candidate material was given a fair chance. Parasitic energy was used to rank the candidates and compare with the current monoethanolamine (MEA) technology. Hundreds of promising candidates were discovered that could reduce the parasitic energy of a coal-fired power plant by 30–40% compared with MEA.

One of the issues arising from high-throughput studies is the accuracy of force fields, especially when the candidate materials are chemically diverse [29]. Kadantsev et al. tackled this issue by parameterizing the electrostatic potential calculated via ab initio [30]. The large and diverse training set contained 543 MOFs combining Zn, Cu, and V building units with 52 different organic carboxylate- and nitrogen-capped ligands and 17 functional groups. This training set was then validated with 693 non-overlapping MOFs. With correlation factors above 0.97, the method was used to improve the prediction of CO_2 uptake and heat of adsorption.

GPU-based calculations were utilized by Kim et al. to screen over 87 000 zeolites based on the simulation of both adsorption and diffusion properties [31]. Channel uniformity was identified as an important textural property to balance both adsorption and diffusion properties. They found zeolites that could outperform the best known zeolite up to a factor of 7. Surprisingly, they also found that some zeolites could do the reverse separation where CO_2 is rejected and other components pass through.

Another screening study of zeolites by Matito-Martos et al. identified candidates for separation and storage of CO_2 [32]. In agreement with Kim et al., channel-type pore topology became an important feature where low pore volumes offer high selective properties while high pore volumes offer high storage properties. Dimensionality of pore space was also used to explain the differences between candidates with similar pore volumes. 1D pores compared to 2D or 3D pores provided a confinement effect where entropy played a role in preferential adsorption. Rigorous testing of the force fields was conducted to ensure the simulations aligned with experimental observations. Overall, many top candidates emerged from the study, along with practical guidelines to design new porous materials with high CO_2 separation properties.

Binary gas mixtures were used to screen ZIF candidates for adsorption- and membrane-based separations by Yilmaz et al. [33]. It became clear that single-gas simulations were not capable of effectively ranking candidate materials because of the large effect of competitive adsorption and diffusion.

Machine learning was adopted by Fernandez et al. to further accelerate the high-throughput procedure [34]. Quantitative structure–property relationships (QSPR) models were trained on a small subset and then tested on 292 050 MOFs: 945 of the top 1000 MOFs were recovered, and only 10% of the whole library required further intensive screening. Overall, an order of magnitude reduction in compute time was achieved, which was an enormous improvement for high-throughput screening.

Virtual screening was applied by Thornton et al. to explore zeolite-based catalysts for CO_2 reduction [35]. It was found that dual adsorption of CO_2 and H_2 could enhance the catalytic potential and efficiency of the catalyst. Optimal cavity sizes of around 6 Å could maximize the change in entropy-enthalpy, and large void spaces >30% could boost the formation of products such as methane, methanol, formaldehyde, and formic acid.

Competitive adsorption with water was identified as an issue in CO_2 capture that is difficult to overcome. Lin et al. performed a large-scale screening study of MOFs to identify MOFs with a higher selectivity toward CO_2 over H_2O [36]. DFT was required to accurately model the specific interactions with each MOF. Electrostatics was found to contribute mostly to the adsorption energy of H_2O, while van der Waals made up around 90% of the CO_2 binding energy.

Finally, Dureckova et al. in 2019 applied robust machine learning models to screen MOF candidates for pre-combustion carbon capture. Pre-combustion flue gas contains roughly 40% CO_2 and 60 % H_2, which means the structure of the adsorbent will be very different compared with those candidates for direct air capture. 358 400 hypothetical structures with the most diverse topologies, including 1166 network topologies, were considered in this study, which is much more than network topologies used previously (< 20). The best models for CO_2 working capacities ($R^2 = 0.944$) and CO_2/H_2 selectivities ($R^2 = 0.872$) were built on a combination of six geometric descriptors and three AP-RDF descriptors. Many top candidates were identified with high CO_2 working capacities greater than 30 mmol g^{-1} and CO_2/H_2 selectivities greater than 250.

5.3 Adsorbent-Based CO₂ Capture

Adsorption is based on a cyclical process, shown in Figure 5.7, in which CO_2 is adsorbed preferentially from a gas stream on to the surface of a solid, typically a mineral zeolite and other materials mentioned in Section 5.5. The gas stream, with most of the CO_2 removed, is then emitted to the atmosphere or utilized for a different purpose. The solid is then purified in stages using differences in either pressure, temperature, or other stimuli to remove the CO_2 and compress it for storage.

Temperature swing adsorption (TSA) is a technique for regenerating a bed of adsorbent that is loaded with the targeted gas. Whereas pressure swing adsorption (PSA) uses changes in pressure to release adsorbed gas, TSA modulates or swings temperature to drive off the adsorbed gas. When PSA operating costs are too high, TSA is often less expensive to operate,

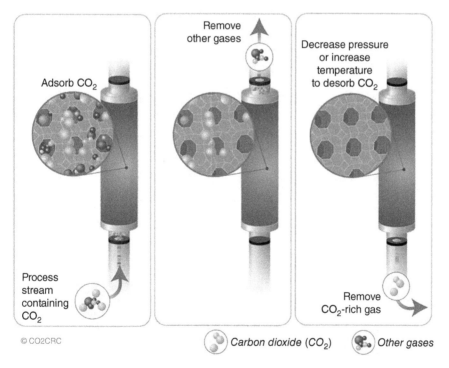

Figure 5.7 Adsorbent-based separation. CO_2 molecules are attracted to the surface of an adsorbent (www.co2crc.com.au/aboutccs/cap_adsorption.html). Source: http://hdl.handle.net/2027.42/90951. Licensed under CCBY 4.0. (*See color plate section for color representation of this figure*).

despite initially costing more to purchase. When high product purities are not achievable with PSA, TSA is usually more suitable. If the temperature to regenerate is more than ~400 °C, then the cost of operating is typically too high, and it would not be beneficial to use TSA. By tuning the adsorption properties of materials, one can reduce the temperature required for regeneration and therefore reduce the energy-cost of carbon capture.

5.3.1 Direct Air Capture

Direct air capture (DAC) of CO_2 has become a promising technology to recycle the global CO_2 supply chain and potentially control atmospheric concentrations. Companies such as Climeworks, Carbon Engineering, and Global Thermostat are leading the way using amine-functionalized materials or aqueous alkali to capture the CO_2. There has been much progress in reducing energy consumption, which is the major cost of the technologies; however, the cost remains inhibitive for large-scale adoption. Adsorbents that utilize the physisorption mechanism require a lower regeneration energy and therefore are promising candidates to further reduce the cost threshold for market uptake. Additional benefits can also include faster kinetics, size selectivity, and lifetime because of the passive nature of the adsorption/desorption cycles compared with chemisorption or caustic approaches.

For DAC with only 0.04% CO_2 in the feed stream, these adsorbents require stronger and selective binding sites than typically found in physisorbents. Some of the promising candidates include hybrid ultramicroporous materials (HUMs) that incorporate hydrophobic pillars such as SiF_6^{2-} (SIFSIX) and TiF_6^{2-} (wTIFSIX) anions within three-dimensional networks [37]. Figure 5.8 depicts the CO_2 molecule trapped within the strong electronegative environment that was simulated by Ziaee et al. [38] and a comparison of selected MOF materials with zeolite 13X in direct CO_2 capture under atmospheric conditions [39]. The simulations also explored the effects of different metals that could be used to fine-tune the pore size and therefore the distance and interactions between CO_2 and the fluorine-decorated pillars.

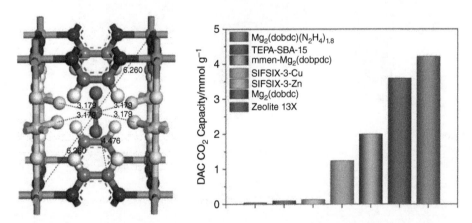

Figure 5.8 (Left) Depiction of a single CO_2 molecule within the pore structure of SIFSIX-3-Cu [38]; (right) comparison of selected MOF materials with zeolite 13X in direct CO_2 capture [39]. Sources: (left) reproduced with the permission of the American Chemical Society; (right) reproduced with the permission of John Wiley & Sons. (*See color plate section for color representation of this figure*).

5.4 Membrane-Based CO_2 Capture

Membrane-based CO_2 separation relies on a material's ability to allow the transport of CO_2 over any competing gases within the mixture. An advantage of this process over adsorbent-based separations is that it is continuous, without any switching of temperature, pressure, or other variables. An upstream pressure is the only energy input required to induce the separation. The performance of a membrane can be characterized by permeability of CO_2 and the selectivity of CO_2 over other gases in the mixture [40]. A Robeson upper bound is a good way to visualize and compare the performance of materials where maximized permeability and selectivity is desired [40]. Methods to calculate permeability and selectivity are provided in Sections 5.2.2 and 5.8.

Mixed matrix membranes (MMMs) combine the advantages of low-cost, mechanically stable polymers and high gas-separation performance of MOFs or inorganic particles and offer an attractive pathway to enhance the performance of CO_2 separation membranes. One of the most difficult challenges in achieving the optimized separation properties using MMM is ensuring a good interface between the polymer and the filler particles, which is difficult to predict [41].

5.5 Candidate Materials

5.5.1 Metal Organic Frameworks

MOFs are a class of porous materials that have potential applications in gas storage, separation processes, and catalysis. The synthesis of MOFs is done by assembling organic linker molecules and metal atoms to form materials with well-defined pores, high surface areas, and desired chemical functionalities (see Figure 5.9). Because of these attractive properties, MOFs are promising candidates for CO_2 capture. Different types of synthesis methods can be applied, including classical hydro (solvo) thermal synthesis, the microwave method, electrochemical synthesis, the diffusion method, and the ultrasonic method.

During synthesis, some of the metal atoms are partially coordinated by solvent molecules. Activating MOFs at an elevated temperature to remove the solvent and open the void space for desired guest molecules is a common practice. All solvent molecules can be removed, including those that are coordinatively bound to framework metal atoms, if the evacuation temperature is high enough. Removing these coordinated solvent molecules leaves coordinatively unsaturated open-metal sites that have been shown to promote high gas uptake and even catalytic activity.

Wilmer et al. pioneered large-scale hypothetical structure development resulting in over 130 000 structures. The work combined geometric principles with known MOF building blocks and was used to discover methane storage [22] followed by a subsequent paper for CO_2 capture [42]. The large-scale CO_2 study revealed structure–property relationships that were difficult to discover with limited experimental data. Pore, surface area, void fraction, and heat of adsorption were among the critical characteristics that could be optimized for CO_2 capture.

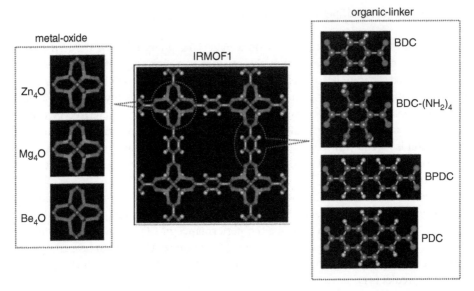

Figure 5.9 Schematic tailoring of the metal oxide and organic linker in IRMOF1. Color code: Zn, green; Mg, cyan; Be, purple; O, red; N, blue; C, gray; H, white. Source: reprinted from Ref. [45] with the permission of the American Chemical Society. (*See color plate section for color representation of this figure*).

5.5.2 Zeolites

Zeolites are hydrated aluminosilicate minerals that form regular, porous structures that act as a molecular sieve. They can be modified to include a large variety of metal cations through a simple ion-exchange process. These modifications lead to large changes in CO_2 sorption capacity, selectivity, and water tolerance.

There has been an immense global impact of zeolites in the refinery industry, which led to about a 30% increase in gasoline yields, thus allowing more efficient utilization of petroleum feedstocks. It is estimated that about five million metric tons per year of natural zeolite is consumed worldwide. For synthetic zeolites, the annual global consumption is approximately 1.8 million metric tons.

For synthetic zeolites, the biggest market (in terms of volume) is in laundry detergents, with 73% of total consumption [43]. Although the detergent industry is still the largest consumer of synthetic zeolites, the consumption of zeolites as detergent builders is expected to decline [43]. Utilization as adsorbents and desiccants represents the first commercial application of synthetic zeolites, and in 2009 this market accounted for 10% of global consumption in terms of tonnage.

Zeolites have promise as CO_2 adsorbents because of their tunable pore sizes and stability. Dehydration of natural gas is a common adsorbent-based separation for zeolites and therefore well suited to large-scale industrial applications. Unfortunately, it is well known that the presence of water significantly decreases the adsorption of CO_2 because water competitively adsorbs on the cations, blocking access for CO_2 [44]. Therefore, zeolites must be tuned further to selectively adsorb CO_2 over water and other competitive components.

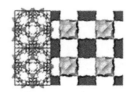

Zn(mIm)₂

Figure 5.10 ZIF structure with building block, topology, and accessible surface area. Source: reprinted from Ref. [15] with the permission of the Royal Society of Chemistry.

5.5.3 Zeolitic Imidiazolate Frameworks

Zeolitic imidazolate frameworks (ZIFs) are a subset of MOFs that share the same topology as zeolites because of the imidazolate-metal-imidazolate coordination angle equivalent to the Si-O-Si coordination angle in zeolites. ZIFs offer promising applications for gas storage and separation. They consist of transition metal ions (Zn^{2+} or Co^{2+}, among others) connected by imidazolate-like linkers; and they have emerged as a novel type of crystalline porous materials that combine highly desirable properties from both zeolites and MOFs, such as microporosity, high surface areas, and exceptional thermal and chemical stability, making them ideal candidates for gas-separation applications. Moreover, by tuning the nature of the organic linker, it is possible to create a huge diversity of structures of controlled pore size and chemical functionality. Figure 5.10 shows the most studied structure of ZIF 8 ($Zn(mIm)_2$).

Large-scale screening of ZIFs for CO_2 capture was led by Lin et al. using zeolites as templates. With over two million hypothetical structures available, the group chose the top thermodynamically accessible structures (0–$30\,kJ\,mol^{-1}$). With hundreds of thousands of structures to explore, the study discovered structures capable of reducing the parasitic energy up to 30–40% of competing technologies. In addition, general guidelines for tuning structural characteristics were given to maximize carbon capture performance.

5.5.4 Mesoporous Carbons

Mesoporous carbons contain pores with diameters between 2 and 50 nm. These structures usually have extremely high diffusivities because of the extraordinary smoothness of the potential-energy surface across the pores. For example, the adsorbate-nanotube potential energy surface can be precomputed on a fine grid and then rapidly evaluated via interpolation during subsequent calculations that reveal enormous diffusivities [14].

A schematic representation of a unit cell of non-doped and nitrogen-doped mesoporous carbons is shown in Figure 5.11 [46] to investigate the effect of nitrogen doping on adsorption capacity and selectivity of CO_2 versus N_2 in a simulated mesoporous carbon model. The results showed that nitrogen doping greatly enhances CO_2 adsorption capacity; with a 7 wt % dopant concentration, the adsorption capacity at 1 bar and 298 K increases from 3 to 12 mmol g⁻¹ (or 48% uptake by weight).

5.5.5 Glassy and Rubbery Polymers

Polymers dominate the gas-separation industry because of their low cost, scale-up, and film versatility within a range of configurations [47]. Simulating glassy polymers is reasonably

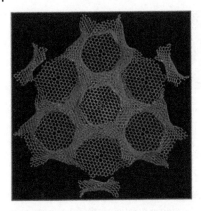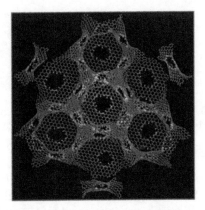

Figure 5.11 Schematic representation of a unit cell of non-doped (left) and nitrogen-doped (right) mesoporous carbons. Color code: C, gray; N, blue; H, white. Source: reprinted from Ref. [46] with the permission of the American Chemical Society.

achievable using a combination of Monte-Carlo and MD [48–50]. It is believed that polymers have reached a maximum performance because they are restricted theoretically to the solution-diffusion model. This belief is difficult to confirm because of the molecular complexity and the continual progress in polymer chemistry to fine-tune structural characteristics [51].

5.6 Porous Aromatic Frameworks

Porous aromatic frameworks (PAFs) and porous polymer networks (PPNs) are among the main candidate materials for gas storage and capture due to their ultrahigh surface area and high physicochemical stability. A practical drawback of MOFs is their limited physicochemical stability. The emergence of PAFs first provides an ideal microporous material combining ultrahigh surface area and high physicochemical stability. Figure 5.12 shows the structure of PAF-302, PAF-303, and PAF-304 [27]. Here, PAF-30X (X = 2, 3, 4) represents

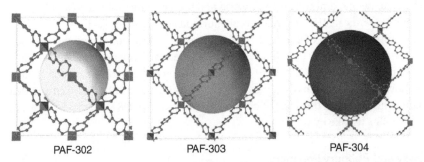

PAF-302   PAF-303   PAF-304

Figure 5.12 Schematic representation of PAF-302, PAF-303, and PAF-304 structures. The colored spheres represent the maximum diameter that can fit inside the porous structures. Source: reprinted from Ref. [27] with the permission of the American Chemical Society.

the number of benzene rings attached to the carbon atom in a three-dimensional structure with uniform pores.

Large-scale screening of PAFs was led by Martin et al. for methane storage [27]. Over 18 000 hypothetical structures were generated and equilibrated using semi-empirical methods. Surface areas up to $10\,000\,m^2\,g^{-1}$ were discovered. This database is a foundation for *in-silico* computational screening for other applications such as carbon capture.

5.7 Covalent Organic Frameworks

Covalent organic frameworks (COFs) are constructed solely from the organic connector (vertex) and linker (edge) building units connected by strong covalent bonds, resulting in various two- and three-dimensional (2D and 3D) porous frameworks. Côté et al. [52] and El-Kaderi et al. [53] reported porous crystalline 2D COFs solely from light organic-linkers covalently bonded with boron-oxide clusters. These 2D COF structures resemble layered graphite composed of graphene sheets. Uribe-Romo et al. [54] synthesized the first 3D crystalline framework (COF-300) constructed solely from C—C and C—N covalent linkages and demonstrated its permanent porosity by studying Ar adsorption at 87 K. Considering that the boron–oxide clusters are hydrolytically unstable, several new 2D COFs with exceptional chemical and thermal stability were recently synthesized, showing potential applications in gas adsorption and separation [55].

Babarao et al. [56] studied for the first time the CO_2 storage capacity of both 2D and 3D COFs using GCMC simulations. [57] They showed that COF-108 has the highest CO_2 storage capacity compared to COF-105 ($82\,mmol\,g^{-1}$) and MOF-177 ($33\,mmol\,g^{-1}$) at 30 bar. Higher CO_2 heat of adsorption at zero loading is found in COF-6 ($32.79\,kJ\,mol^{-1}$) compared to other COFs. Their results showed that CO_2 adsorption capacity has good correlation with structural properties such as pore volume, porosity, density, and surface area. Yang et al. [58] performed adsorption studies in 2D COF-8 and COF-10 structures and found stepped behavior in adsorption isotherms. They showed that the CO_2-CO_2 interaction, CO_2-COF interaction, pore size, and temperature play a crucial role in the stepped behavior of adsorption isotherm. In another study by Cao et al. [59], the effect of doping the metal ions in 3D COFs was reported: they found the highest adsorption of $9.28\,mmol\,g^{-1}$ at 1 bar and 298 K, which is much higher than any COFs and many MOFs.

5.8 Criteria for Screening Candidate Materials

Evaluation criteria are required in order to target or discard candidates for carbon capture. As discussed earlier, a number of predictions can be made, and there are a number of properties important for optimizing carbon capture. Bae and Snurr [60] proposed evaluation criteria that are described here in addition to other important factors including CO_2 uptake, working capacity, regenerability, break-through time, and selectivity.

5.8.1 CO_2 Uptake

Predicted CO_2 uptake within an adsorbent or a membrane is an important factor to consider for carbon-capture applications. Table 5.1 lists candidate materials and their predicted CO_2

Table 5.1 Adsorption of pure CO_2 by adsorbents at different conditions.

Material	CO_2 uptake (wt %)	Condition	Reference
MOF 200	71	50 bar and 300 K	[61]
MOF 210	71	50 bar and 300 K	[61]
MIL 101	63.7	50 bar and 303 K	[61]
MOF 177	60	35 bar and 298 K	[61]
IRMOF 1	46.8	35 bar and 298 K	[62]
IRMOF 6	44.1	35 bar and 298 K	[62]
MIL 100	44.1	50 bar and 303 K	[61]
IRMOF 3	42.7	35 bar and 298 K	[62]
Cu3(BTC)2	39.7	35 bar and 298 K	[62]
MOF 74	30.5	35 bar and 298 K	[62]
Mg/DOBDC	26	0.1 bar and 300 K	[28]
Ni/DOBDC	18	0.1 bar and 300 K	[28]
CO/DOBDC	13	0.1 bar and 300 K	[28]
Zi/DOBDC	6	0.1 bar and 300 K	[28]
Pd(2-pymo)2	4	0.1 bar and 300 K	[28]
HKUST-1	3	0.1 bar and 300 K	[28]
UMCM-150	2	0.1 bar and 300 K	[28]
UMCM-150(N)2	2	0.1 bar and 300 K	[28]

uptake at different conditions. MOF-200 tops the chart in CO_2 capture capacity. In general, the CO_2 uptake in any porous material increases with pressure and tends toward a saturation level. It has been discovered that there are some materials with different slopes but an identical saturation level; hence it is difficult to compare all materials across a wide range of operating conditions. Nonetheless, CO_2 uptake serves as a preliminary indicator for capture performance.

5.8.2 Working Capacity

The working capacity is the difference in adsorption capacity and desorption capacity. For a PSA process, the material adsorbs CO_2 at high pressure and desorbs at the low pressure, as shown in Figure 5.13. Working capacity can then be calculated as

$$\Delta N = N_{ads} - N_{des}$$

where N_{ads} and N_{des} are the CO_2 uptake at adsorption and desorption pressures, respectively. This calculation can also be used for TSA, where working capacity is the difference between uptake at the adsorption and desorption temperatures. Other switching processes such as light-switching or electric field-switching can also adopt this criteria. Working capacity can also be used to calculate the regeneration energy (or parasitic energy) by equating a swing in pressure and temperature with the associated energy cost.

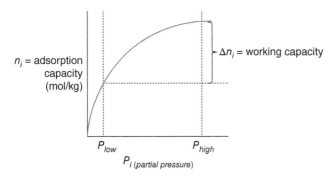

Figure 5.13 Working capacity extracted from predicted isotherm between a desorption pressure (P_{low}) and an adsorption pressure (P_{high}).

5.8.3 Selectivity

Selectivity is a measure for describing the amount of CO_2 separated over other gas components, or the separation purity. For adsorbent-based separations, the selectivity can be defined as the uptake of CO_2 over the uptake of the competing components – typically, N_2, CH_4, H_2O, H_2, and others, depending on the source of emissions and method of separation. For membrane-based separations, the selectivity should be calculated as the permeation (or permeability) of CO_2 divided by the permeation of the competing gases, which is a combination of adsorption and diffusion as indicated earlier in Section 5.4.

There is usually a trade-off between uptake and selectivity. Na-J (Barrer and White) (JBW) has the highest selectivity for the CO_2/CH_4 mixture, but its adsorption capacity is less than many others. Bae and Snurr describe selectivity as follows [60]:

$$\alpha_{12}^{ads} = \frac{N_1^{ads} * Y_2}{N_2^{ads} * Y_1}$$

Here, N is the adsorbed amount and y is the molar fraction in the gas phase. Subscripts 1 and 2 indicate the strongly adsorbed component (CO_2) and the weakly adsorbed component (CH_4, N_2, or H_2), respectively. Superscript *ads* refers to adsorption conditions. This evaluation criteria can be used to quickly evaluate materials for use in real pressure or vacuum-swing adsorption processes.

5.8.4 Diffusivity

Diffusivity is important for predicting the speed at which gas pass through the materials for both adsorbents and membranes. The diffusivity (or diffusion coefficient) is defined as the displacement of individual molecules, while the transport diffusivity (Fickian or collective diffusivity) is defined as the net molecular flux that is generated by a concentration gradient [63]. For any guest species, the diffusivity values vary by several orders of magnitude depending on the pore size and topology. Diffusivity is also dependent on the loading or concentration within the pores.

5.8.5 Regenerability

Regenerability can also be an important parameter that captures the amount of adsorption sites regenerated at each cycle [60]. This can be calculated as the ratio of working capacity over absolute capacity, R (%):

$$R = \frac{\Delta N_1}{N_1^{ads}} * 100$$

Regenerability is likely to change over multiple cycles depending on the mechanism of adsorption and operating conditions. However, this equation gives a reasonable indication of how well the material is regenerated, at least at the initial pristine condition. Bae and Snurr found a negative correlation between regenerability and the heat of adsorption, which will be discussed further in Section 5.8.7.

5.8.6 Breakthrough Time in PSA

Breakthrough time is a specific measure for PSA. In the literature, it has been found that MOFs with very high selectivity for a particular gas shows very low CO_2 uptake; and some materials with very high working capacity show low uptake. This is because there should be an optimum cutoff between the selectivity and working capacity. Breakthrough time is the combination of both of them.

A packed bed adsorber is used to measure breakthrough characteristics that has been demonstrated with Mg-MOF-74 and UTSA-16, and matched excellently with simulation as shown by Bloch et al. [64]. In addition, the study by Xiang et al. [65] using breakthrough experiments showed that UTSA-16 displays high uptake ($160\,cm^3\,cm^{-3}$) of CO_2 at ambient conditions, making it a potentially useful adsorbent material for post-combustion CO_2 capture and biogas stream purification.

5.8.7 Heat of Adsorption

Heat of adsorption can be directly calculated by GCMC simulations or indirectly calculated through DFT calculations and other methods. The higher heat of adsorption implies that the structure has a higher affinity for the gas of interest. Along with other parameters discussed, heat of adsorption can be used to assess the capture feasibility of a material. This parameter is closely related to regenerability, where the heat of adsorption can approximately reflect the amount of energy required to regenerate the material. For example, most MOFs demonstrate a CO_2 heat of adsorption between 20 and $50\,kJ\,mol^{-1}$.

5.9 *In-Silico* Insights

5.9.1 Effect of Water Vapor

It is well-known that water vapor is omnipresent. It is often present in gaseous mixtures and can influence the CO_2 adsorption and structure stability of MOFs. For example, MOF-177 decomposes under humid conditions [66], and MOF-200 and MOF-210 are unstable when in direct contact with water. In addition, the CO_2 uptake on the hydrated

Mg-MOF-74 dramatically decreases as compared to that under dry conditions [67]. MIL-101 and MIL-100 have very good water stability [68] and high CO_2 capacity [69] and are hydrophilic [70]. The CO_2 working capacity and selectivity from a CO_2/CH_4 feed mixture by MIL-100(Fe) significantly increased (about 150%) in the presence of water vapor due to formation of more adsorptive sites toward CO_2, but it badly weakened (decrease by 44%) those of MIL-101(Cr) due to H_2O competitive adsorption.

In Cu-BTC, the adsorption behavior can be tuned by the presence of water molecules coordinated to open metal sites. The interaction between the quadruple moment of CO_2 and the electric field created by water molecules leads to enhanced CO_2 uptake and CO_2 selectivity over N_2 and CH_4 [71]. In a study by Yazaydin et al. [72], the predicted adsorption isotherms in Cu-BTC by GCMC simulation included 2×10^7 step equilibration period followed by a 2×10^7 step production run. DFT calculations were used to optimize the position of the hydrogen atoms of the water molecules and to calculate the partial charges of the Cu-BTC framework atoms (Figure 5.14).

Water affects post-combustion CO_2 capture in MgMOF-74, which was investigated using a combination of GCMC and DFT by Yu and Balbuena [73]. There was a decrease in CO_2 adsorption capacity in MgMOF-74. This effect was a consequence of the reduction of binding energy between CO_2 and the water-coordinated MgMOF-74 framework. DFT calculations indicate that such binding energy is much lower than that without coordinated water (see Figure 5.15).

The coordinated water molecules in HKUST-1, however, lead to improved CO_2 uptake. The comparison of binding energies between CO_2-Cu and CO_2-coordinated water in this system indicated that CO_2 adsorption is stronger in the hydrated HKUST-1 than that in the dry framework, which was ascribed to the enhanced Coulombic interaction between CO_2 and the coordinated water [73]. In another study, they observed that water coordinating to Mg-HKUST-1 lowers CO_2 uptake drastically, whereas it increases the CO_2 adsorption in Zn-, Co-, and Ni-HKUST-1, respectively [74].

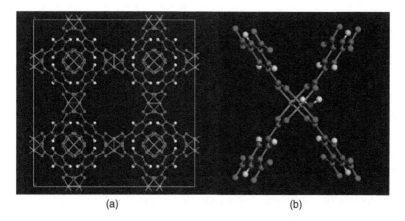

(a)   (b)

Figure 5.14 (a) Dry Cu-BTC unit cell; (b) hydrated Cu-BTC (4 wt %) with coordinated water molecule from density functional theory (DFT). Color code: Cu, orange; O, red; C, gray; H, white. The oxygen atom of the coordinated water molecule is shown in blue. Source: reprinted from Ref. [72] with the permission of the American Chemical Society. (*See color plate section for color representation of this figure*).

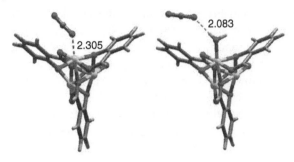

Figure 5.15 Optimized structures of the Mg-MOF-74 cluster with (right) and without (left) water coordination interacting with CO_2. Color code: Mg, green; O, red; C, gray; H, lavender. Source: reprinted from Ref. [74] with the permission of the American Chemical Society. (*See color plate section for color representation of this figure*).

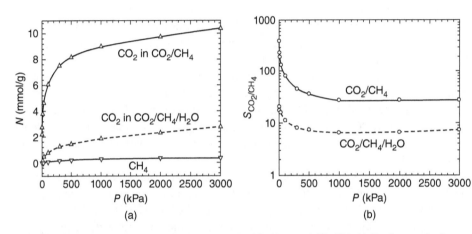

Figure 5.16 (a) Isotherms and (b) selectivities for CO_2/CH_4 and $CO_2/CH_4/H_2O$ mixtures in rho-ZMOF. The bulk composition is 50:50 for CO_2/CH_4, and 50:50:0.1 for $CO_2/CH_4/H_2O$. Source: reprinted from Ref. [76] with the permission of the Royal Society of Chemistry.

In other cases, the effect of water, say, in ionic MOFs has a much larger effect on CO_2 uptake and selectivity. It can be either detrimental or beneficial depending on the topology of the framework. For example, the selectivity in cationic soc-MOF increases at low pressures even in the presence of trace amounts of water, particularly for CO_2/CH_4 mixtures [75]. Similarly, the selectivity of CO_2/CH_4 in MIL-101 is enhanced by the presence of terminal water molecules. However, in anionic rho-ZMOF, the presence of water has a huge effect on the adsorption capacity and selectivity of CO_2/CH_4 mixtures (Figure 5.16). A similar effect is observed for CO_2/H_2 and CO_2/N_2 mixtures in anionic rho-ZMOF. In some MOFs, a small amount of water has a negligible effect on the separation of gas mixtures: for example, the CO_2/H_2 and CO_2/N_2 mixtures in bio-MOF-11, [18] CO_2/CH_4 mixture in Zn(BDC)(TED)0.5 [19], and CO_2/CH_4 and CO_2/N_2 mixtures in ZIF-68 and ZIF-69 [76].

Therefore, the effect of a trace amount of water on the adsorption selectivity of CO_2 in the CO_2/CH_4 mixture is MOF dependent. When the interaction between water molecules and MOF is large enough, the effect becomes significant, as the electrostatic interactions between the framework and fluids play a crucial role. When the interaction between water molecules and MOF is weak enough to allow the water molecules to move freely in the materials, the effect on selectivity is negligible. However, at moderate interaction strength, the effect can be either negligible or significant, depending on the competition of the water molecules to the adsorption of CO_2.

Seda Keskin and her group did a high-throughput computational screening on the CSD MOF database to identify the top MOF adsorbent and membrane for CO_2/N_2 and CO_2/H_2 separation. Both GCMC and MD simulations were performed in several MOFs, and the effect of H_2O on mixed-gas separation was studied on the top-performing MOFs. In this way, the number of promising MOF adsorbents and membranes to be investigated for flue gas and syngas CO_2 separation in future experimental studies was narrowed down from thousands to tens [77, 78].

5.9.2 Effect of Metal Exchange

Lau et al. [79] explored the post-synthetic metal exchange approach to increase CO_2 uptake in a UiO-66 framework; see Figure 5.17. The CO_2 uptake of Ti-exchanged UiO-66 obtained from molecular modeling showed that the CO_2 uptake almost doubles compared to the parent UiO-66. This is further confirmed in experimental work where the CO_2 uptake of UiO-66 (Zr) is enhanced by nearly 81% via post-synthetic exchange with Ti (IV) ions. The smaller pores in Ti-exchanged UiO-66 enhanced the CO_2 isosteric heat Qst and, consequently, increased the CO_2 uptake. Additionally, the lower framework density of Ti-exchanged UiO-66 also contributed to higher gravimetric CO_2 uptake. As the theoretical enhancement in CO_2 gravimetric uptake of UiO-66(Ti_{100}) is ~19%, other factors contribute to drastically enhanced CO_2 uptake in Ti-exchanged UiO-66. Meanwhile, the random nature of exchange sites precludes a close match of simulated and experimental results for the heterometallic samples; yet the trend of enhanced CO_2 uptake with increased Ti loading is clearly observed in both simulated and experimental data. Notable is that no significant increase for CO_2 uptake is observed in simulations for a theoretical homometallic Ti-based MOF.

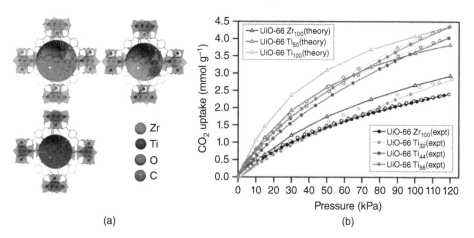

(a)   (b)

Figure 5.17 (a) The Zr-based metal organic framework (MOF) UiO-66 with postsynthetic exchange with Ti(IV) to deliver heterometallic MOFs, with decreased octahedral cages sizes; (b) CO_2 uptake as a function of pressure and Ti concentration. Source: reprinted from Ref. [80] with the permission of the Royal Society of Chemistry. (*See color plate section for color representation of this figure*).

5.9.3 Effect of Ionic Exchange

One of the approaches to increase the binding energy of CO_2 with the framework is the creation of a strong electrostatic field within the cavities based on the existence of extra-framework ions that can be ion-exchanged. MOFs with extra-large cavities can accommodate ions with different shapes and polarities. In this regard, zeolite-like metal-organic framework (ZMOF) materials with RHO and SOD topologies attracted huge attention in creating charged frameworks that are then ion-exchangeable. The presence of extra-framework ions creates strong electric fields that favors the adsorption of guest molecules such as CO_2 much more strongly compared to other gases with less interaction.

Babarao and Jiang [13] reported the high selectivity of CO_2 over gases such as CH_4, N_2, and H_2 in Na-exchanged rho-ZMOFs using molecular simulation. Later, Nalaparaju et al. predicted the adsorption properties of CO_2 and N_2 in monovalent Na-, divalent Mg-, and trivalent Al-exchanged rho-ZMOFs, and the effect of cations on the adsorption properties was discussed [80]. Figure 5.18 shows the equilibrium location of cations obtained through the simulated annealing method.

The effect of cations on adsorption of CO_2 and N_2 was further predicted using Monte Carlo simulation. CO_2 uptake is significantly affected by the type of cation and increases on the order of the charge-to-diameter ratio. The reason is that the electrostatic interactions between CO_2 and cations as well as the ionic framework increase with an increasing charge-to-diameter ratio. CO_2 is more strongly adsorbed than N_2 due to its quadrupolar moment. Also, N_2 exhibits a linear isotherm and is far away from saturation. In the three rho-ZMOFs studied, the N_2 uptake follows the order of increasing dispersion interaction with the cation (Figure 5.19).

5.9.4 Effect of Framework Charges

Framework charges play a critical role in the purification of gas mixtures including CO_2 from different point sources [81, 82]. The contribution is shown to be more significant at low pressures and less important with increasing pressure [83]. For applications around

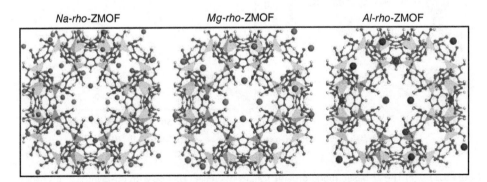

Na-rho-ZMOF   *Mg-rho-ZMOF*   *Al-rho-ZMOF*

Figure 5.18 Atomic structures of rho-ZMOFs. Color code: In, cyan; N, blue; C, gray; O, red; H, white; Na⁺, purple; Mg²⁺, pink; Al³⁺, dark blue. Source: reprinted from Ref. [81] with the permission of Elsevier. (*See color plate section for color representation of this figure*).

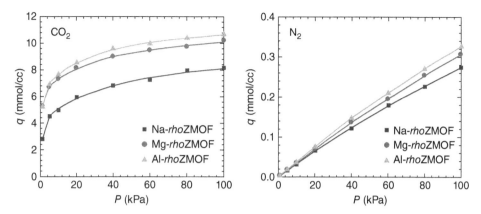

Figure 5.19 Adsorption of isotherms of CO_2 and N_2 in Na-, Mg-, and Al-rho-ZMOF. The symbols are from simulations, and the lines are fit using the dual-site Langmuir model. Source: reprinted from Ref. [81] with the permission of Elsevier.

atmospheric pressure, such as flue gas purification, the effect of framework charges cannot be ignored unless the pores are large (>3.3 nm).

Figure 5.20 shows the contribution of charges calculated by Zheng et al. [83]. When the pore size is larger than 3.3 nm, there is a linear relationship with pore size, and the contribution becomes smaller than 10%. With increasing pressure, the framework charge contribution decreases rapidly and becomes less than 10% at pressures higher than 2.0 MPa.

At low pressure, the contribution of framework charges to CO_2 uptake is large. The electrostatic interaction between the framework and CO_2 plays a dominant role in the adsorption capacity, as the absolute values of the charges in the metal clusters are highest in the frameworks; for materials with the same primitive topologies, the electrostatic contribution increases with decreasing pore size at 0.1 MPa. The contribution of the framework charges is less than 10% at 0.1 MPa, when the pore size is larger than 3.3 nm. The effects of the framework charges in MOFs with large pores can be ignored for the initial screening of MOF materials for CO_2 capture applied at atmospheric or higher pressures, such as in the flue gas purification process.

With increasing pressure, CO_2 molecules begin to adsorb mainly around organic units. At moderate pressure, the framework charge contribution decreases rapidly with increasing pressure. The charges in the organic linkers are smaller than those in the metal clusters. The CO_2 molecules tend to gather in the center of the pore with a further increase in pressure (2.0 MPa), which explains the quick decrease in the contribution of framework charges with increasing pressure. The contribution of framework charges can be ignored for large-scale initial computational screening of MOF materials without introducing large errors, such as in the natural gas upgrading process.

In Figure 5.20, it is interesting to note that, for materials with the same primitive topologies, the electrostatic contribution increases with decreasing pore size at 0.1 MPa (Figure 5.20a for IRMOFs and Figure 5.20b for ZIFs and PCNs). For ZIF-69, ZIF-3, and ZIF-10, CO_2 molecules are mainly adsorbed in the small pores; and for the IRMOFs, the CO_2 molecules first occupy the corner regions of the larger pores.

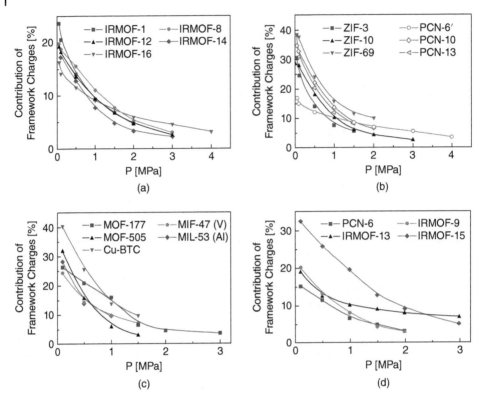

Figure 5.20 Contribution of framework charges to CO_2 uptake at 298 K: (a–c) in MOFs without catenation; (d) in MOFs with catenation. Source: reprinted from Ref. [3] with the permission of the American Chemical Society.

5.9.5 Effect of High-Density Open Metal Sites

In a study by Karra et al. [3], the role of open metal sites in Cu-BTC through GCMC simulation showed that MOFs with open metal sites have the potential for enhancing adsorption separations of molecules of differing polarities. But the pore size relative to the sorbate size also played a significant role. The simulations show that electrostatic interactions between the CO dipole and the partial charges on the MOF atoms dominate the adsorption mechanism. Binary simulations show that Cu-BTC is quite selective for CO over hydrogen and nitrogen for all three mixture compositions at 298 K. The removal of CO from a 5% mixture with methane is slightly enhanced by the electrostatic interactions of CO with the copper sites. In general, Cu-BTC exhibits no significant selectivity for CO over methane for the equimolar and 95% mixture.

Similar conclusions are derived for CO_2: at low pressure, MOFs possessing a high density of open metal sites are found to adsorb significant amounts of CO_2. There is also a strong correlation between the heat of adsorption and the number of open metal sites [84]. Experimental CO_2 uptake at 0.1 bar (the anticipated partial pressure of CO_2 in flue gas) and room temperature for 14 MOFs was reported. MOFs with a large capacity for CO_2 at high pressures often do not perform well at low pressure. Changing the metal from Zn in

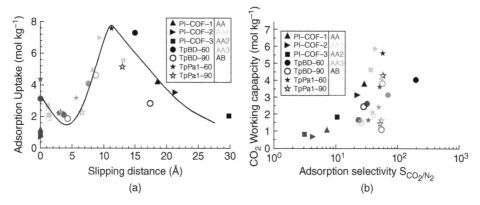

Figure 5.21 (a) Non-monotonous variation of CO_2 adsorption uptake (at 1 bar) with the interlayer slipping distance; (b) variation of CO_2 working capacity (difference between uptake at 1 bar adsorption and 0.1 bar desorption pressure) with adsorption selectivity for a gas mixture of CO_2:N_2 in 15 : 85 ratio, at 1 bar. All results are at 298 K.

M\DOBDC (M-MOF-74) to Mg, Co, or Ni provides big changes in CO_2 uptake. M\DOBDC (where M refers to either Zn, Mg, Ni, or Co; and DOBDC is dioxybenzenedicarboxylate) MOFs have open metal sites that can interact with adsorbate molecules, and Mg\DOBDC performs particularly well. In another study, the metal in M\DOBDC was substituted with other first transition metal elements, which showed that Ti- and V-MOF-74 have greater affinity for CO_2 than Mg-MOF-74 by 6–9 kJ mol^{-1} [85]. The best-performing materials are the various forms of M\DOBDC, all of which have high density of open metal sites.

5.9.6 Effect of Slipping

Sharma et al. [86] reported for the first time the effect of interlayer slipping on CO_2 adsorption and CO_2:N_2 separation in microporous TpPa1 and mesoporous TpBD and PI-COFs using quantum mechanical and GCMC simulations. They found that the slipping affects the number of adsorption sites in the COF structures, resulting in approximately three times higher CO_2 working capacity (or delivery capacity) and CO_2:N_2 selectivity as compared to an eclipsed structure. The highest CO_2 working capacity of 5.8 mol kg^{-1} and CO_2:N_2 separation selectivity of 197 (at 1 bar and 298 K) was observed for slipped PI-COF-2 and TpBD COFs, respectively (Figure 5.21). The molecular insight presented in their work is qualitatively applicable to other similar slipped COFs and is useful for the development of COFs for enhanced gas storage and separation applications. Similarly, in another work, they observed that slipping increases the CO_2: CH_4 selectivity, whereas functionalization results in a decrease in selectivity [87].

References

1 Frenkel, D. and Smit, B. (2002). *Understanding Molecular Simulation from Algorithms to Applications*, 2e. Elsevier.

2 Duren, T., Bae, Y.S., and Snurr, R.Q. (2009). *Using molecular simulation to characterise metal-organic frameworks for adsorption applications. Chemical Society Reviews* 38 (5): 1237–1247.

3 Karra, J.R. and Walton, K.S. (2008). *Effect of open metal sites on adsorption of polar and nonpolar molecules in metal-organic framework Cu-BTC. Langmuir* 24 (16): 8620–8626.

4 Yang, Q. and Zhong, C. (2006). *Electrostatic-field-induced enhancement of gas mixture separation in metal-organic frameworks: a computational study. ChemPhysChem* 7 (7): 1417–1421.

5 Walton, K.S., Milllward, A.R., Dubbeldam, D. et al. (2008). *Understanding inflections and steps in carbon dioxide adsorption isotherms in metal-organic frameworks. Journal of the American Chemical Society* 130 (2): 406–407.

6 Honicke, I.M., Senkovska, I., Bon, V., et al., *Balancing mechanical stability and ultrahigh porosity in crystalline framework materials.* Angewandte Chemie International Edition, 2018. 57(42): p. 13780–13783.

7 Gomez-Gualdron, D.A., Moghadam, P.Z., Hupp, J.T. et al. (2016). *Application of consistency criteria to calculate BET areas of micro- and mesoporous metal-organic frameworks. Journal of the American Chemical Society* 138 (1): 215–224.

8 Duren, T., Millange, F., Férey, G. et al. (2007). *Calculating geometric surface areas as a characterization tool for metal-organic frameworks. Journal of Physical Chemistry C* 111 (42): 15350–15356.

9 Walton, K.S. and Snurr, R.Q. (2007). *Applicability of the BET method for determining surface areas of microporous metal-organic frameworks. Journal of the American Chemical Society* 129 (27): 8552–8556.

10 Sarkisov, L. and Harrison, A. (2011). *Computational structure characterisation tools in application to ordered and disordered porous materials. Molecular Simulation* 37 (15): 1248–1257.

11 Rouquerol, J., Llewellyn, P., and Rouquerol, F. (2006). Is the BET equation applicable to microporous adsorbents? In: *Characterization of Porous Solids Vii - Proceedings of the 7th International Symposium on the Characterization of Porous Solids* (ed. P.L. Llewellyn et al.), 49–56. Elsevier.

12 Vuong, T. and Monson, P.A. (1996). *Monte Carlo simulation studies of heats of adsorption in heterogeneous solids. Langmuir* 12 (22): 5425–5432.

13 Babarao, R. and Jiang, J.W. (2009). *Unprecedentedly high selective adsorption of gas mixtures in rho zeolite-like metal-organic framework: a molecular simulation study. Journal of the American Chemical Society* 131 (32): 11417–11425.

14 Yilmaz, G. and Keskin, S. (2014). *Molecular modeling of MOF and ZIF-filled MMMs for CO2/N2 separations. Journal of Membrane Science* 454 (0): 407–417.

15 Thornton, A.W., Dubbeldam, D., Liu, M.S. et al. (2012). *Feasibility of zeolitic imidazolate framework membranes for clean energy applications. Energy & Environmental Science* 5 (6): 7637–7646.

16 Keskin, S. and Sholl, D.S. (2010). *Selecting metal organic frameworks as enabling materials in mixed matrix membranes for high efficiency natural gas purification. Energy & Environmental Science* 3 (3): 343–351.

17 Bouma, R.H.B., Checchetti, A., Childichimo, G., and Drioli, E. (1997). *Permeation through a heterogeneous membrane: the effect of the dispersed phase. Journal of Membrane Science* 128 (2): 141–149.

18 Bánhegyi, G. (1986). *Comparison of electrical mixture rules for composites. Colloid and Polymer Science* 264 (12): 1030–1050.

19 Chen, Y.F., Babarao, R., Sandler, S.I., and Jiang, J.W. (2010). *Metal–organic framework MIL-101 for adsorption and effect of terminal water molecules: from quantum mechanics to molecular simulation. Langmuir* 26 (11): 8743–8750.

20 Babarao, R., Hu, Z., Jiang, J.W. et al. (2007). *Storage and separation of CO2 and CH4 in Silicalite, C168 Schwarzite, and IRMOF-1: a comparative study from Monte Carlo simulation. Langmuir* 23 (2): 659–666.

21 Curtarolo, S., Hart, L.W., Nardelli, M.B. et al. (2013). *The high-throughput highway to computational materials design. Nature Materials* 12: 191.

22 Wilmer, C.E., Leaf, M., Lee, C.Y. et al. (2011). *Large-scale screening of hypothetical metal–organic frameworks. Nature Chemistry* 4: 83.

23 Chung, Y.G., Camp, J., Haranczyk, M. et al. (2014). *Computation-ready, experimental metal–organic frameworks: a tool to enable high-throughput screening of Nanoporous crystals. Chemistry of Materials* 26 (21): 6185–6192.

24 Deem, M.W., Pophale, R., Cheeseman, P.A., and Earl, D.J. (2009). *Computational discovery of new zeolite-like materials. The Journal of Physical Chemistry C* 113 (51): 21353–21360.

25 Baerlocher, C. and McCusker, L.B. *Database of Zeolite Structures.* Structure Commission of the International Zeolite Association. http://www.iza-structure.org/databases.

26 Lin, L.-C., Berger, A.H., Martin, R.L. et al. (2012). *In silico screening of carbon-capture materials. Nature Materials* 11: 633.

27 Martin, R.L., Simon, C.M., Smit, B., and Haranczyk, M. (2014). *In silico Design of Porous Polymer Networks: high-throughput screening for methane storage materials. Journal of the American Chemical Society* 136 (13): 5006–5022.

28 Yazaydın, A.Ö., Snurr, R.Q., Park, T.H. et al. (2009). *Screening of metal–organic frameworks for carbon dioxide capture from flue gas using a combined experimental and modeling approach. Journal of the American Chemical Society* 131 (51): 18198–18199.

29 McDaniel, J.G., Li, S., Tylianakis, E. et al. (2015). *Evaluation of force field performance for high-throughput screening of gas uptake in metal–organic frameworks. The Journal of Physical Chemistry C* 119 (6): 3143–3152.

30 Kadantsev, E.S., Boyd, P.G., Daff, T.D. et al. (2013). *Fast and accurate electrostatics in metal organic frameworks with a robust charge equilibration parameterization for high-throughput virtual screening of gas adsorption. The Journal of Physical Chemistry Letters* 4 (18): 3056–3061.

31 Kim, J., Abouelnasr, M., Lin, L.-C., and Smit, B. (2013). *Large-scale screening of zeolite structures for CO2 membrane separations. Journal of the American Chemical Society* 135 (20): 7545–7552.

32 Matito-Martos, I., Martin-Calvo, A., Gutiérrez-Sevillano, J.J. et al. (2014). *Zeolite screening for the separation of gas mixtures containing SO2, CO2 and CO. Physical Chemistry Chemical Physics* 16 (37): 19884–19893.

33 Yilmaz, G., Ozcan, A., and Keskin, S. (2015). *Computational screening of ZIFs for CO2 separations. Molecular Simulation* 41 (9): 713–726.

34 Fernandez, M., Boys, P.G., Daff, T.D. et al. (2014). *Rapid and accurate machine learning recognition of high performing metal organic frameworks for CO2 capture. The Journal of Physical Chemistry Letters* 5 (17): 3056–3060.

35 Thornton, A.W., Winkler, D.A., Liu, M.S. et al. (2015). *Towards computational design of zeolite catalysts for CO2 reduction. RSC Advances* 5 (55): 44361–44370.

36 Li, S., Chung, Y.G., and Snurr, R.Q. (2016). *High-throughput screening of metal–organic frameworks for CO2 capture in the presence of water. Langmuir* 32 (40): 10368–10376.

37 Kumar, A., Hua, C., Madden, D.G. et al. (2017). *Hybrid ultramicroporous materials (HUMs) with enhanced stability and trace carbon capture performance. Chemical Communications* 53 (44): 5946–5949.

38 Ziaee, A., Chovan, D., Lusi, M. et al. (2016). *Theoretical optimization of pore size and chemistry in SIFSIX-3-M hybrid Ultramicroporous materials. Crystal Growth & Design* 16 (7): 3890–3897.

39 Hu, Z.G., Wang, Y., Shah, B.B., and Zhao, D. (2019). CO_2 capture in metal-organic framework adsorbents: an engineering perspective. *Advanced Sustainable Systems* 3 (1): 1800080.

40 Robeson, L.M. (2008). *The upper bound revisited. Journal of Membrane Science* 320 (1–2): 390–400.

41 Chung, T.-S., Jiang, L.Y., Kulprathipanjs, S. et al. (2007). *Mixed matrix membranes (MMMs) comprising organic polymers with dispersed inorganic fillers for gas separation. Progress in Polymer Science* 32 (4): 483–507.

42 Wilmer, C.E., Farha, O.K., Bae, Y.-S. et al. (2012). *Structure-property relationships of porous materials for carbon dioxide separation and capture. Energy & Environmental Science* 5 (12): 9849–9856.

43 Davis, S. (2009). *Chemical Economics Handbook*. SRI Consulting.

44 Yilmaz, B., Trukhan, N., and Müller, U. (2012). *Industrial outlook on zeolites and metal organic frameworks. Chinese Journal of Catalysis* 33 (1): 3–10.

45 Babarao, R. and Jiang, J. (2008). *Molecular screening of metal–organic frameworks for CO2 storage. Langmuir* 24 (12): 6270–6278.

46 Babarao, R., Dai, S., and Jiang, D.-e. (2012). *Nitrogen-doped mesoporous carbon for carbon capture – a molecular simulation study. The Journal of Physical Chemistry C* 116 (12): 7106–7110.

47 Baker, R.W. (2004). *Membrane Technology and Applications*. West Sussex, UK: Wiley.

48 Greenfield, M.L. and Theodorou, D.N. (1993). *Geometric analysis of diffusion pathways in glassy and melt atactic polypropylene. Macromolecules* 26 (20): 5461–5472.

49 Gray-Weale, A.A., Henchman, R.H., Gilbert, R.G. et al. (1997). *Transition-state theory model for the diffusion coefficients of small penetrants in glassy polymers. Macromolecules* 30: 7296–7306.

50 Jiang, Y., Willmore, F.T., Sanders, D. et al. (2011). *Cavity size, sorption and transport characteristics of thermally rearranged (TR) polymers. Polymer* 52 (10): 2244–2254.

51 Thornton, A.W., Doherty, C.M., Falcaro, P. et al. (2013). *Architecturing Nanospace via thermal rearrangement for highly efficient gas separations. The Journal of Physical Chemistry C* 117 (46): 24654–24661.

52 Cote, A.P., El-Kaderi, H.M., Furukawa, H. et al. (2007). *Reticular synthesis of micro-porous and mesoporous 2D covalent organic frameworks. Journal of the American Chemical Society* 129 (43): 12914–12915.

53 El-Kaderi, H.M., Hunt, J.R., Medoza-Cortés, J.L. et al. (2007). *Designed synthesis of 3D covalent organic frameworks. Science* 316 (5822): 268–272.

54 Uribe-Romo, F.J., Doonan, C.J., Furukawa, H. et al. (2011). *Crystalline covalent organic frameworks with Hydrazone linkages. Journal of the American Chemical Society* 133 (30): 11478–11481.

55 Zeng, Y.F., Zou, R.Q., and Zhao, Y.L. (2016). *Covalent organic frameworks for CO2 capture. Advanced Materials* 28 (15): 2855–2873.

56 Babarao, R. and Jiang, J.W. (2008). *Exceptionally high CO2 storage in covalent-organic frameworks: atomistic simulation study. Energy & Environmental Science* 1 (1): 139–143.

57 Babarao, R., Custelcean, R., Hay, B.P. et al. (2012). *Computer-aided Design of Interpene-trated Tetrahydrofuran-Functionalized 3D covalent organic frameworks for CO2 capture. Crystal Growth & Design* 12 (11): 5349–5356.

58 Yang, Q.Y. and Zhong, C.L. (2009). *Molecular simulation study of the stepped behaviors of gas adsorption in two-dimensional covalent organic frameworks. Langmuir* 25 (4): 2302–2308.

59 Lan, J.H., Cao, D., Wang, W., and Smit, B. (2010). *Doping of alkali, alkaline-earth, and transition metals in covalent-organic frameworks for enhancing CO2 capture by first-principles calculations and molecular simulations. ACS Nano* 4 (7): 4225–4237.

60 Bae, Y.S. and Snurr, R.Q. (2011). *Development and evaluation of porous materials for carbon dioxide separation and capture. Angewandte Chemie-International Edition* 50 (49): 11586–11596.

61 Xian, S.K., Peng, J., Zhang, Z. et al. (2015). *Highly enhanced and weakened adsorption properties of two MOFs by water vapor for separation of CO2/CH4 and CO2/N-2 binary mixtures. Chemical Engineering Journal* 270: 385–392.

62 Millward, A.R. and Yaghi, O.M. (2005). *Metal-organic frameworks with exceptionally high capacity for storage of carbon dioxide at room temperature. Journal of the American Chemical Society* 127 (51): 17998–17999.

63 Krishna, R. and van Baten, J.M. (2010). *In silico screening of zeolite membranes for CO2 capture. Journal of Membrane Science* 360 (1–2): 323–333.

64 Bloch, E.D., Queen, W.L., Krishna, R. et al. (2012). *Hydrocarbon separations in a metal-organic framework with open iron(II) coordination sites. Science* 335 (6076): 1606–1610.

65 Xiang, S.C., He, Y., Zhang, Z. et al. (2012). *Microporous metal-organic framework with potential for carbon dioxide capture at ambient conditions. Nature Communications* 3: 9.

66 Chae, H.K., Siberio-Pérez, D.Y., Kim, J. et al. (2004). *A route to high surface area, poros-ity and inclusion of large molecules in crystals. Nature* 427 (6974): 523–527.

67 DeCoste, J.B., Peterson, G.W., Schindler, B.J. et al. (2013). *The effect of water adsorp-tion on the structure of the carboxylate containing metal-organic frameworks Cu-BTC, Mg-MOF-74, and UiO-66. Journal of Materials Chemistry A* 1 (38): 11922–11932.

68 Férey, G., Mellot-Draznieks, C., Serre, C. et al. (2005). *A chromium terephthalate-based solid with unusually large pore volumes and surface area. Science* 309 (5743): 2040–2042.

69 Zhang, Z., Hunag, S., Xian, S. et al. (2011). *Adsorption equilibrium and kinetics of CO2 on chromium terephthalate MIL-101. Energy & Fuels* 25 (2): 835–842.

70 Küsgens, P., Rose, M., Senkovska, I. et al. (2009). *Characterization of metal-organic frameworks by water adsorption. Microporous and Mesoporous Materials* 120 (3): 325–330.

71 Peng, X., Lin, L.-C., Sun, W., and Smite, B. (2015). *Water adsorption in metal-organic frameworks with open-metal sites. AIChE Journal* 61 (2): 677–687.

72 Yazaydin, A.O., Benin, A.I., Faheem, S.A. et al. (2009). *Enhanced CO2 adsorption in metal-organic frameworks via occupation of open-metal sites by coordinated water molecules. Chemistry of Materials* 21 (8): 1425–1430.

73 Yu, J.M. and Balbuena, P.B. (2013). *Water effects on Postcombustion CO2 capture in Mg-MOF-74. Journal of Physical Chemistry C* 117 (7): 3383–3388.

74 Yu, J.M., Wu, Y.F., and Balbuena, P.B. (2016). *Response of metal sites toward water effects on Postcombustion CO2 capture in metal-organic frameworks. ACS Sustainable Chemistry & Engineering* 4 (4): 2387–2394.

75 Babarao, R. and Jiang, J. (2009). *Upgrade of natural gas in rho zeolite-like metal–organic framework and effect of water: a computational study. Energy & Environmental Science* 2 (10): 1088–1093.

76 Huang, H.L., Zhang, W., Liu, D., and Zhong, C. (2012). *Understanding the effect of trace amount of water on CO2 capture in natural gas upgrading in metal-organic frameworks: a molecular simulation study. Industrial & Engineering Chemistry Research* 51 (30): 10031–10038.

77 Avci, G., Velioglu, S., and Keskin, S. (2018). *High-throughput screening of MOF adsorbents and membranes for H-2 purification and CO2 capture. ACS Applied Materials & Interfaces* 10 (39): 33693–33706.

78 Daglar, H. and Keskin, S. (2018). *Computational screening of metal-organic frameworks for membrane-based CO2/N-2/H2O separations: best materials for flue gas separation. Journal of Physical Chemistry C* 122 (30): 17347–17357.

79 Lau, C.H., Babarao, R., and Hill, M.R. (2013). *A route to drastic increase of CO2 uptake in Zr metal organic framework UiO-66. Chemical Communications* 49 (35): 3634–3636.

80 Nalaparaju, A., Khurana, M., Farooq, S. et al. (2015). *CO2 capture in cation-exchanged metal-organic frameworks: holistic modeling from molecular simulation to process optimization. Chemical Engineering Science* 124: 70–78.

81 Hamad, S., Balestra, S.R.G., Bueno-Perez, R. et al. (2015). *Atomic charges for modeling metal-organic frameworks: why and how. Journal of Solid State Chemistry* 223: 144–151.

82 Sharma, A., Huang, R., Malani, A., and Babarao, R. (2017). *Computational materials chemistry for carbon capture using porous materials. Journal of Physics D: Applied Physics* 50 (46).

83 Zheng, C.C., Liu, D., Yang, Q. et al. (2009). *Computational study on the influences of framework charges on CO2 uptake in metal-organic frameworks. Industrial & Engineering Chemistry Research* 48 (23): 10479–10484.

84 Caskey, S.R., Wong-Foy, A.G., and Matzger, A.J. (2008). *Dramatic tuning of carbon dioxide uptake via metal substitution in a coordination polymer with cylindrical pores. Journal of the American Chemical Society* 130 (33): 10870–10871.

85 Park, J., Kim, H., Han, S.S., and Jung, Y. (2012). *Tuning metal-organic frameworks with open-metal sites and its origin for enhancing CO2 affinity by metal substitution. Journal of Physical Chemistry Letters* 3 (7): 826–829.

86 Sharma, A., Malani, A., Medhekar, N.V., and Babarao, R. (2017). *CO2 adsorption and separation in covalent organic frameworks with interlayer slipping. CrystEngComm* 19 (46): 6950–6963.

87 Sharma, A., Babarao, R., Medhekar, N.V., and Malani, A. (2018). *Methane adsorption and separation in slipped and functionalized covalent organic frameworks. Industrial & Engineering Chemistry Research* 57 (14): 4767–4778.

6

Ultrathin Membranes for Gas Separation

Ziqi Tian[1], Song Wang[1], Sheng Dai[2,3] and De-en Jiang[1]

[1]*Department of Chemistry, University of California, Riverside, CA, USA*
[2]*Chemical Sciences Division, Oak Ridge National Laboratory, Oak Ridge, TN, USA*
[3]*Department of Chemistry, The University of Tennessee, Knoxville, TN, USA*

6.1 Introduction

Gas separation is an important process with a wide range of industrial applications, including air separation, hydrogen production, natural gas purification, and post- and pre-combustion carbon dioxide capture. In comparison with traditional distillation and sorbent-based separation methods, membrane-based technology is more energy-efficient, thus providing an attractive alternative in solving some critical environmental problems such as carbon capture. In the past three decades, commercial polymeric and inorganic membranes have been developed for a variety of practical applications [1–3].

Materials for Carbon Capture, First Edition. Edited by De-en Jiang, Shannon M. Mahurin and Sheng Dai.
© 2020 John Wiley & Sons Ltd. Published 2020 by John Wiley & Sons Ltd.

The solution-diffusion mechanism is the most employed theory to understand gas-separation phenomena across polymeric and liquid membranes [4]. Governed by this mechanism, the separation performance of a membrane material is constrained by the Robeson upper bound, which states that there is a trade-off between selectivity and permeability [5]. However, membrane productivity is measured by its permeance, not the thickness-normalized permeability. Gas permeance is inversely proportional to membrane thickness [6]. By this argument, a one-atom-thin membrane would offer the highest permeance. If the one-atom-thin membrane has micropores with size between diameters of a targeted gas pair, high selectivity can be achieved through a molecular-sieving separation [7–9]. Therefore, ultrathin membranes such as porous graphene have the potential to offer high selectivity and high permeance for gas separation.

Two-dimensional (2D) materials offer an exciting starting point to design novel ultra-thin membranes [10]. Graphene, a single-atom-thick sheet composed of sp^2 hybridized carbon, is the simplest 2D material. Nanoporous graphene and its derivatives have been investigated both theoretically and experimentally, revealing new opportunities for next-generation membrane technology [11–13].

These atom-thin porous materials separate gases by a molecular sieving effect. Therefore, it is necessary to compare the size of CO_2 with other gases such H_2 and N_2 in the light of pre- and post-combustion carbon capture, respectively. Empirical kinetic diameters of small gases are generally used [14]. We have shown that quantum mechanical diameters determined by the calculated iso-electronic density surfaces at small values are in quite good agreement with empirical kinetic diameters (Figure 6.1) [15]. One can see that to separate CO_2 from N_2, the pore size should be about 3.4 Å, while to separate CO_2 from H_2, the pore size should be about 3.0 Å.

In this chapter, we review and discuss the progress of graphene-based ultrathin membranes for gas separation, especially for CO_2 capture, from both theoretical design papers and experimental efforts. First, we examine porous graphene, the first proposed ultrathin porous membranes for gas separation, including the original concept and the experimental confirmation. Then, we cover other systems inspired by porous graphene, including poly-phenylene membrane, 2D porous organic polymers, porous carbon nanotubes, and porous porphyrins. Finally, we offer an outlook for ultrathin membranes for carbon capture.

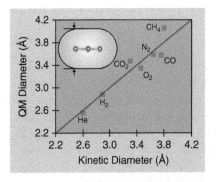

Figure 6.1 Quantum mechanical (QM) diameters versus the experimental kinetic diameters. Source: reproduced from Ref. [15] with permission from the American Chemical Society.

6.2 Porous Graphene

Since its breakthrough discovery in 2004 [16], graphene has attracted increasing attention in both science and technology, because of its ideal proving ground for theoretical prediction and great potential applications, involving surface catalysis [17, 18], nanoelectronics [19–21], and energy storage [22, 23]. However, the pristine graphene sheet is impermeable to all the gas molecules, due to the dense delocalized π-electron cloud in the center of the aromatic ring [24]. To selectively separate gases, nanopores with appropriate sizes have to be created in graphene.

6.2.1 Proof of Concept

In 2009, we first demonstrated the potential capability of porous graphene for gas separation using density functional theory (DFT) calculations [25]. Subnanometer-sized pores were designed on the perfect graphene sheet, as shown in Figure 6.2. Two neighboring six-membered rings were removed to create nanopores. At the same time, eight unsaturated carbon atoms were generated. There were two approaches to passivate the porous structure. First, one could saturate these sp^2 hybridized carbon atoms with hydrogen atoms (Figure 6.2a). Determined from electron density isosurface at a relatively small isovalue of 0.003 a.u., the pore was approximately rectangular in shape with a width of 2.5 Å and

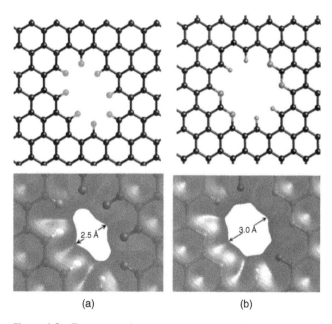

(a)   (b)

Figure 6.2 Two approaches to create nanopores in a graphene sheet. (a) An all-hydrogen saturated pore (C, black; H, grey) and its electron density isosurface (isovalue of 0.003 a.u.); (b) a nitrogen-functionalized pore (C, black; H, smaller grey; N, bigger grey) and its electron density isosurface (isovalue of 0.003 a.u.). Source: reproduced from Ref. [25] with permission from the American Chemical Society.

length of 3.8 Å. For the second approach, the four carbon atoms on the zigzag edges of the nanopore were substituted with nitrogen atoms, and the other four carbon atoms were passivated with hydrogen atoms (Figure 6.2b). This pore was a little larger than the first one, with dimensions of 3.0×3.8 Å.

The potential energy surfaces of hydrogen and methane molecules passing through these two nanopores were mapped with DFT calculations. To consider dispersion interactions, which should be critical to the interaction between neutral molecules and graphene, van der Waals density functional (vdW-DF) theory was employed with periodic boundary conditions. At this level, energy barriers of H_2 and CH_4 crossing the all-hydrogen saturated pore were 0.22 and 1.60 eV, respectively. In comparison, for the larger nitrogen-functionalized pore, they were 0.04 and 0.51 eV, respectively. Evidently, these two nanopores were both too small for methane to pass through. Based on the Arrhenius equation, the ideal selectivity of H_2/CH_4 could be estimated from the calculated diffusion barriers. At room temperature (300 K), H_2/CH_4 selectivities of hydrogen-saturated and nitrogen-functionalized pores were on the order of 10^8 and 10^{23}, respectively, which were much higher than commercial polymer and silica membranes [26].

6.2.2 Experimental Confirmation

Although the designed porous graphene showed excellent performance on hydrogen separation from theoretical calculations, it could be challenging to introduce uniform and controlled nanopores on graphene. Two methods can introduce nanopores with controllable sizes: one was to punch holes on pristine graphene layer (the top-down method); the other was to construct 2D frameworks from small molecules as building blocks (bottom-up method). With the development of nanotechnology, several graphene-derived porous materials have been prepared by the former approach [27–35]. They had remarkable gas-separation performance, confirming our computational proof of concept.

Bunch et al. first introduced size-controlled nanopores in pristine graphene by ultraviolet-induced oxidative (UV/ozone) etching [29]. Bilayer graphene was selected as precursor because previous experiments had shown more controlled etching and increased stability of pores in bilayer graphene compared to single layer [36]. As shown in Figure 6.3a, initially, pristine graphene membranes were mechanically exfoliated and fabricated over wells that had been etched in silicon oxide. These wells were filled with certain gas species by slow diffusion through SiO_2 substrate. After removing the gas outside the membrane, the higher pressure inside the well caused the membrane to bulge upward. The maximum deflection of the membrane, δ, was measured by atomic force microscopy (AFM) to characterize the leak rates of various filled gases before and after etching (Figure 6.3b). The rate of change of δ with time $(-d\delta/dt)$ demonstrates molecular selectivity of porous graphene membrane. After etching, $-d\delta/dt$ of H_2 and CO_2 increased by two orders of magnitude. In comparison, those of Ar and CH_4 were almost unchanged, implying that H_2 and CO_2 passed through the porous membrane, but Ar and CH_4 are too large to cross these etched pores. The rate $(-d\delta/dt)$ was further depicted versus kinetic diameters of gas molecules in Figure 6.3c, suggesting that the pore size was close to the diameter of Ar (3.4 Å). This work strongly supported our theoretical prediction. The measured H_2 leak rate was estimated on the order of 10^{-23} mol s^{-1} Pa^{-1}, which was much smaller than computational estimation of hydrogen-saturated pore in our work

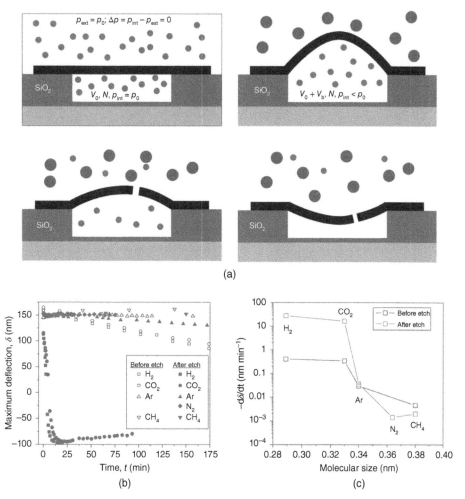

Figure 6.3 (a) Schematic of pristine and porous graphene membranes on silicon oxide substrates at different pressures; (b) maximum membrane deflection, δ, versus time, t, for various gases before and after introducing pores into graphene; (c) average rate of changes of δ with time (−dδ/dt) versus molecular size before and after etching. Source: Reproduced from Ref. [29] with permission from Nature Publishing Group.

(10^{-20} mol s^{-1} Pa^{-1}), probably because the experimental pore had a lower density and a higher passing-through energy barrier than the computational model. In addition, this membrane could distinguish CO_2 from CH_4 and N_2 based on molecular size, showing outstanding potential for carbon capture.

Porous graphene can also be produced with an ion/electron beam. Golovchenko et al. reported an approach for creating nanopores in graphene with atomic precision with an energetic argon ion beam [30]. They created nanopores with sharp pore-size distribution of 3 Å, while no gas permeation experiment was further performed. Later on, Park et al. developed a facile method to efficiently introduce a large number of nanopores with controllable sizes [34]. They placed two chemical-vapor-deposition (CVD) graphene layers on a punctured SiN_x substrate and then used a He-based focused ion beam (FIB) to drill apertures of <10 nm in diameter. Although the average pore diameter was slightly larger than both the pore size of the theoretical model and kinetic diameters of the separated gas molecules, this porous graphene still showed good separation selectivity. In this case, an effusion process might govern gas transport. The mean free paths of various gas molecules rather than the molecular size determined gas selectivity.

6.2.3 More Realistic Simulations to Obtain Permeance

Bunch et al.'s experiment not only confirmed the gas-separating ability of our proposed molecular-sieving-like porous graphene but also indicated that membranes with pore size of 3.4 Å have outstanding CO_2/N_2 separation property. Due to the high computational cost and limited accessible timescale, our previous DFT-based MD simulation was performed for only about tens of picoseconds for a very high pressure (over 200 bar) and a high temperature (600 K) without a distinction of the feed and permeate sides. Inspired by Bunch et al.'s experiment and to simulate more realistic membrane-separation conditions, we then performed classical molecular dynamics (CMD) to estimate the permeances and selectivity of various gas molecules [37–39].

A sandwich-like model was set up to model gas permeation through porous graphene, as shown in Figure 6.4a. A porous membrane with pore density of $0.04\,nm^{-2}$ was fixed in the center of the system, and studied gas molecules were placed initially in the upper chamber at different initial pressures. We focused on a nitrogen-functionalized nanopore (4N4H nanopore; Figure 6.4a, same as in Figure 6.2b), which had dimension similar to that created in the experiment. CMD simulations were performed in the canonical (NVT) ensemble with 2D periodic boundary conditions.

The CO_2/N_2 separation process was simulated first. We tracked the rate of the single component gas diffusing through the pores to the other side. At an initial pressure of 10 atm, CO_2 could readily pass through the porous membrane, while N_2 never permeated in 16 ns. This result was in good agreement with the Bunch experiment [29], mainly because the size of the nitrogen-functionalized pore in the simulation closely resembled

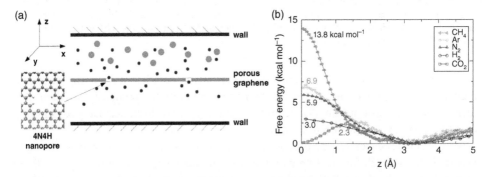

Figure 6.4 (a) Schematic of sandwich-like simulation model and the structure of the nanopore. Source: reproduced from Ref. [38] with permission from Royal Society of Chemistry. (b) Free-energy profiles of various gas permeations as a function of the absorption heights. Source: reproduced from Ref. [39] with permission from Elsevier.

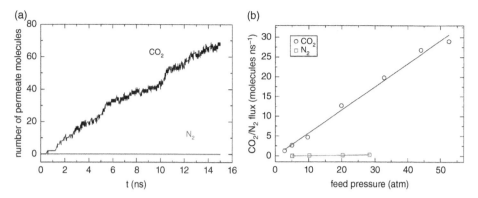

Figure 6.5 (a) CO_2 and N_2 permeations at 10 atm; (b) their fluxes as a function of pressure. Source: reproduced from Ref. [38] with permission from the Royal Society of Chemistry.

that in the experiment (after etch in Figure 6.3c), and the separation mechanism followed the size-sieving model. We found a nearly linear correlation between the numbers of permeate CO_2 molecules and simulation time (Figure 6.5a). The gas permeation rate could be estimated from the slope of this linear equation. For CO_2, permeance was about 2.9×10^5 GPU (1 GPU = 3.35×10^{-10} mol m^{-2} s^{-1} Pa^{-1}).

Pressure influence on gas permeance was further considered by performing simulations over various loadings of CO_2 or N_2. Similarly, the flux at each specific pressure could be obtained from the rate of number of permeate molecules versus time. CO_2 flux also showed good linear dependence on pressure (Figure 6.5b). However, because there were almost no permeate nitrogen molecules, nitrogen permeance was difficult to estimate from the MD trajectory: at relatively high pressure of 20 atm, only one N_2 passed through the porous membrane in a 40 ns dynamic simulation. To reasonably estimate the selectivity, a free-energy computation was performed using the umbrella sampling method. Free-energy profiles of various gas molecules are depicted in Figure 6.4b. The energy barrier for N_2 was located at the pore center, implying that the kinetic diameter of N_2 (3.64 Å) was larger than the dimension of the nanopore (3.4 Å). On the other hand, the free-energy profile and snapshot of a passing-through event showed that, when a CO_2 molecule was above the membrane, it should overcome the free-energy barrier to reorient itself from parallel to perpendicular to the nanopore. Once CO_2 approached the pore center, it would permeate readily, because the nitrogen atoms with negative charge on the pore rim enhanced the attraction between CO_2 and the nanopore. Interaction was energetically favorable when CO_2 was located at the pore center. The free-energy barrier of N_2 passing through the nanopore was estimated to be 5.9 kcal mol^{-1}, which was greater than that of CO_2 (2.3 kcal mol^{-1}). These free-energy barriers were relatively independent of pressure in a range from 2 to 30 atm. Using the free-energy barriers with kinetic theory, the CO_2 and N_2 fluxes at 10 atm were predicted as about 2.7 and 0.01 ns^{-1}, respectively, corresponding to a selectivity of 300. This simulation reproduced the outstanding performance of porous graphene for CO_2/N_2 separation in experiment. The excellent CO_2 permeance on the order of magnitude of 10^5 GPU and high CO_2/N_2 selectivity of 300 offered an attractive opportunity for post-combustion CO_2 capture.

Table 6.1 Free-energy barriers (ΔG) of five gas molecules passing through the 3.4-Å nanopore (Figure 6.4a), and their theoretical and experimental selectivity versus CO_2 (S_{calc} and S_{exp}, respectively) [39].

	H_2	CO_2	Ar	N_2	CH_4
Diameter/Å	2.89	3.30	3.40	3.64	3.80
ΔG/kcal/mol	3.0	2.3	6.9	5.9	13.8
S_{calc}	1.6:1	1:1	1:1400	1:300	$1:6.6 \times 10^7$
S_{exp}	1.7:1	1:1	1:500	1:7000	1:9000

Next, permeances of other gas molecules such as H_2, Ar, and CH_4 were simulated to elucidate selectivity trends. H_2 is the smallest diatomic molecule. It could easily cross the nitrogen-functionalized nanopore in CMD simulation. The permeance of H_2 obtained from CMD trajectories was on the order of magnitude of 10^5 GPU. Unlike H_2 and CO_2, the molecular diameters of argon and methane are larger than the nitrogen-functionalized nanopore, so they were difficult to pass through the nanopore, as confirmed in the CMD simulation. With the same simulation conditions, the trend of flux was that $H_2 > CO_2 >> N_2 > Ar > CH_4$, generally following the sequence of kinetic diameters. To discuss selectivity of different gas pairs, free-energy barriers of Ar and CH_4 across the nanopore were calculated with the previously mentioned umbrella sampling method. From the potential energy curves in Figure 6.4b, we list corresponding energy barriers for the five gases (H_2, CO_2, N_2, Ar, and CH_4) in Table 6.1. From energy barriers, the ideal selectivities were calculated to compare with the experimental ones. The sequence of energy barrier ($CH_4 > Ar > N_2 > H_2 > CO_2$) was generally in accord with that of simulated flux. Although the calculated free-energy barrier of H_2 permeation was higher than that of CO_2, H_2 flux was in fact higher than CO_2 flux because of the much lighter mass of hydrogen. The predicted CO_2/Ar and CO_2/N_2 selectivities were in qualitative agreement with the experimental measurements; however, simulation predicted a much higher selectivity for CO_2/CH_4 pair. The difference could be due to two reasons: (i) experimental leak rates of Ar, N_2 and CH_4 were too small to be measured accurately; and (ii) the relaxation of the porous membrane during the gas-transport event was ignored in CMD simulation.

Based on the simulation, nanoporous graphene showed excellent capability for both post-combustion (CO_2/N_2) carbon capture, in good agreement with experiment. In 2010, the US Department of Energy projected a target cost of CO_2 capture of $20–25/ton of CO_2 [40]. Merkel et al. reported that a membrane with a CO_2 permeance of 4000 GPU and CO_2/N_2 selectivity of 40 corresponded to a cost of CO_2 capture of $15/ton [41]. Evidently, graphene-based ultrathin membranes can achieve this target.

6.2.4 Further Simulations of Porous Graphene

Besides our studies, many other researchers have also explored various single-layered porous graphene models theoretically [42–51]. Chen et al. investigated the influence of nanopores on the mechanical properties of graphene [52]. Membrane strength was determined by pore size, shape, and porosity. The increasing of pore size and porosity led to a decrease in strength. This work is helpful to the design of porous graphene in practice. Zhao et al. found that more permeation events of nitrogen were observed than those of hydrogen when the nanopores were slightly larger than the kinetic diameter of N_2, because nitrogen had stronger vdW interactions with graphene than hydrogen, which accumulated nitrogen on the surface of the membrane [42]. Schwerdtfeger et al. showed that nanoporous graphene could provide suitable barriers for separation of helium isotopic mixture containing fermionic ^{3}He and bosonic ^{4}He [43, 44]. At a temperature of 10 K, the ^{3}He/^{4}He ratio was estimated to be 19 with an acceptable flux of 10^{-9} mol cm^{-2} s^{-1}. Xue et al. reported that some fluorine-modified porous graphene had excellent selectivity for CO_2/N_2 separation [49].

To understand the separation mechanism of a porous single layer, Strano et al. developed a model for gas permeation through ultrathin membranes [53]. It was considered that gas molecules diffused through a single-atom-thick membrane via five steps: adsorption to the surface, association to a pore, translocation through a pore, dissociation from the pore, and desorption from the surface. Various expressions were derived under different rate-limiting steps, providing predictions and limitations for the following theoretical and experimental works. Similarly, Hadjiconstantinou et al. assumed that there were two types of gas fluxes crossing the membrane [54]. One was the direct flux, in which the molecule directly crossed the nanopore from the bulk phase of one side to the other side of the membrane. The other was the surface flux, in which molecule crossed after being adsorbed on membrane. CMD simulations showed that, for gases that didn't adsorb on graphene, such as He and H_2, significant contribution came from direct flux. In contrast, for N_2, CH_4, and other absorbable gases, surface flux might be much larger than direct flux. To increase the contribution of surface flux, Aluru et al. introduced a water slab between a gas mixture and porous graphene [55]. Based on water-solubility, the selectivity ratio could be enhanced. Covered by the water slab, graphene with large nanopores showed high CO_2/O_2 and CO_2/N_2 selectivities, thus having the potential for CO_2 separation from air.

6.2.5 Effect of Pore Density on Gas Permeation

Pore density is an important factor dictating gas separations through one-atom-thin porous graphene membranes. Although the role of pore density has been alluded to in several works [37, 56–59], how exactly the pore density affects permeation has not been fully addressed, especially in the light of the direct gas-phase pathway versus the indirect adsorbed phase or surface pathway. Based on the 4N4H nanopore we constructed (Figure 6.2b and Figure 6.4a), we designed a series of porous graphene membranes with different pore densities from 0.01 to 1.28 nm^{-2}, as shown in Figure 6.6a [60]. CO_2 and He molecules were tested, which represent two different types of gas molecule with strong and weak adsorption onto the nanoporous graphene membrane surface, respectively. From the initial permeation rate by MD simulations, we computed the flux and flux per pore as a function of the pore density (Figure 6.6b and c). One can see that for He the flux increases almost linearly and the flux per pore is nearly a constant with the pore density. On the other hand, the flux of CO_2 shows a similar linear increase when the pore density is less than 0.3 nm^{-2}, but the increase greatly slows down after 0.3 nm^{-2}. And correspondingly, a

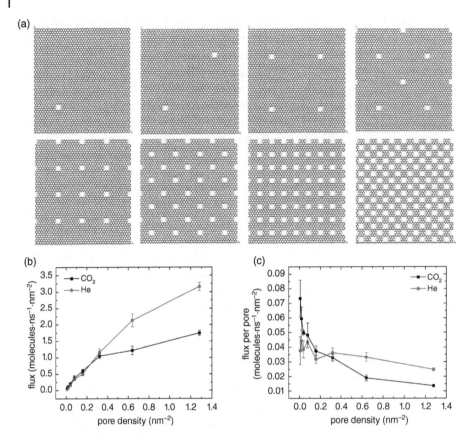

Figure 6.6 (a) The 10×10 nm^2 porous-graphene membrane with different pore density of the same pore (3.0 Å $\times 3.8$ Å in size): 0.01 nm^{-2}; 0.02 nm^{-2}; 0.04 nm^{-2}; 0.08 nm^{-2}; 0.16 nm^{-2}; 0.32 nm^{-2}; 0.64 nm^{-2}; 1.28 nm^{-2}. Initial flux vs. pore density of graphene membranes for CO_2 and He permeation: (b) flux per unit membrane area; (c) flux per pore. Source: reproduced from Ref. [60] with permission from the Royal Society of Chemistry.

roughly exponential decay of the flux per pore with the pore density for CO_2 is displayed. This distinct and interesting difference between CO_2 and He begs a detailed analysis.

According to previous kinetic analysis of gas permeation through a porous graphene membrane [53, 54], the total flux is composed of direct flux and surface flux. For He, due to weak adsorption, we found that its total flux is dominated by the direct flux, especially at high pore densities. Since the direct flux scales with the permeable area, which in turn scales with the pore density, a net flux linear with the pore density and a constant per-pore flux should be obtained, as seen in Figure 6.6b and c. For CO_2, due to strong adsorption, the surface flux dominates the total flux, as the direct flux is minor for all pore densities (Figure 6.7a). Since the surface flux is relevant to surface adsorption, we analyzed the effect of adsorption behavior. In Figure 6.7b, one can see that the per-pore coverage on the feed side is much greater than on the permeate side. Furthermore, we found that the decreasing trend of the per-pore coverage of CO_2 on the feed side can well explain the decreasing per-pore flux of CO_2 with the pore density in Figure 6.6c. So far, our simulations showed

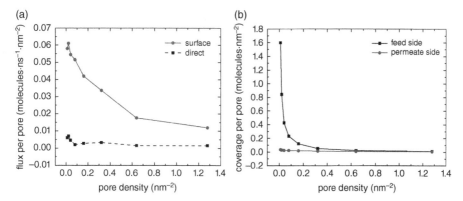

Figure 6.7 (a) The surface flux and the direct flux of CO_2, across graphene membranes of different pore densities; (b) per-pore coverage of CO_2 on feed and permeate sides of the graphene membrane at different pore densities. Source: reproduced from Ref. [60] with permission from the Royal Society of Chemistry.

that the higher the pore density, the greater the flux. More important, since the adsorption greatly modulates the dependence of the flux on the pore density for a strongly adsorbing gas such as CO_2, making the membranes asymmetric by creating dissimilar surfaces is a promising strategy to lead to more interesting permeation behavior.

6.3 Graphene-Derived 2D Membranes

Creating holes in graphene to make porous graphene represents the top-down approach to achieve a one-atom-thin membrane. But the other approach is a bottom-up method of assembling chemical building blocks into a 2D porous membrane. There are several systems designed or synthesized according to this approach that are closely related to porous graphene. We review them here.

6.3.1 Poly-phenylene Membrane

Poly-phenylene has been produced by a surface-promoted aryl-aryl coupling of hexaiodo-substituted cyclohexa-m-phenylene [61]. The synthetic route is shown in Figure 6.8a. At first, macrocycle molecules containing six phenyl rings were deposited on well-defined Ag(111) surface. At room temperature, dehalogenation readily occurred. Subsequently, the temperature increased above 575 K, resulting in intermolecular coupling and formation of a honeycomb network. This fabricated 2D poly-phenylene had a single pore size and nanometer-scale periodicity.

Zhou et al. explored the structural properties of poly-phenylene and its capability for H_2 separation from CO_2 and CH_4 via DFT computation [63]. The unit cell of polyphenylene contained two phenyl rings, as shown in Figure 6.8b. Based on calculated electron-density isosurfaces with isovalue of 0.003 a.u. (Figure 6.8b), hexagonal pores were found with a width of approximately 2.48 Å, which was a little smaller than the kinetic diameter of H_2 (2.89 Å). As a result, hydrogen had a moderate permeation barrier of 0.61 eV, while the

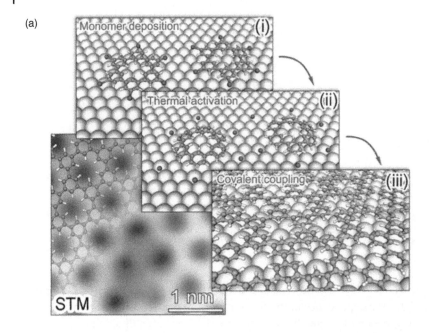

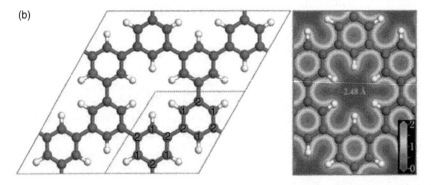

Figure 6.8 (a) Synthetic route of the surface-promoted aryl-aryl coupling to fabricate poly-phenylene and its STM image supported on a Ag(111) surface. Source: reproduced from Ref. [62] with permission from 2010, John Wiley and Sons. (b) The 2 × 2 supercell of poly-phenylene and the electron density of a nanopore. A unit cell was indicated by dash line in the supercell. Color code: C, green; H, white. Source: reproduced from Ref. [63] with permission from the Royal Society of Chemistry. (*See color plate section for color representation of this figure*).

computed diffusion barriers for CO_2, CO, and CH_4 were 2.21, 2.35, and 5.19 eV, respectively, indicating that there was no passing-through event of these molecules under normal conditions. According to the Arrhenius equation, selectivities for H_2/CO_2, H_2/CO, and H_2/CH_4 at 300 K were estimated on the order of magnitude of 10^{26}, 10^{29}, and 10^{76}, respectively. Herein, the polyphenylene membrane showed remarkably high performance for hydrogen purification and pre-combustion CO_2 separation.

Based on high-level ab initio calculations, Shrier further extended the potential application of poly-phenylene membranes to noble gas separation [64]. Compared with the low barrier height for He, Ne and CH_4 would be strongly rejected by this porous membrane, indicating that the poly-phenylene membrane could efficiently separate He from natural gas feedstock. Almost at the same time, Blankenburg et al. systematically studied diffusion barriers through the polyphenylene membrane for a series of atmospheric gas molecules, including H_2, He, Ne, O_2, CO_2, NH_3, CO, N_2, and Ar [62]. From energy barriers calculated by a dispersion-corrected DFT approach, the selectivity was estimated for various molecular combinations at 300 K, leading to the same conclusion as Zhou and Shrier's studies that this 2D poly-phenylene membrane had high hydrogen and helium selectivities.

Later on, some modified poly-phenylene systems were further investigated theoretically [65–71]. Lu et al. demonstrated that boron doping in poly-phenylene could improve selective H_2 separation [67]. From grand canonical Monte Carlo simulations, cation-decorated membranes showed high H_2 storage capacities. Nitrogen doping could enlarge the nanopore and dramatically reduce the energy barrier for oxygen passing through [70]. Besides heteroatom-doping, Limpijumnong et al. showed that tensile stress could deform the structure of poly-phenylene and then effectively increase the permeation rate of H_2, O_2, and CO_2 [69]. As a result, permeation rate and selectivity of various gases were able to be controlled by applying tensile stress. If tensile stress was appropriate, large gas molecules, such as CO_2, would pass through this porous membrane. Kan et al. demonstrated that the pore size would enlarge after an inter-layer-connection of two layers of poly-phenylene [71]. This double-layer structure had both improved permeability and selectivity for hydrogen.

6.3.2 Graphyne and Graphdiyne Membranes

Graphyne and graphdiyne are graphene derivatives, with respect to their structures and properties [72]. Phenyl rings are connected with one and two acetenyl chains in the graphyne and graphdiyne structures, respectively, as depicted in Figure 6.9. Square-centimeter-scale graphdiyne film has been successfully fabricated via cross-coupling of hexaethynylbenzene on a copper surface at relatively mild conditions (60 °C, nitrogen atmosphere) [73]. Transmission electron microscope (TEM) images and X-ray diffraction (XRD) patterns further confirmed its regular structure. Because of its well-defined nanopores and single-layer structure, graphdiyne is supposed to be an experimentally available ultrathin membrane for gas separation.

Smith et al. first demonstrated the potential application of graphdiyne as a superior separation membrane for hydrogen purification [74]. The pore size defined by the vdW surface was about 3.8 Å, which was between the vdW diameter of H_2 and CH_4. The energy barrier of H_2 passing through nanopores was 0.10 eV, while the calculated energy barriers of CH_4 and CO passing through the pore were 0.72 and 0.33 eV, respectively. Thus graphdiyne was supposed to be an ideal membrane for H_2 separation. Following this work, Buehler et al. report a reactive molecular dynamics investigation on H_2 separation from syngas [75]. Determined by the Arrhenius equation, the H_2 permeation barrier was consistent with DFT results. Qiao et al. demonstrated that nitrogen-doping on the graphdiyne nanomesh successfully reduced the H_2 permeation barrier and increased the CH_4 and CO barriers simultaneously, enhancing membrane performance on hydrogen purification [76]. Luo et al. compared the hydrogen separation performances of three porous membranes, involving graphyne,

graphene graphyne graphdiyne

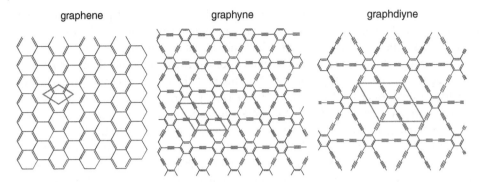

Figure 6.9 The structures of graphene, graphyne, and graphdiyne. Their unit cells are indicated by the rhombuses. Source: reproduced from Ref. [72] with permission from the Royal Society of Chemistry.

graphdiyne, and rhombic-graphyne [77]. The rhombic-graphyne that possesses the middle pore size allowed highly selective H_2 separation with acceptable permeability.

6.3.3 Graphene Oxide Membranes

Ultrathin porous graphene oxide was also synthesized as a state-of-the-art membrane for high-performance gas separation. For few-layered graphene oxide membranes, gas passed through both pores and tiny channels. Li et al. prepared 1.8-nm-thick graphene oxide by a facile filtration process [32]. H_2 could be separated from CO_2 and other larger gas molecules through the selective structural defects. Park et al. achieved selective gas diffusion by introducing well-controlled nanopores and gas-flow channels via different stacking methods [31]. In their work, CO_2 could be selectively removed from natural gas. Not only pore size, but also channel structure played a significant role in tunable gas-transport behaviors of these graphene oxide laminates. The permeation process of graphene oxide was more complicated than that of porous graphene.

Some CMD simulations focused on the performance of graphene-oxide membranes for gas separation. Xue et al. reported that thanks to intense electrostatic interactions, hydroxyl-functionalization of the graphene surface enhanced the CO_2 affinity, and nitrogen-doping of the nanopore rim increased the CO_2/N_2 selectivity [78]. Xu et al. showed that based on few-layered porous graphene oxide, a high-performance membrane for gas-separation could be established by varying the interlayer space and decorating the surface with hydroxyl groups [79]. Furthermore, similar to the graphene structure, Yang et al. theoretically designed a porous silicone membrane and showed that it had high hydrogen and helium selectivities superior to carbon-based ultrathin membranes [80, 81].

6.3.4 2D Porous Organic Polymers

From the structures of porous graphene and graphene derivatives, several organic 2D polymers have been designed as novel porous membranes. Shrier et al. constructed a series of 2D polymer by linking phenyl rings with various planar groups involving phenyl and ethenyl [82–84]. These membranes had various pore sizes, leading to different gas selectivity.

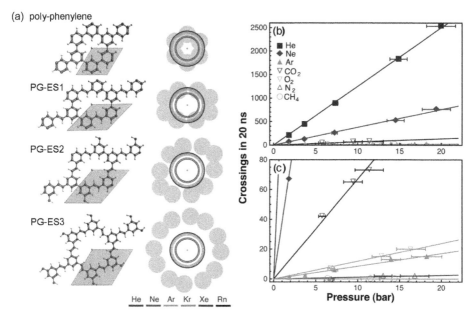

Figure 6.10 (a) Structures and vdW radii overlaps of poly-phenylene, PG-ES1, PG-ES2 and PG-ES3; (b) PG-ES1 crossings during classical molecular dynamics (CMD) simulations; (c) the same data with an emphasis on large noble gas species. Source: reproduced from Ref. [83] with permission from the American Chemical Society.

The unit cell of poly-phenylene had three hexagonal directions. One could replace the biphenyl-like unit with E-stilbene-like groups along one, two, or three directions, as shown in Figure 6.10a. PG-ES1 (porous graphene-E-stilbene-1) was extended with an E-stilbene-like unit in one direction. Its pore size defined by the vdW surface was larger than in poly-phenylene. The nanopore still had a high diffusion energy barrier for the spherical CH_4 molecule. It was only feasible for the linear molecules to pass through, including CO_2, N_2, and O_2. Adsorption and transport processes of these four gases were studied with CMD simulation at 325 K, which was the temperature for post-combustion CO_2 capture [82]. Because at simulation conditions, adsorption equilibrium constants were very small and coverage rates were far from saturation, each gas molecule adsorbed on the surface and crossed the nanopore independently. Pure gas systems were studied instead of gas mixtures. According to the probability density distribution, CO_2 was much more intensely adsorbed on the surface than the other species. All three adsorbed linear molecules tended to be parallel to the surface. 20 ns CMD simulations were performed at various pressures. Even at pressures higher than 15 atm, only a few N_2 and O_2 molecules crossed the porous membrane, compared to nearly hundreds of CO_2 molecules passing through under the same conditions. No CH_4 passed through the pore even during 50 ns. The trend of flux was consistent with ordering of the kinetic diameters: $CO_2 < O_2 < N_2 < CH_4$. It was also found that linear molecules lingered on the surface, allowing them to adjust orientation before passing through the pore. The permeances of various gases were constant with respect to pressure. The simulated CO_2, O_2, and N_2 permeances were reported as 3×10^5, 6×10^4, and 6×10^3 GPU, respectively. CO_2 permeance was much higher than commercial membranes (for instance, 45 nm thick Polyactive films had 2×10^3 GPU CO_2 permeance [85]). The conservatively estimated selectivities of CO_2/N_2 and CO_2/O_2 were 60 and 5, respectively. These results showed that PG-ES1 achieved a combination of high CO_2 permeance and selectivity, enabling economical CO_2 capture.

This series of two 2D materials was also studied for noble gas separation [83]. Classical force-field parameters were adjusted according to dispersion-corrected DFT energy curves to compute the noble gas permeances (Figure 6.10b). The combination of high permeance and selectivity for He made PG-ES1 a promising ultrathin membrane for separation of He from various sources. Subsequently, Xue et al. showed that PG-ES1 was also a good material to separate H_2 from a gas mixture [86].

The biphenyl-like group in one direction of 2D polyphenylene could be replaced with a p-terphenyl (TP)-like unit to enlarge the pore size, denoted as PG-TP1 [84]. The distance between two opposite phenyl rings was about 7 ± 0.2 Å. The separation of a series of small hydrocarbons through this porous membrane was studied by CMD simulation. At practical conditions, the permeances of ethane and short olefins were on the order of magnitude of 10^6 GPU. In comparison, the best existing membranes had permeances of 1–100 GPU [87, 88]. These results indicated that the PG-TP1 membrane was a promising material for high-performance hydrocarbon gas separation.

There have also been some other simulations on 2D organic polymers for gas separation [89–91]. Similar to benzene, sym-triazine is another possible subunit for construction of 2D polymers. Poly-triazine, which is usually denoted as g-C_3N_3 (graphite C_3N_3), can be seen as nitrogen-doped poly-phenylene. Zhou and Li et al. showed that g-C_3N_3 had high hydrogen permeance and selectivity among other atmospheric gases [89, 90]. In addition, a new porous C2N monolayer was synthesized via a simple wet-logical reaction [92]. Based on DFT and ab initio MD calculations, it was considered to possess superior hydrogen separation performance, demonstrated by Xu et al. [91].

6.4 Porous Carbon Nanotube

To move porous one-atom-thin membranes toward practical application, one can change the morphology of the nano-material. A carbon nanotube (CNT) is a close analog to a nanoscale hollow fiber. CNT has many unique applications in nanofluidics, which benefit from the fast transport of molecules in the cavity of CNT [56, 93, 94]. Rolling porous graphene membranes into porous CNTs may combine the molecular-sieve and nanofluidic properties of these two novel materials. In practice, some porous CNT membranes have been prepared [95, 96]. Furthermore, an array of aligned CNTs could be assembled in a polymer substrate to form a nanoporous membrane with high-flux transport of fluid [97, 98].

We proposed windowed CNT with nitrogen-functionalized pores for CO_2 separation from natural gas [99]. The CNT model in a CMD simulation is shown in Figure 6.11a and b, which contained two coaxial single-wall CNTs. The inner tube was windowed (14,0) CNT to provide the gas-separating ability, while the outer one was a perfect CNT without a window, as a boundary of simulation. When the distance between the two tubes was larger than about 6 Å, the outer tube had no influence on gas permeation. With this model, the pressure drop could be estimated as the gas molecules passed through the nanopores in the inner tube.

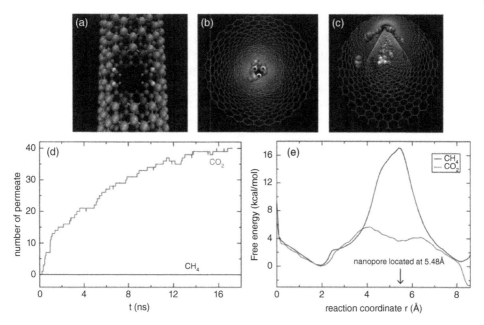

Figure 6.11 (a) 4N4H windows on the carbon nanotube (CNT) wall; (b) initial configuration of classical molecular dynamics (CMD) simulation, where gas mixture is inside the tube; (c) snapshot around t = 16 ns; (d) number of permeate molecules with time. The gas mixture is composed of 80% CO_2 and 20% CH_4 at initial pressure of 175 atm. (e) Free-energy profiles of gas passing through with the radial distance from the axis of CNT. The gas mixture is composed of 50% CO_2 and 50% CH_4 at initial pressure of 88 atm. Source: reproduced from Ref. [99] with permission from the American Chemical Society.

The 4N4H nanopore of the inner tube had the same structure as the nitrogen-functionalized pore constructed in Figure 6.2b and Figure 6.4a. The pore size was estimated to be 3.4 Å, which was between the kinetic diameters of CH_4 (3.8 Å) and CO_2 (3.3 Å). Pairs of pores were created along the tube, which were opposite to each other. Pore density was about one pair per 5 nm of the tube length. CMD simulation at room temperature was performed to study the permeation processes of CO_2 and CH_4. At first, the gas mixture was filled into the inner tube. The number of molecules passing through the nanopores was monitored to calculate flux and permeance, which are plotted in Figure 6.11d. Apparently, CO_2 could readily permeate into the interspace, while no CH_4 crossed the pore in 16 ns. CO_2 permeation was very fast without obvious binding with the pore rim. Before CO_2 successfully passed through the pore, it had to orientate itself from parallel absorption configuration to a perpendicular configuration. In contrast, CH_4 could not pass through this small pore. The fact that CO_2 easily passed through the window implied that attraction from pyridinic nitrogen improved CO_2 permeation and CO_2/CH_4 separation. Free-energy profiles of CH_4 and CO_2 permeation are shown in Figure 6.11e. The transition state for CH_4 diffusion was at the center of the pore with a barrier height of 17 kcal mol^{-1}. In comparison, a permeation barrier for CO_2 permeation was 5.5 kcal mol^{-1}, relating to a configuration that CO_2 was 1.5 Å above the center of the pore. CO_2 reorientation before permeation led to this energy barrier. Based on the Arrhenius equation, CO_2/CH_4 selectivity was estimated on the order of 10^8, which was independent of the pressure. Simultaneously, CO_2 permeance ranged from 10^7 to 10^5 GPU, which was much greater than that of advanced polymer membranes (about 100 GPU) [100]. The flux of the retentate (CH_4 in this study) was another important factor for evaluation of membrane performance. CMD simulation also showed that although nanopores on the inner wall hindered the diffusion of CH_4, the diffusion rate was still one order of magnitude greater than that in normal layered materials, such as zeolites [101]. To conclude, the windowed CNT was suggested as a superior platform for removing CO_2 from natural gas.

Johnson et al. independently studied the ability of a very similar single-wall porous CNT to separate H_2/CH_4 and CO_2/CH_4 mixtures [102]. Besides DFT calculation and CMD simulation with a fixed porous CNT model, they also considered the structure relaxation of porous CNT in molecular dynamics with the reactive empirical bond order (REBO) potential [103]. To avoid rotation and translation of CNT in a flexible model, some atoms in CNT were frozen. The potential energy surface of each gas passing through was calculated with DFT (the Perdew-Burke-Ernzerhof [PBE] functional), dispersion corrected DFT (PBE-D2), and classical potentials with rigid and flexible CNT models. Similar to previous work, there was no permeation barrier for H_2 or CO_2. For CH_4, all four calculations showed a barrier that kept CH_4 from moving through the pore. PBE-D2 and the classical flexible model yielded similar permeation barrier heights for CH_4, which were much lower than those obtained by PBE and the classical rigid models. The obvious difference between PBE and PBE-D2 results indicated that the vdW correction made a significant contribution to the potential energy surfaces for CH_4 and CO_2. Moreover, the similarity between PBE-D2 and classical flexible model results implied that the classical flexible model was a good approximation for considering non-bonded interaction accurately and effectively in dynamics simulation. In CMD simulations with both the classical rigid and flexible models, configurations of gas mixture outside the CNT were constructed as initial states. For a $1:1$ $H_2:CH_4$ mixture corresponding to an initial pressure of 340 atm, dynamics simulations with both rigid and flexible models at room temperature showed almost-infinite H_2/CH_4 selectivity (Figure 6.12b). Although the flexible and rigid models had the same equilibrium H_2 loadings in porous CNT, the former model could reach the equilibrium state more quickly than the latter model (100 ps vs. 300 ps). For a $1:1$ $CO_2:CH_4$ mixture with a pressure of 200 atm, porous CNT could also effectively separate CO_2 from CH_4 with nearly infinite selectivity, as shown in Figure 6.12d. In the flexible model, equilibration took about 10 times longer time than for the H_2/CH_4 mixture. Additionally, they found that CO_2 and CH_4 tended to adsorb on the surface of CNT, generating ring-like layers. This work drew the same conclusion as our study, and provided a new flexible model for accurate dynamics simulation.

Recently, Kral et al. studied selective absorption, transport, and separation of other fluids in porous CNT using CMD simulation [104]. They showed that porous CNT could be used in separation of organic mixture containing alcohol and benzene. After charging, CNT was able to separate different ions from solution, which extended the application of porous CNTs.

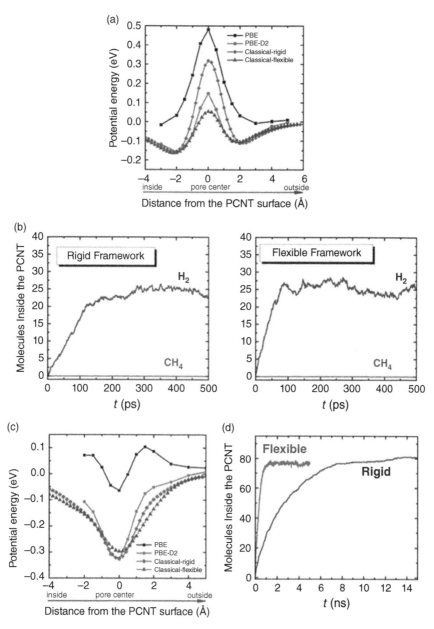

Figure 6.12 (a) Potential energy profile for CH_4 across carbon nanotube (CNT) pore computed from different methods; (b) number of permeate H_2 and CH_4 with time for rigid and flexible CNT models from a gas mixture composed of 50% H_2 and 50% CH_4; (c) potential energy profile for CO_2 across CNT pore computed from different methods; (d) number of permeate CO_2 with time for rigid and flexible CNT models. Source: reproduced from Ref. [102] with permission from the American Chemical Society.

6.5 Porous Porphyrins

Based on graphene and its derivatives, theoretical chemists have studied many nanoporous 2D structures as molecule-sieving membranes. But until now, it is difficult to introduce numerous uniform nanopores into a large area of impermeable surface through a top-down approach. On the other hand, the proposed 2D polymers, which are polymerized from a benzene-like subunit through a bottom-up method, often have relatively small pores. For example, only H_2 and He can pass through poly-phenylene and PG-ES1 (Figure 6.10). One can extend the pore size by inserting linear groups, such as acetenyl, ethenyl, and phenyl, while the enlarged pore will weaken the mechanical properties of the membrane. In the past 10 years, a large number of 2D covalent organic frameworks (COFs) have been synthesized experimentally [105–107], offering a new opportunity for porous membrane construction. Recently, we designed some expanded porphyrin-based membranes for selective CO_2 separation [108].

Porphyrins are common in biochemistry and host-guest chemistry, and often used as subunits for constructions of 2D COFs [107]. They are planar macrocycle molecules. Basic porphyrin, also known as porphine, is composed of four pyrrole rings. Various divalent metal cations can coordinate with porphyrins in the center of molecule, implying that the pore diameter is too small (<1.6 Å) for any gas to cross. However, expanded porphyrins composed of more than four heterocycles have larger pores [109]. It was reported that some expanded porphyrins could selectively coordinate with SO_4^{2-}, which has a kinetic diameter of about 5–6 Å [110]. Moreover, there are nitrogen or oxygen atoms around the pore rim. Negative charge on these heteroatoms can selectively attract CO_2 molecules. As a result, 2D membranes assembled with expanded porphyrins can have appropriate pore sizes and intense CO_2 affinity, resulting in high performance for CO_2 separation.

We screened a series of expanded porphyrin derivatives, mainly including three classes: sapphyrin, amethyrin, and cyclo[8]pyrrole [109]. The labels and structures are shown in Figure 6.13a. To enlarge the pore in the center, some pyrrolic nitrogen atoms and their neighbor hydrogen atoms were replaced with oxygen atoms. In other words, some pyrrole rings were substituted by furans. We determined the pore diameter by the smallest distance between two opposite atoms on the pore rim. Taking into account the vdW radius, we found that the cyclo[8]pyrrole derivatives had appropriate pore sizes for CO_2 permeation. Based on the electron-density iso-value surface at relatively small iso-values, we could draw the same conclusion.

We mapped the potential energy surfaces of CO_2 and N_2 crossing the expanded porphyrins at BLYP-D3/def2-QZVPP level. The potential-energy curves in Figure 6.13b showed that the interaction between the quadrupole moment of CO_2 and negative charges on the heteroatoms significantly decreased the steric hindrance of the small pores. For the three compounds labeled sap-3O, ame-2O, and ame-4O, the CO_2 permeation barriers were 0.277, 0.381, and 0.120 eV, respectively. CO_2 molecules could overcome these barriers under relatively mild experimental conditions. In comparison, both CO_2 and N_2 could readily pass through the pores of cyclo[8]pyrrole derivatives. Membranes composed of these subunits might have low selectivity for CO_2 over N_2. From the energy barriers, the estimated ideal

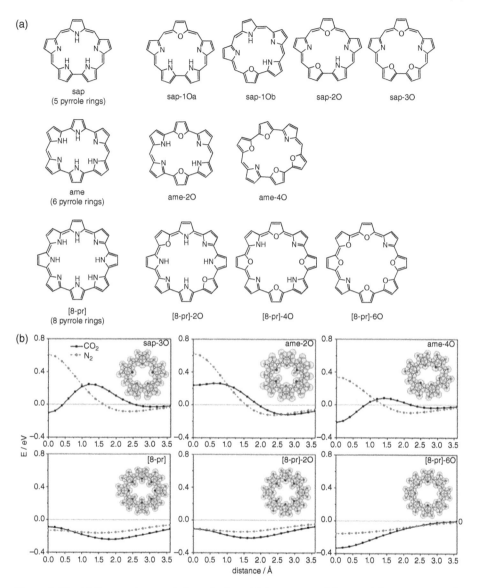

Figure 6.13 (a) Screened expanded porphyrins with large pores; (b) potential energy curves for CO_2 and N_2 passing through selected porous porphyrins computed at the BLYP-D3/def2-QZVPP level. Source: reproduced from Ref. [108] with permission from the American Chemical Society.

CO_2/N_2 selectivities of sap-3O, ame-2O, and ame-4O were on the order of magnitude of 10^5 to 10^7.

Subsequently, we constructed 2D membranes with ame-2O and ame-4O as the building blocks. Following to a meso-meso linked approach [111] and strategy for sequential covalent linking [112], we proposed synthetic routes for the 2D membranes. CMD simulations were performed (Figure 6.14), to validate the high selectivity and yield gas permeance. Estimated CO_2 permeances of the ame-2O and ame-4O membranes were 8.0×10^3 and 7.2×10^4

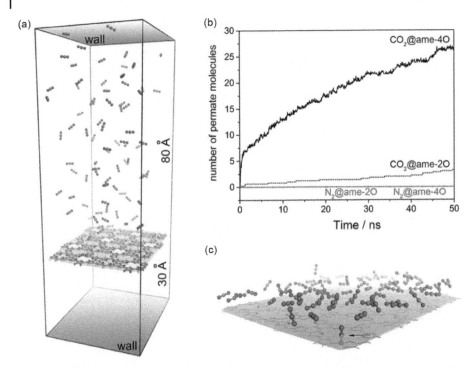

Figure 6.14 (a) A model for molecular dynamic simulation; (b) pure CO_2 and N_2 permeation through two porphyrin-based membranes at an initial pressure of 20 atm; (c) snapshot of CO_2 passing through the pore on an ame-2O membrane. Source: reproduced from Ref. [108] with permission from the American Chemical Society.

GPU, respectively. In contrast, no N_2 molecules passed through the porous membrane in 50 ns. Therefore, these expanded porphyrin-based membranes should have potential as an ultrathin material for high-efficiency post-combustion CO_2 capture.

6.6 Flexible Control of Pore Size

Pore size is a crucial factor impacting gas separation, but to precisely control the pore size down to 3–5 Å proves challenging [15]. And it is even harder to continuously tune the pore size due to the molecular construction forming the pores that non-continuously [54]. To overcome the challenges in pore-size control of one-atom-thin membranes, we proposed two promising strategies as follows.

6.6.1 Ion-Gated Porous Graphene Membrane

In 2017, we designed a composite membrane composed a monolayer (less than 5 Å) of ionic liquid (IL) atop a porous graphene, which contains hydrogen-terminated hexagonal pores with a diameter of 6.0 Å (Figure 6.15). The IL [emim][BF$_4$] was chosen because its wettability on graphene had been investigated and it could form a single adsorbed layer film

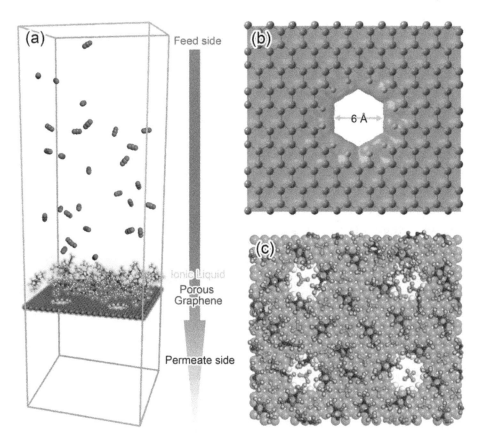

Figure 6.15 (a) A bichamber system for simulation of gas permeance through the composite membrane; top and bottom sides of the box are impermeable. (b) Structure of a hydrogen-terminated, 6 Å pore inside the graphene layer with an electronic density contour. (c) Top view of the composite membrane from the feed side: a monolayer of the [emim][BF$_4$] IL on the porous graphene. Color code in (c): C, gray; H, white; N, blue; B, red; F, green. Cations are highlighted with a blue edge, anions with red. Source: reproduced from Ref. [113] with permission from 2017, American Chemical Society. (*See color plate section for color representation of this figure*).

on a graphene sheet [113]. Figure 6.16 shows the results of CO_2, N_2, and CH_4 permeation and adsorption. Because the pore size is much greater than the kinetic diameters of CO_2 (3.30 Å), N_2 (3.64 Å), and CH_4 (3.80 Å), without the IL layer, all three gases could pass through the porous graphene membrane without any hindrance and achieve equilibrium quickly (Figure 6.16a). However, with the IL layer, one can see that the membrane is highly selective for CO_2 permeation, while CH_4 permeation is reduced the most (Figure 6.16b). Based on linear regression, a high CO_2 permeance value of 1.39×10^5 GPU and an impressive pure-gas selectivity of 42 for CO_2/CH_4 are achieved.

Further studies of gas adsorption showed that adsorption selectivities are 4.6 and 2.7 for CO_2/N_2 and CO_2/CH_4, respectively. These results revealed that CO_2/N_2 permselectivity of 5.2 is mainly due to adsorption selectivity because of a more favorable interaction of the IL layer with CO_2 than with N_2. In contrast, the CO_2/CH_4 permselectivity of 42 is

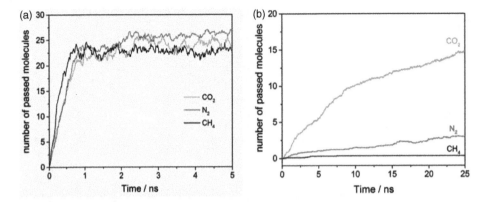

Figure 6.16 Pure gas permeation through the 6.0-Å porous graphene with an initial feed pressure of 10 atm at 298 K: (a) without the ionic liquid layer; (b) with the ionic liquid layer. Source: reproduced from Ref. [113] with permission from the American Chemical Society.

mainly due to diffusivity selectivity, because the IL layer modulates the pore size. By checking the snapshots, we found that the pore is gated by the anion [BF_4], which possesses strong interaction with electropositive hydrogen termination. If the pore rim is functionalized with electronegative dopants such as nitrogen, the cations can be also used to modulate the pore size. In this work, we used porous graphene as an ultrathin membrane due to its simple structure, while the porous substrate can also be extended to other 2D materials with nanopores, such as MoS_2, 2D COFs, and 2D polymers. Moreover, the relatively larger pore size can be more easily created experimentally than the precisely controlled smaller pores.

6.6.2 Bilayer Porous Graphene with Continuously Tunable Pore Size

To continuously tune the pore size, we proposed another strategy containing a bilayer porous graphene, as shown in Figure 6.17. The same pore with all-hydrogen termination was employed (Figure 6.17a). The distance between the two layers of graphene sheets was fixed at 3.4 Å, similar to the interlayer spacing in graphite, so there is no adsorption between the two layers. Figure 6.17b,c shows how the effective pore size and shape are continuously tuned by the offset, which represents the relative position of two layers of graphene sheet in a lateral direction [114].

Then, we used CMD simulations to quantify the permeance and selectivity as a function of offset (Figure 6.18). One can see that the selectivity of CO_2/CH_4 (Figure 6.18a) increases as the offset increases or the effective pore size decreases. When the effective pore size is about 3.6 Å, selectivity reaches 100 with a high CO_2 permeance of 6×10^5 GPU. A much higher selectivity can be achieved when one further decreases the effective pore size, because the permeation of CH_4 molecule is blocked. Since a too-small effective pore size also causes the decrease of CO_2 permeance, a trade-off between permeance and selectivity suggests that control of the effective pore size between 3.6 and 3.7 Å affords the optimal balance. Likewise, the selectivity of N_2/CH_4 reaches 68 while N_2 permeance remains 4×10^5 GPU when the effective pore size is about 3.6 Å (Figure 6.18b).

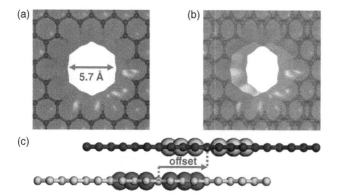

Figure 6.17 Construction of a porous bilayer graphene membrane: (a) the single-layer graphene with a larger pore than the sizes of small gas molecules, such as CO_2; (b) the bilayer nanoporous graphene membrane from stacking two single-layer membranes from (a); (c) side view of the bilayer nanoporous graphene membrane that shows the offset of the pore centers. Isosurfaces of electron density are shown in (a) and (b) to define the pore shape, and larger spheres in (c) indicate the pore rims. Source: reproduced from Ref. [114] with permission from the American Chemical Society. (*See color plate section for color representation of this figure*).

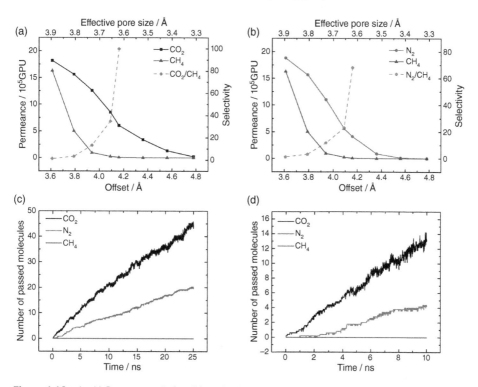

Figure 6.18 (a–b) Permeance (left axis) and selectivity (right axis) as a function of the offset (bottom axis) or the effective pore size (top axis) in bilayer nanoporous graphene membranes: (a) CO_2 vs. CH_4 and (b) N_2 vs. CH_4. (c–d) Permeation of CO_2, N_2, and CH_4 with time through the bilayer porous graphene membranes: (c) pore size of 10.4 Å and (d) pore size of 25.2 Å. Source: reproduced from Ref. [114] with permission from the American Chemical Society.

More bilayer membranes with larger pore sizes in the single layer have also been studied. For the single-layer graphene with pore size of both 10.4 Å and 25.2 Å, through tuning offset, the effective pore size of 3.6 Å could be obtained. They also present very high selectivities of CO_2/CH_4 and N_2/CH_4 (Figure 6.18c and d). To experimentally realized the desired offset, we suggest that one prepare many parallel random samples of porous graphene bilayer membranes with different offsets and then transfer them on a porous support for characterization of the offset and test of membrane performance. To prepare such a bilayer membrane, one can simply stack one porous-graphene layer on another or fold a porous single-layer porous graphene into a bilayer, which has been reported by Ruoff et al. recently [115]. Hence, our work suggests a promising direction to achieve advanced control in pore sizes for selective gas separations via ultrathin membranes.

6.7 Summary and Outlook

In this chapter, we have reviewed recent progress on ultrathin membranes for gas separation. Due to their intrinsic nature of atomic thickness, these 2D membranes are expected to be superior membranes in various applications. Porous graphene was theoretically designed as the starting point. Many simulations have been performed to design novel 2D materials and to evaluate their separation capabilities, showing that ultrathin membranes exhibit extraordinary permeance and selectivity simultaneously. Several strategies have been demonstrated to be able to improve the separation performance of porous one-atom-thick membrane. For example, in graphene-based and organic membranes, nitrogen, oxygen, and fluorine usually possess negative charge. These heteroatom dopants on the pore rim dramatically decrease the permeation barrier of CO_2 and then enhance selective CO_2 capture. Gas adsorption on the surface also plays a critical role in the permeation process, and thus surface modification may influence selectivity. In addition, one can also change the topologies of materials to create ultrathin membranes for practical applications. One available approach is rolling the porous single-layer membrane into a nanotube.

Until now, most of the studies of these ultrathin membranes were based on theoretical simulations of ideal models. Although CMD and highly accurate ab initio calculations have provided good estimations of permeance and selectivity of proposed materials, it is still challenging to realize these structures. In recent years, great progress in experiments has been achieved. With the development of nanotechnology, some porous membranes based on graphene have been successfully synthesized, and indeed had high permeance and selectivity at the same time. On the other hand, a bottom-up approach opens the door to construction of a leak-free thin film with uniform pores from small molecules. Several porous 2D polymers, such as poly-phenylene and graphdiyne, have been prepared.

There are remaining challenges and opportunities. 2D membranes could lack the mechanical strength that polymeric membranes enjoy. Hence, one strategy is to create membranes on the order of 1 nm in thickness based on carbon molecular sieves. This should provide a much higher mechanical strength for making composite membranes, e.g., with ionic liquids. Moreover, advanced characterization techniques are needed to combine with computation to understand deeper the separation mechanisms in such systems. Further, there are still many difficulties related to the preparation of large-area ultrathin films and introduction of size-controlled nanopores. To realize the applications of porous ultrathin membranes, large-scale fabrication and controllable-nanopore creation techniques have to be further developed.

Acknowledgments

This work was supported by the Division of Chemical Sciences, Geosciences and Biosciences, Office of Basic Energy Sciences, US Department of Energy. This research used resources of the National Energy Research Scientific Computing Center, a DOE Office of Science User Facility supported by the Office of Science of the US Department of Energy under Contract DE-AC02-05CH11231.

References

1 Pandey, P. and Chauhan, R.S. (2001). Membranes for gas separation. *Prog. Polym. Sci.* 26: 853–893.

2 Baker, R.W. (2002). Future directions of membrane gas separation technology. *Ind. Eng. Chem. Res.* 41: 1393–1411.

3 Bernardo, P., Drioli, E., and Golemme, G. (2009). Membrane gas separation: a review/state of the art. *Ind. Eng. Chem. Res.* 48: 4638–4663.

4 George, S.C. and Thomas, S. (2001). Transport phenomena through polymeric systems. *Prog. Polym. Sci.* 26: 985–1017.

5 Robeson, L.M. (2008). The upper bound revisited. *J. Membr. Sci.* 320: 390–400.

6 Oyama, S.T., Lee, D., Hacarlioglu, P., and Saraf, R.F. (2004). Theory of hydrogen permeability in nonporous silica membranes. *J. Membr. Sci.* 244: 45–53.

7 de Vos, R.M. and Verweij, H. (1998). High-selectivity, high-flux silica membranes for gas separation. *Science* 279: 1710–1711.

8 Carta, M., Malpass-Evans, R., Croad, M. et al. (2013). An efficient polymer molecular sieve for membrane gas separations. *Science* 339: 303–307.

9 Peng, Y., Li, Y.S., Ban, Y.J. et al. (2014). Metal-organic framework nanosheets as building blocks for molecular sieving membranes. *Science* 346: 1356–1359.

10 Zhao, Y., Xie, Y., Liu, Z. et al. (2014). Two-dimensional material membranes: an emerging platform for controllable mass transport applications. *Small* 10: 4521–4542.

11 Jiao, Y., Du, A., Hankel, M., and Smith, S.C. (2013). Modelling carbon membranes for gas and isotope separation. *Phys. Chem. Chem. Phys.* 15: 4832–4843.

12 Liu, G., Jin, W., and Xu, N. (2015). Graphene-based membranes. *Chem. Soc. Rev.* 44: 5016–5030.

13 Xu, Q., Xu, H., Chen, J. et al. (2015). Graphene and graphene oxide: advanced membranes for gas separation and water purification. *Inorg. Chem. Front.* 2: 417–424.

14 Breck, D.W. (1974). *Zeolite Molecular Sieves: Structure, Chemistry and Use*. New York: Wiley.

15 Mehio, N., Dai, S., and Jiang, D.E. (2014). Quantum mechanical basis for kinetic diameters of small gaseous molecules. *J. Phys. Chem. A* 118: 1150–1154.

16 Novoselov, K.S., Geim, A.K., Morozov, S.V. et al. (2004). Electric field effect in atomically thin carbon films. *Science* 306: 666–669.

17 Pan, X.L. and Bao, X.H. (2011). The effects of confinement inside carbon nanotubes on catalysis. *Acc. Chem. Res.* 44: 553–562.

18 Kong, X.K., Chen, C.L., and Chen, Q.W. (2014). Doped graphene for metal-free catalysis. *Chem. Soc. Rev.* 43: 2841–2857.

19 Schwierz, F. (2010). Graphene transistors. *Nat. Nanotechnol.* 5: 487–496.

20 Bao, Q.L. and Loh, K.P. (2012). Graphene photonics, plasmonics, and broadband optoelectronic devices. *ACS Nano* 6: 3677–3694.

21 Osada, M. and Sasaki, T. (2012). Two-dimensional dielectric nanosheets: novel nanoelectronics from nanocrystal building blocks. *Adv. Mater.* 24: 210–228.

22 Stoller, M.D., Park, S.J., Zhu, Y.W. et al. (2008). Graphene-based ultracapacitors. *Nano Lett.* 8: 3498–3502.

23 Lightcap, I.V. and Kamat, P.V. (2013). Graphitic design: prospects of graphene-based nanocomposites for solar energy conversion, storage, and sensing. *Acc. Chem. Res.* 46: 2235–2243.

24 Berry, V. (2013). Impermeability of graphene and its applications. *Carbon* 62: 1–10.

25 Jiang, D.E., Cooper, V.R., and Dai, S. (2009). Porous graphene as the ultimate membrane for gas separation. *Nano Lett.* 9: 4019–4024.

26 Ockwig, N.W. and Nenoff, T.M. (2007). Membranes for hydrogen separation. *Chem. Rev.* 107: 4078–4110.

27 Fischbein, M.D. and Drndic, M. (2008). Electron beam nanosculpting of suspended graphene sheets. *Appl. Phys. Lett.* 93: 113107.

28 O'Hern, S.C., Stewart, C.A., Boutilier, M.S.H. et al. (2012). Selective molecular transport through intrinsic defects in a single layer of CVD graphene. *ACS Nano* 6: 10130–10138.

29 Koenig, S.P., Wang, L., Pellegrino, J., and Bunch, J.S. (2012). Selective molecular sieving through porous graphene. *Nat. Nanotechnol.* 7: 728–732.

30 Russo, C.J. and Golovchenko, J.A. (2012). Atom-by-atom nucleation and growth of graphene nanopores. *Proc. Natl. Acad. Sci. U. S. A.* 109: 5953–5957.

31 Kim, H.W., Yoon, H.W., Yoon, S.M. et al. (2013). Selective gas transport through few-layered graphene and graphene oxide membranes. *Science* 342: 91–95.

32 Li, H., Song, Z.N., Zhang, X.J. et al. (2013). Ultrathin, molecular-sieving graphene oxide membranes for selective hydrogen separation. *Science* 342: 95–98.

33 Boutilier, M.S.H., Sun, C.Z., O'Hern, S.C. et al. (2014). Implications of permeation through intrinsic defects in graphene on the design of defect-tolerant membranes for gas separation. *ACS Nano* 8: 841–849.

34 Celebi, K., Buchheim, J., Wyss, R.M. et al. (2014). Ultimate permeation across atomically thin porous graphene. *Science* 344: 289–292.

35 Xie, G., Yang, R., Chen, P. et al. (2014). A general route towards defect and pore engineering in graphene. *Small* 10: 2280–2284.

36 Liu, L., Ryu, S.M., Tomasik, M.R. et al. (2008). Graphene oxidation: thickness-dependent etching and strong chemical doping. *Nano Lett.* 8: 1965–1970.

37 Liu, H., Dai, S., and Jiang, D.E. (2013). Permeance of H_2 through porous graphene from molecular dynamics. *Solid State Commun.* 175–176: 101–105.

38 Liu, H., Dai, S., and Jiang, D.E. (2013). Insights into CO_2/N_2 separation through nanoporous graphene from molecular dynamics. *Nanoscale* 5: 9984–9987.

39 Liu, H., Chen, Z., Dai, S., and Jiang, D.E. (2015). Selectivity trend of gas separation through nanoporous graphene. *J. Solid State Chem.* 224: 2–6.

40 Alivisatos, A.P. and Buchanan, M. V.. 2010. Basic research needs for carbon capture: Beyond 2020. Office of Science, US Department of Energy. https://science.energy.gov/~/media/bes/pdf/reports/files/Basic_Research_Needs_for_Carbon_Capture_rpt.pdf.

41 Merkel, T.C., Lin, H.Q., Wei, X.T., and Baker, R. (2010). Power plant post-combustion carbon dioxide capture: an opportunity for membranes. *J. Membr. Sci.* 359: 126–139.

42 Du, H., Li, J., Zhang, J. et al. (2011). Separation of hydrogen and nitrogen gases with porous graphene membrane. *J. Phys. Chem. C* 115: 23261–23266.

43 Hauser, A.W., Schrier, J., and Schwerdtfeger, P. (2012). Helium tunneling through nitrogen-functionalized graphene pores: pressure- and temperature-driven approaches to isotope separation. *J. Phys. Chem. C* 116: 10819–10827.

44 Hauser, A.W. and Schwerdtfeger, P. (2012). Nanoporous graphene membranes for efficient $^3He/^4He$ separation. *J. Phys. Chem. Lett.* 3: 209–213.

45 Hauser, A.W. and Schwerdtfeger, P. (2012). Methane-selective nanoporous graphene membranes for gas purification. *Phys. Chem. Chem. Phys.* 14: 13292–13298.

46 Qin, X., Meng, Q., Feng, Y., and Gao, Y. (2013). Graphene with line defect as a membrane for gas separation: design via a first-principles modeling. *Surf. Sci.* 607: 153–158.

47 Ambrosetti, A. and Silvestrelli, P.L. (2014). Gas separation in nanoporous graphene from first principle calculations. *J. Phys. Chem. C* 118: 19172–19179.

48 Lei, G., Liu, C., Xie, H., and Song, F. (2014). Separation of the hydrogen sulfide and methane mixture by the porous graphene membrane: effect of the charges. *Chem. Phys. Lett.* 599: 127–132.

49 Wu, T., Xue, Q., Ling, C. et al. (2014). Fluorine-modified porous graphene as membrane for CO_2/N_2 separation: molecular dynamic and first-principles simulations. *J. Phys. Chem. C* 118: 7369–7376.

50 Azamat, J., Khataee, A., and Joo, S.W. (2015). Molecular dynamics simulation of trihalomethanes separation from water by functionalized nanoporous graphene under induced pressure. *Chem. Eng. Sci.* 127: 285–292.

51 Nieszporek, K. and Drach, M. (2015). Alkane separation using nanoporous graphene membranes. *Phys. Chem. Chem. Phys.* 17: 1018–1024.

52 Liu, Y. and Chen, X. (2014). Mechanical properties of nanoporous graphene membrane. *J. Appl. Phys.* 115: 034303.

53 Drahushuk, L.W. and Strano, M.S. (2012). Mechanisms of gas permeation through single layer graphene membranes. *Langmuir* 28: 16671–16678.

54 Sun, C., Boutilier, M.S., Au, H. et al. (2014). Mechanisms of molecular permeation through nanoporous graphene membranes. *Langmuir* 30: 675–682.

55 Lee, J. and Aluru, N.R. (2013). Water-solubility-driven separation of gases using graphene membrane. *J. Membr. Sci.* 428: 546–553.

56 Holt, J.K., Park, H.G., Wang, Y.M. et al. (2006). Fast mass transport through sub-2-nanometer carbon nanotubes. *Science* 312: 1034–1037.

57 Sircar, S., Golden, T., and Rao, M. (1996). Activated carbon for gas separation and storage. *Carbon* 34: 1–12.

58 Gin, D.L. and Noble, R.D. (2011). Designing the next generation of chemical separation membranes. *Science* 332: 674–676.

59 Yuan, Z., Govind Rajan, A., Misra, R.P. et al. (2017). Mechanism and prediction of gas permeation through sub-nanometer graphene pores: comparison of theory and simulation. *ACS Nano* 11: 7974–7987.

60 Wang, S., Tian, Z., Dai, S., and Jiang, D.E. (2018). Effect of pore density on gas permeation through nanoporous graphene membranes. *Nanoscale* 10: 14660–14666.

61 Bieri, M., Treier, M., Cai, J.M. et al. (2009). Porous graphenes: two-dimensional polymer synthesis with atomic precision. *Chem. Commun.*: 6919–6921.

62 Blankenburg, S., Bieri, M., Fasel, R. et al. (2010). Porous graphene as an atmospheric nanofilter. *Small* 6: 2266–2271.

63 Li, Y., Zhou, Z., Shen, P., and Chen, Z. (2010). Two-dimensional polyphenylene: experimentally available porous graphene as a hydrogen purification membrane. *Chem. Commun.* 46: 3672–3674.

64 Schrier, J. (2010). Helium separation using porous graphene membranes. *J. Phys. Chem. Lett.* 1: 2284–2287.

65 Schrier, J. (2011). Fluorinated and nanoporous graphene materials as sorbents for gas separations. *ACS Appl. Mater. Interfaces* 3: 4451–4458.

66 Schrier, J. and McClain, J. (2012). Thermally-driven isotope separation across nanoporous graphene. *Chem. Phys. Lett.* 521: 118–124.

67 Lu, R., Rao, D., Lu, Z. et al. (2012). Prominently improved hydrogen purification and dispersive metal binding for hydrogen storage by substitutional doping in porous graphene. *J. Phys. Chem. C* 116: 21291–21296.

68 Hankel, M., Jiao, Y., Du, A. et al. (2012). Asymmetrically decorated, doped porous graphene as an effective membrane for hydrogen isotope separation. *J. Phys. Chem. C* 116: 6672–6676.

69 Jungthawan, S., Reunchan, P., and Limpijumnong, S. (2013). Theoretical study of strained porous graphene structures and their gas separation properties. *Carbon* 54: 359–364.

70 Lu, R., Meng, Z., Rao, D. et al. (2014). A promising monolayer membrane for oxygen separation from harmful gases: nitrogen-substituted polyphenylene. *Nanoscale* 6: 9960–9964.

71 Huang, C., Wu, H., Deng, K. et al. (2014). Improved permeability and selectivity in porous graphene for hydrogen purification. *Phys. Chem. Chem. Phys.* 16: 25755–25759.

72 Inagaki, M. and Kang, F. (2014). Graphene derivatives: graphane, fluorographene, graphene oxide, graphyne and graphdiyne. *J. Mater. Chem. A* 2: 13193–13206.

73 Li, G.X., Li, Y.L., Liu, H.B. et al. (2010). Architecture of graphdiyne nanoscale films. *Chem. Commun.* 46: 3256–3258.

74 Jiao, Y., Du, A., Hankel, M. et al. (2011). Graphdiyne: a versatile nanomaterial for electronics and hydrogen purification. *Chem. Commun.* 47: 11843–11845.

75 Cranford, S.W. and Buehler, M.J. (2012). Selective hydrogen purification through graphdiyne under ambient temperature and pressure. *Nanoscale* 4: 4587–4593.

76 Jiao, Y., Du, A., Smith, S.C. et al. (2015). H2purification by functionalized graphdiyne – role of nitrogen doping. *J. Mater. Chem. A* 3: 6767–6771.

77 Zhang, H., He, X., Zhao, M. et al. (2012). Tunable hydrogen separation in sp–sp^2 hybridized carbon membranes: a first-principles prediction. *J. Phys. Chem. C* 116: 16634–16638.

78 Shan, M., Xue, Q., Jing, N. et al. (2012). Influence of chemical functionalization on the CO_2/N_2 separation performance of porous graphene membranes. *Nanoscale* 4: 5477–5482.

79 Jiao, S. and Xu, Z. (2015). Selective gas diffusion in graphene oxides membranes: a molecular dynamics simulations study. *ACS Appl. Mater. Interfaces* 7: 9052–9059.

80 Hu, W., Wu, X., Li, Z., and Yang, J. (2013). Helium separation via porous silicene based ultimate membrane. *Nanoscale* 5: 9062–9066.

81 Hu, W., Wu, X., Li, Z., and Yang, J. (2013). Porous silicene as a hydrogen purification membrane. *Phys. Chem. Chem. Phys.* 15: 5753–5757.

82 Schrier, J. (2012). Carbon dioxide separation with a two-dimensional polymer membrane. *ACS Appl. Mater. Interfaces* 4: 3745–3752.

83 Brockway, A.M. and Schrier, J. (2013). Noble gas separation using PG-ESX(X= 1, 2, 3) nanoporous two-dimensional polymers. *J. Phys. Chem. C* 117: 393–402.

84 Solvik, K., Weaver, J.A., Brockway, A.M., and Schrier, J. (2013). Entropy-driven molecular separations in 2D-nanoporous materials, with application to high-performance paraffin/olefin membrane separations. *J. Phys. Chem. C* 117: 17050–17057.

85 Yave, W., Car, A., Wind, J., and Peinemann, K.V. (2010). Nanometric thin film membranes manufactured on square meter scale: ultra-thin films for CO_2 capture. *Nanotechnology* 21: 395301.

86 Tao, Y., Xue, Q., Liu, Z. et al. (2014). Tunable hydrogen separation in porous graphene membrane: first-principle and molecular dynamic simulation. *ACS Appl. Mater. Interfaces* 6: 8048–8058.

87 Grinevich, Y., Starannikova, L., Yampol'skii, Y. et al. (2011). Membrane separation of gaseous C1-C4 alkanes. *Pet. Chem.* 51: 585–594.

88 Ma, X.L., Lin, B.K., Wei, X.T. et al. (2013). Gamma-alumina supported carbon molecular sieve membrane for propylene/propane separation. *Ind. Eng. Chem. Res.* 52: 4297–4305.

89 Ma, Z., Zhao, X., Tang, Q., and Zhou, Z. (2014). Computational prediction of experimentally possible g-C_3N_3 monolayer as hydrogen purification membrane. *Int. J. Hydrog. Energy* 39: 5037–5042.

90 Chen, Z., Li, P., and Wu, C. (2015). A uniformly porous 2D CN (1 : 1) network predicted by first-principles calculations. *RSC Adv.* 5: 11791–11796.

91 Xu, B., Xiang, H., Wei, Q. et al. (2015). Two-dimensional graphene-like C2N: an experimentally available porous membrane for hydrogen purification. *Phys. Chem. Chem. Phys.* 17: 15115–11516.

92 Mahmood, J., Lee, E.K., Jung, M. et al. (2015). Nitrogenated holey two-dimensional structures. *Nat. Commun.* 6: 6486.

93 Majumder, M., Chopra, N., Andrews, R., and Hinds, B.J. (2005). Nanoscale hydrodynamics enhanced flow in carbon nanotubes. *Nature* 438: 44.

94 Kim, S., Jinschek, J.R., Chen, H. et al. (2007). Scalable fabrication of carbon nanotube/polymer nanocomposite membranes for high flux gas transport. *Nano Lett.* 7: 2806–2811.

95 Kim, M., Sohn, K., Kim, J., and Hyeon, T. (2003). Synthesis of carbon tubes with mesoporous wall structure using designed silica tubes as templates. *Chem. Commun.*: 652–653.

96 Rodriguez, A.T., Chen, M., Chen, Z. et al. (2006). Nanoporous carbon nanotubes synthesized through confined hydrogen-bonding self-assembly. *J. Am. Chem. Soc.* 128: 9276–9277.

97 Yoon, D., Lee, C., Yun, J. et al. (2012). Enhanced condensation, agglomeration, and rejection of water vapor by superhydrophobic aligned multiwalled carbon nanotube membranes. *ACS Nano* 6: 5980–5987.

98 Hinds, B.J., Chopra, N., Rantell, T. et al. (2004). Aligned multiwalled carbon nanotube membranes. *Science* 303: 62–65.

99 Liu, H., Cooper, V.R., Dai, S., and Jiang, D.E. (2012). Windowed carbon nanotubes for efficient CO_2 removal from natural gas. *J. Phys. Chem. Lett.* 3: 3343–3347.

100 Park, H.B., Jung, C.H., Lee, Y.M. et al. (2007). Polymers with cavities tuned for fast selective transport of small molecules and ions. *Science* 318: 254–258.

101 Skoulidas, A.I., Ackerman, D.M., Johnson, J.K., and Sholl, D.S. (2002). Rapid transport of gases in carbon nanotubes. *Phys. Rev. Lett.* 89: 185901.

102 Bucior, B.J., Chen, D.-L., Liu, J., and Johnson, J.K. (2012). Porous carbon nanotube membranes for separation of H_2/CH_4 and CO_2/CH_4 mixtures. *J. Phys. Chem. C* 116: 25904–25910.

103 Brenner, D.W., Shenderova, O.A., Harrison, J.A. et al. (2002). A second-generation reactive empirical bond order (REBO) potential energy expression for hydrocarbons. *J. Phys. Condens. Matter* 14: 783–802.

104 Yzeiri, I., Patra, N., and Kral, P. (2014). Porous carbon nanotubes: molecular absorption, transport, and separation. *J. Chem. Phys.* 140: 104704.

105 Furukawa, H. and Yaghi, O.M. (2009). Storage of hydrogen, methane, and carbon dioxide in highly porous covalent organic frameworks for clean energy applications. *J. Am. Chem. Soc.* 131: 8875–8883.

106 Dawson, R., Cooper, A.I., and Adams, D.J. (2012). Nanoporous organic polymer networks. *Prog. Polym. Sci.* 37: 530–563.

107 Feng, X., Ding, X.S., and Jiang, D.L. (2012). Covalent organic frameworks. *Chem. Soc. Rev.* 41: 6010–6022.

108 Tian, Z., Dai, S., and Jiang, D.E. (2015). Expanded porphyrins as two-dimensional porous membranes for CO_2 separation. *ACS Appl. Mater. Interfaces* 7: 13073–13079.

109 Saito, S. and Osuka, A. (2011). Expanded porphyrins: intriguing structures, electronic properties, and reactivities. *Angew. Chem. Int. Ed.* 50: 4342–4373.

110 Seidel, D., Lynch, V., and Sessler, J.L. (2002). Cyclo[8]pyrrole: a simple-to-make expanded porphyrin with no meso bridges. *Angew. Chem. Int. Ed.* 41: 1422–1425.

111 Mori, H., Tanaka, T., Lee, S. et al. (2015). meso-meso linked porphyrin-[26]hexaphyrin-porphyrin hybrid arrays and their triply linked tapes exhibiting strong absorption bands in the NIR region. *J. Am. Chem. Soc.* 137: 2097–2106.

112 Lafferentz, L., Eberhardt, V., Dri, C. et al. (2012). Controlling on-surface polymerization by hierarchical and substrate-directed growth. *Nat. Chem.* 4: 215–220.

113 Tian, Z., Mahurin, S.M., Dai, S., and Jiang, D.E. (2017). Ion-gated gas separation through porous graphene. *Nano Lett.* 17: 1802–1807.

114 Wang, S., Dai, S., and Jiang, D.E. (2019). Continuously tunable pore size for gas separation via a bilayer nanoporous graphene membrane. *ACS Appl. Nano Mater.* 2: 379–384.

115 Wang, B., Huang, M., Kim, N.Y. et al. (2017). Controlled folding of single crystal graphene. *Nano Lett.* 17: 1467–1473.

7

Polymeric Membranes

Jason E. Bara and W. Jeffrey Horne

Department of Chemical & Biological Engineering, University of Alabama, Tuscaloosa, AL, USA

7.1 Introduction

7.1.1 Overview of Post-Combustion CO_2 Capture

The United States Department of Energy (DOE) calls for post-combustion carbon dioxide (CO_2) capture and sequestration (CCS) processes to be retrofitted at existing coal-fired power plants that will remove at least 90% of CO_2 emissions while incurring no more than a 35% increase in the cost of electricity (COE) [1, 2]. CCS processes operate by separating CO_2 from power plant flue gas (via several possible techniques) [3, 4]. The captured CO_2 is subsequently compressed for direct injection into geologic formations for permanent sequestration or beneficial uses such as enhanced oil recovery (EOR) [1, 2]. While coal-based electric power generation accounts for nearly 40% of the electricity produced in the United States, it also the largest source of CO_2 emissions [2]. Reliance on coal (as a % of electricity generation) is even greater in China and India [5].

Growing concerns over the contributions of CO_2 and other greenhouse gases (GHGs) to global climate change have prompted major efforts to use fossil fuels (especially coal) in a more responsible manner. A number of diverse process technologies are currently at various

Materials for Carbon Capture, First Edition. Edited by De-en Jiang, Shannon M. Mahurin and Sheng Dai.
© 2020 John Wiley & Sons Ltd. Published 2020 by John Wiley & Sons Ltd.

stages of evaluation as potential solutions that can contribute to meeting the challenge of energy efficient CCS [2, 3, 5, 6]. DOE targets a 2020–2030 timeline for full-scale technology demonstrations and commercialization at operating power plants, with support for R&D efforts for the foreseeable future [1, 2]. Energy-efficient capture of CO_2 from flue gas is an especially challenging engineering process due to gas-stream conditions [2, 4]. A typical flue gas stream leaves the stack at ~40–60 °C and just above atmospheric pressure (~15 psia), containing 10–14% CO_2 at a partial pressure of ~2 psia. Figure 7.1 presents a block flow diagram incorporating a CO_2 capture process into the power plant emissions stream [2].

Chemical solvent (i.e. aqueous amine) processes are the leading candidates for initial deployment due to their proliferation, large knowledge base, and long history of use for CO_2 and H_2S removal in the natural gas industry [3, 6–8]. Significant work has been done to identify amine molecules and process configurations that reduce the overall energy penalty compared to the aq. monoethanolamine (MEA) baseline [6]. Although pilot plant campaigns have made progress in recent years, an amine solvent process has yet to be deployed at the full scale of a "typical" 550 MW coal-fired power plant [1]. Furthermore, challenges relating to amine lifetimes and emissions associated with their volatility and degradation due to flue gas components such as SO_2, NO_x, and other species must still be surmounted [9–11].

Membrane-based processes can offer advantages for post-combustion CO_2 capture as they possess intrinsic benefits such as lower energy consumption, a much more stable separation medium (e.g. polymer versus liquid), and minimal maintenance requirements [1, 2, 12–18]. Membranes account for a small but growing portion of commercial acid gas removal systems in the natural gas industry [13, 14] and are currently at small-scale pilot demonstrations for post-combustion CO_2 capture at coal-fired power plants (i.e. CO_2/N_2 separation) [2, 12, 19], and also for pre-combustion CO_2 capture at integrated gasification combined cycle (IGCC) plants [2, 20–22]. Although a variety of material types

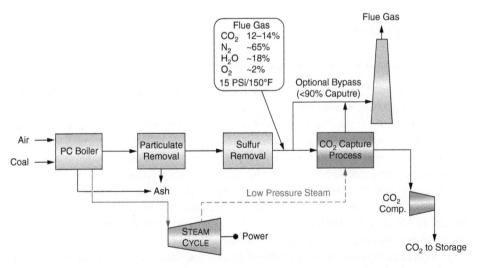

Figure 7.1 Post-combustion CO_2 capture block flow diagram. Source: reprinted with permission from Ref. [2].

such as metals, ceramics, carbons, immobilized liquids, and polymers can be used as membranes to separate CO_2 from other gases [14], this chapter will specifically focus on the relationships between polymer structure and membrane performance with respect to CO_2/N_2 separations.

7.1.2 Polymer Membrane Fundamentals and Process Considerations

Most polymer membranes operate via the solution–diffusion (S–D) mechanism, wherein the permeability (P_i) of the polymer to a given gas is the product of the individual species' solubility (S_i) and diffusivity (D_i) within the material (Eq. (7.1)) [23, 24]. Permeability is defined as a relationship between flux (J_i) normalized by partial pressure driving force (Δp_i) and membrane thickness (l) (Eq. (7.2)). Permeability of a membrane to a gas is typically expressed in the non-SI unit barrer, named after Richard Maling Barrer (1 barrer = 10^{-10} (cm^3 [STP]).cm.cm^{-2}.s^{-1}.(cm Hg)) [25]. The separation factor (selectivity) for a gas pair ($\alpha_{i,j}$) is defined as the ratio of permeabilities (P_i/P_j) that can be separated into solubility and diffusivity contributions (Eq. (7.3)).

$$P_i = S_i \cdot D_i \tag{7.1}$$

$$P_i = \frac{J_i \cdot l}{\Delta p_i} \tag{7.2}$$

$$\alpha_{i,j} = \frac{P_i}{P_j} = \frac{S_i}{S_j} \cdot \frac{D_i}{D_j} \tag{7.3}$$

Within the natural gas industry, membrane processes have found commercial adoption under conditions with low flow rates and high CO_2 partial pressures (Figure 7.2) [13]. As such, membranes have been more competitive with physical solvent processes (e.g. Selexol, Rectisol) than with amine systems to date. Membranes can be combined with amine "polishing" processes at higher concentrations and higher flow rates [13]. However, as is apparent in Figure 7.2, amine processes dominate at low concentrations and high flow rates, i.e. conditions reflective of a post-combustion CO_2 capture process. For membranes to become more competitive with amines at low pressures, advanced polymer materials and perhaps even polymer materials that feature some reversible chemical reaction with CO_2 may be needed to advance the viability of polymer membranes for post-combustion CO_2 capture.

CO_2 is present at low partial pressures (~2 psia) in the flue gas of coal-fired power plants, which results in a very small driving force available S-D driven transport (Eq. (7.2)), and significant compression of the flue gas to increase the transmembrane pressure differential is not economically or energetically feasible [2]. As previously stated, membranes are already used commercially within the natural gas industry to separate CO_2 from CH_4; however, the partial pressure of CO_2 in that application is often ≫100 psia, making it relatively easy to achieve high CO_2 flux and removal from CH_4 [13, 14, 26]. Thus, highly permeable membranes that enable maximum CO_2 flux at a minimal driving force are needed to efficiently capture CO_2 from flue gas.

Since dense polymer membranes follow the S-D mechanism of gas transport, which is governed by Fick's Law [23], the flux through a polymer membrane is inversely proportional to membrane thickness. Thinner membranes will be more productive per unit area,

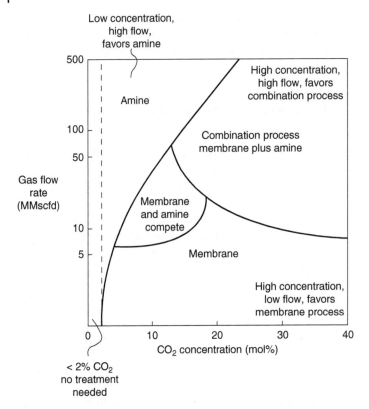

Figure 7.2 Process conditions favoring amine- and/or membrane-based removal of CO_2 in natural gas treating. Source: reprinted from Ref. [13] with permission from the American Chemical Society.

which will lead to a smaller process footprint (i.e. less membrane area to process a given volume of gas). Thus, a true measure of a membrane's operational performance is its permeance to a gas, rather than permeability. Gas permeance units (gpu) are calculated simply as the membrane's permeability in barrers (Eq. (7.2)) divided by membrane thickness (in μm). Models indicate that membrane permeances (of >1000 gpu and CO_2/N_2 selectivities >40 can present favorable economics for post-combustion CO_2 capture [12]. Analysis of these models leads to a clear conclusion that increased permeance is more desirable than increased selectivity (provided $\alpha CO_2/N_2 > 40$) in determining the cost of CO_2 capture (Figure 7.3) [12].

In order to achieve the high throughputs necessary for treating the large volumes of flue gas in post-combustion CO_2 capture, very thin membranes (<500 nm) contained in modules with high surface area to volume ratios (e.g. bundles of hollow fibers) will ultimately be required [2, 12]. Thus, to increase CO_2 permeance, either polymer films can be made thinner (smaller denominator) or their permeability to CO_2 can be increased (larger numerator).

Theoretically, any polymer with the requisite CO_2/N_2 selectivity could be a viable candidate membrane material for post-combustion CO_2 capture if it were to be cast as a sufficiently thin, defect-free film (Figure 7.3). For example, a polymer with a CO_2 permeability

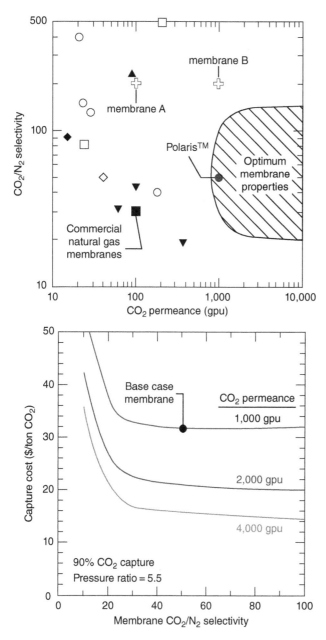

Figure 7.3 Performance (left) and economic (right) models for membrane-based post-combustion CO_2 capture. Source: reprinted from Ref. [12] with permission from Elsevier.

of 1000 barrers need only be cast as 1 μm thick film to achieve a permeance of 1000 gpu, and would reach 2000 gpu at 500 nm thick. However, materials with permeabilities of 100 and 10 barrers would need to be cast as 100 nm and 10 nm films, respectively, to achieve permeance of 1000 gpu. As it cannot be assumed that all polymer materials are amenable to forming such ultrathin films, laboratory research efforts may be more fruitful if directed at increasing membrane permeability.

A valuable and intuitive visual guide to understanding the state of the art in polymer membranes is through the use of *Robeson plots*, which graph experimental data for various materials examined for a given separation [27–30]. Robeson plots illustrate the relationship between experimentally observed gas pair selectivities as determined by the ratio of their permeabilities (Figure 7.3) and the permeability of the more permeable species. Figure 7.4 presents a Robeson plot for CO_2/N_2 with data as of 2008 [28].

As can be seen in Figure 7.4, a "flux-selectivity tradeoff" is shown to exist, wherein membrane permeability increases at the expense of CO_2/N_2 selectivity as illustrated by the negative slope of the data cluster. Such flux-selectivity tradeoffs are also observed in the other separations analyzed by Robeson including CO_2/CH_4, O_2/N_2, H_2/CO_2, etc. [28, 30]. Freeman has shown that the basis of the flux-selectivity trade-off for a given gas pair is their relative sizes and condensabilities [29].

In Figure 7.4, an obvious clustering exists of materials with CO_2 permeabilities in the range of 1–100 barrer and CO_2/N_2 selectivities of 10–40. Thus, when reconciling Robeson plots for CO_2/N_2 separation with the performance requirements presented in Figure 7.3, there are relatively few candidate polymer materials that appear capable of enabling cost- and energy-efficient post-combustion CO_2 capture.

This chapter will focus on summarizing several of the most promising classes of polymer materials for post-combustion CO_2 capture using membrane technologies. Where possible,

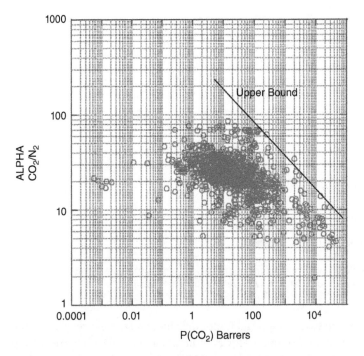

Figure 7.4 Robeson plot as of 2008 for polymeric membranes in CO_2/N_2 separation. Source: reprinted from Ref. [28] with permission from the (2008) Elsevier.

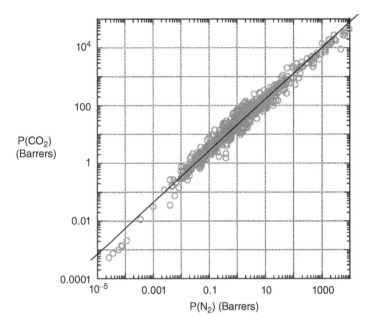

$P(CO_2)$ (Barrers)

$P(N_2)$ (Barrers)

Figure 7.5 Robeson's correlation for CO_2 and N_2 permeabilities in polymer membranes. Source: reprinted from Ref. [27] with permission from Elsevier.

data for CO_2 and N_2 will be presented, although many studies on new materials have tended to focus on the existing commercial CO_2/CH_4 separation market. Nonetheless, even in the absence of direct measurements, expected ranges of N_2 permeabilities and, in turn, CO_2/N_2 separation factors can be roughly estimated for such materials from Robeson's correlation (Figure 7.5) [27].

In addition to the polymer materials on the Robeson plot that follow the S-D mechanism of transport, materials that make use of reversible chemical reactions that hold the potential for greatly enhancing CO_2 permeability via facilitated transport will also be examined. A brief discussion of the use of polymers used in membrane contactor devices for CO_2 capture will also be presented. Finally, we will provide our perspective on potential advancements that might be achieved through the design of advanced polymers.

7.2 Polymer Types

7.2.1 Poly(Ethylene Glycol)

Materials formed from poly(ethylene glycol) (PEG) (also known as poly(ethylene oxide) [PEO]) all share a common —(CH_2CH_2O)— repeat unit within their structures. The ether functionality within the repeat unit is polar, and the lone pairs exhibit affinity for the highly electropositive carbon and the quadrupolar character of the CO_2 molecule [31]. There are a number of strategies by which to produce PEG-based materials for membranes, including acid- or base-initiated polymerization of ethylene oxide (Figure 7.6, top) and thermally

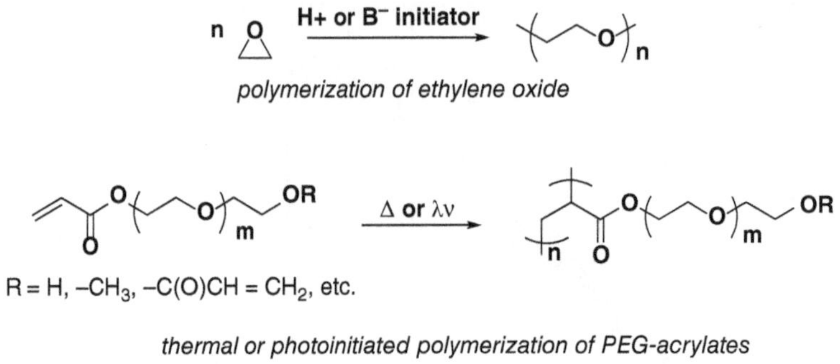

polymerization of ethylene oxide

R = H, –CH$_3$, –C(O)CH = CH$_2$, etc.

thermal or photoinitiated polymerization of PEG-acrylates

Figure 7.6   Polymerization mechanisms by which PEG-based materials are commonly produced.

or photoinitiated polymerizations of PEG-(meth)acrylate/di(meth)acrylate monomers (Figure 7.6, bottom).

PEG produced from the polymerization of ethylene oxide tends to crystallize as the molecular weight (i.e. number of repeat units) grows [31]. Although these types of PEG materials have been shown to have CO_2/N_2 selectivities of ~50 due to favorable ether-CO_2 interactions, CO_2 permeability has been measured at only 12 barrers, due the glassy nature of the material owed to strong chain packing that reduces fractional free volume (FFV) [31]. FFV is a measure of the relative amount of void space within a polymer material, and increased/decreased FFV is strongly correlated with increased/decreased permeability [32].

Alternatively, thermally or photoinitiated radical polymerization of PEG-acrylate species results in more rubbery materials that exhibit greatly increased CO_2 permeability (100–600 barrers) while retaining high CO_2/N_2 selectivity (α = 40–60) [12, 31, 33–37]. Interestingly, the inclusion of PEG-diacrylate crosslinkers in forming PEG-based membranes does not necessarily result in reductions in gas permeability or FFV, and a balance of PEG-acrylate and PEG-diacrylate species can be used to control membrane properties and stability [33, 34, 37]. Since PEG is a hydrophilic polymer, it may be susceptible to swelling in the presence of water vapor present in the flue gas, and some crosslinking may thus be required to maintain membrane integrity.

CO_2/N_2 selectivity in PEG-based membranes is primarily derived from large solubility differences between these two gases (i.e. $S_{CO2} \gg S_{N2}$) [31]. Yet the ratio of their diffusivities (D_{CO2}/D_{N2}) in PEG-based materials is likely to be ≤ 1, which is surprising since CO_2 is effectively smaller than N_2 based on their respective kinetic diameters, 3.30 and 3.64 Å [31]. Thus, the interactions between CO_2 and the ether linkages create a very favorable solubility contribution to permeability selectivity, but this is somewhat offset by reduced diffusivity (Figure 7.3).

Based on their very good performance for CO_2/N_2 separation, low cost, and ease of fabrication, PEG-based membranes are leading candidates for post-combustion CO_2 capture to be deployed at operating power plants [12]. Although MTR has not revealed the nature of the polymer used in the Polaris membranes (cf. Figure 7.3), it is probable that they are derived from PEG-based materials [38].

Many other design strategies also utilize PEG (or PEG-like segments) to produce CO_2-selective membranes, including: random and block copolymers [38–40], blending of

PEG with other polymers [41–43], and PEG-inorganic hybrids [38]. There are simply too many reports in which PEG has been used as a membrane material for CO_2 separation to be thoroughly covered within the scope of this chapter. Readers are referred to a recent review paper by Shao and co-workers (and references therein) for much greater detail on PEG-containing membranes for CO_2 removal [38].

7.2.2 Polyimides and Thermally Rearranged Polymers

Polyimides are among the most desirable polymers for gas-separation membranes due to their highly desirable physical properties [44, 45]. Polyimides such as Matrimid and Kapton were developed for use in microelectronics and as thermal insulators. Wholly aromatic polyimides such as these are conventionally synthesized via an initial condensation of a diamine with a dianhydride at near-ambient temperature, followed by thermal imidization at higher temperature. The resultant polyimides are typically glassy polymers. Matrimid and Upilex are two of the most relevant polyimides for gas separations in established commercial gas-separation applications (e.g. H_2/CH_4, CO_2/CH_4, O_2/N_2, etc.) based on their ability to be processed into high-quality films and fibers, even though their performance in these separations is well below Robeson's upper bounds [44, 46]. For a comprehensive review of polyimides, readers are referred to a review article by Paul and co-workers [45].

Yet increasing CO_2 permeability in polyimides is a priority if they are to find utility as membrane materials for post-combustion CO_2 capture [44]. Recent developments in polyimide design have revolved around the use of the fluorinated dianhydride 6-FDA with bulky aromatic diamines (Figure 7.7) [44]. These materials have been observed to have much higher CO_2 permeabilities (500–700 barrer) due to the large disruptions in chain packing and increased FFV caused by $-CF_3$ groups and multiple $-CH_3$ groups present on aromatic diamines such as durene diamine [44]. The inclusion of very bulky, three-dimensional triptycene linkages within the polyimide backbone (Figure 7.7) has also been shown to increase performance of polyimides with CO_2 permeabilities approaching 3000 barrers [47–49]. These materials are found at or near Robeson's 2008 upper bound for CO_2/CH_4 separation [28] and must also be viewed as among the most promising for

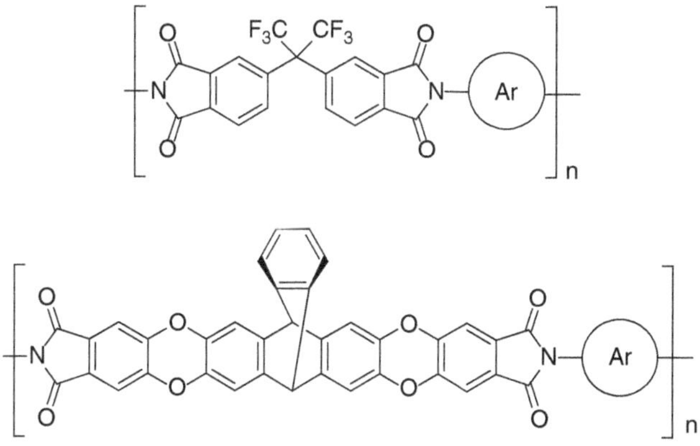

Figure 7.7 Structures of polyimides synthesized from 6-FDA (top) and triptycene moieties (bottom) with generic aromatic (Ar) diamines.

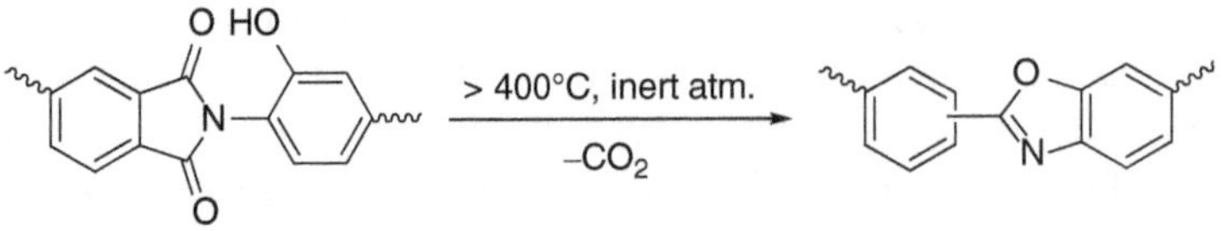

Figure 7.8 Thermal rearrangement from polyimide to poly(benzoxazole).

advancing the state of the art in CO_2/N_2 separation. Disadvantages cited for these materials are increased cost, plasticization, aging, and difficult synthetic methods associated with obtaining bulkier aromatic diamines [44].

In applications treating natural gas, polyimides tend to suffer from plasticization due to swelling induced by exposure to high-pressure CO_2 (10–35 bar), resulting in increased permeability of both CO_2 and CH_4, and reduced CO_2 selectivity [13, 14, 44]. This is unlikely to be a concern in post-combustion CO_2 capture, as the total pressure of the flue gas stream is ~1 bar and it contains only 12–15 vol% CO_2. However, because glassy polymers are not at equilibrium, polyimides also suffer from physical aging over time, rearrangement of polymer chains, loss of free volume, and reduced permeability (and increased selectivity) [44, 45]. It may be possible to somewhat alleviate physical aging in polyimides through ambient temperature crosslinking, but research is needed to identify appropriate crosslinking agents and/or polyimide structural components [45].

Thermally rearranged (TR) polymers (produced from *o*-hydroxy aromatic diamines) are a promising and emerging variant of conventional polyimides [44]. These materials are formed by a rearrangement of the polyimide to a poly(benzoxazole) at temperatures above 400 °C, with this transition shown in Figure 7.8. The resultant membrane can exhibit CO_2 permeabilities as high as 2000 barrers, with CO_2/CH_4 selectivity of ~40 without plasticization. However, there are still significant research efforts needed to advance the viability of TR polymers such as reducing the rearrangement temperature, improving mechanical properties, and understanding the effect of the *ortho* substituent on the final material gas-separation performance [44].

7.2.3 Polymers of Intrinsic Microporosity (PIMs)

PIMs are another group of condensation polymers that has shown promise as a platform for highly permeable gas-separation membranes [44, 50–54]. The PIM backbone features a spiro linkage that contorts the rigid structure, resulting an open structure with high free volumes (Figure 7.9). Brunauer–Emmett–Teller (BET) surface area analysis [55] with

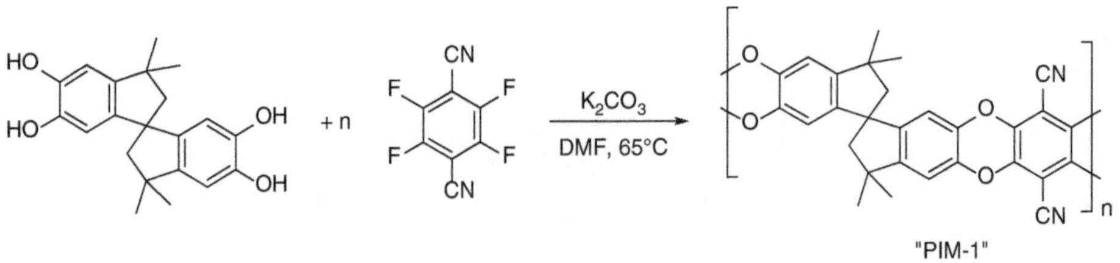

Figure 7.9 Synthesis and structure of PIM-1 reported by Budd and co-workers.

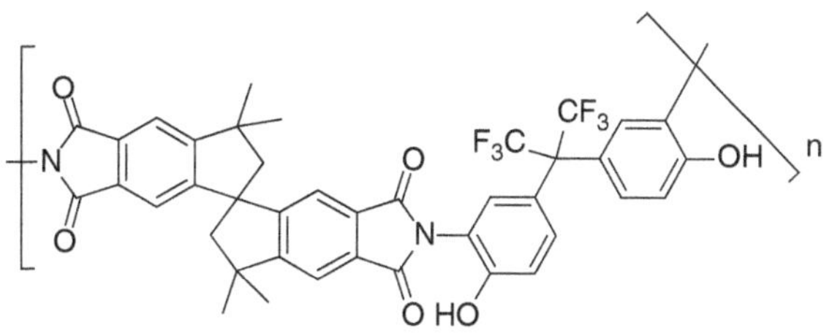

Figure 7.10 Example of a thermally rearrangable PIM-polyimide. Source: reprinted from Ref. [60] with permission from the American Chemical Society.

N_2 has shown that PIMs possess specific surface areas of 600–900 m^2 g^{-1} [53, 54]. The best-performing PIMs have exhibited CO_2 permeabilites on the order of 1000–4500 barrers, and can surpass the upper bound of Robeson plots for CO_2/N_2 and CO_2/CH_4 [53, 54].

The main precursors for the synthesis of PIMs, aromatic tetrols and activated aromatic halogens, are certainly rarer, more expensive, and likely to offer fewer opportunities for structural tailoring compared to the dianhydrides and diamines that are used to form polyimides [44]. Furthermore, not all PIMs synthesized are necessarily capable of forming defect-free thin films that can be tested as membranes [44, 53, 54].

Recent efforts have focused on improving improve the processability and performance of PIMs with an emphasis on new structural subunits within PIMs [50, 51, 56–58], crosslinking [59], blends with PEG [43], and hybrid thermally rearrangable PIM-polyimides (Figure 7.10) [52, 60, 61].

7.2.4 Poly(Ionic Liquids)

Polymerized ionic liquids (poly[ILs]) offer a vast range of possibilities for the design of materials for polymeric gas-separation membranes [62–67], owed largely to the tailorable nature of the imidazolim cations that possesses five points of possible functionalization/derivatization [68]. Poly(IL) materials are typically fabricated through photoinitiated radical polymerization of an IL wherein a styrene, vinyl, or acrylate group is appended to the cation (Figure 7.11), resulting in a polymer with pendant cations bound to the polymer backbone with "free" anions [70–72]. Poly(IL) materials produced from ILs with polymerizable anions are also possible [70, 71, 73, 74].

The development of poly(IL) membranes stemmed from their initial use in supported ionic liquid membranes (SILMs) [65, 75–79]. Scovazzo et al. showed that SILMs could have CO_2 permeabilities as high as 1000 barrer and were more promising for CO_2/N_2 separation ($\alpha_{ij} = 20$–40) than for CO_2/CH_4 compared to conventional polymer membranes on Robeson plots [28, 30, 75, 76, 80, 81]. However, a limitation of SILMs containing ILs is the stability

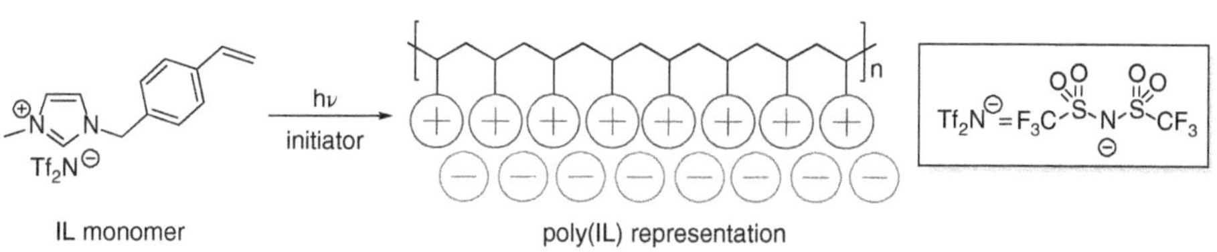

IL monomer   poly(IL) representation

Figure 7.11 Generalized concept of a photoinitiated radical polymerization of an IL. Source: reprinted from Ref. [69] with permission from the American Chemical Society.

of the membrane, as the IL is held in place via weak capillary forces, and it can be easily "blown out" at pressure differentials well below 1 atm [67, 82].

Bara et al. focused on the design of poly(IL) membranes wherein a polymerizable IL monomer was used in the SILM and then photopolymerized to form a mechanically robust polymer membrane [83–85]. These poly(IL) membranes retained, or even improved upon, the CO_2/N_2 and CO_2/CH_4 selectivities exhibited by analogous SILMs, although CO_2 permeability was reduced to the range of 10–40 barrer, since a large volume fraction of poly(styrene) was now present in the material, which hindered gas diffusion [83–85]. To remedy this reduction in permeability, Bara and co-workers then put forth the concept of poly(IL)-IL composite membranes as a means to use the poly(IL) as a type of support framework to hold non-polymerizable "free" IL in the membrane [62, 86–88]. By incorporating just 20 mol% free IL within a poly(IL) matrix, CO_2 permeability improved ∼400% compared to the neat poly(IL), while CO_2/N_2 selectivity improved 35% [86–88]. The coulombic interactions between ions in the poly(IL)-IL composite prevented any blow out of the free IL due to pressure, as originally suggested by Lodge [67]. Further work has shown that the poly(IL)-IL concept is highly tunable and logical structure–property relationships can be used to understand membrane performance [89–95]. Furthermore, the free IL content in the composite can exceed 50 mol% while maintaining a mechanically stable material, and permeabilities of the composites can begin to approach those of SLMs [89–91].

Although a thin-film composite membrane with an active layer consisting of a poly(IL)-IL composite with CO_2 permeance of 6100 gpu and CO_2/N_2 selectivity of 22 has been reported [96], progress in improving CO_2 permeability of poly(IL) membranes for CO_2/N_2 has appeared to become incremental in recent years. It seems the intrinsic CO_2 permeability of SILMs represents an apparent "speed limit" that cannot be exceeded by poly(IL)-IL composites. While a number of other membrane materials, such as block copolymers, have incorporated ILs into their structures [97, 98], there have not been major breakthroughs in terms of greatly improved permeability or selectivity. However, the use of imidazolium cations as a building block for polymers remains an attractive approach to membrane design because it presents a modular architecture that can be tailored in many ways [66, 99]. The use of imidazolium cations in new and perhaps drastically different approaches to poly(IL) design must be explored to further advance the progress of these materials. Recent examples include the synthesis and study of imidazolium ionenes [100] and polyimide-ionenes [101, 102] as gas-separation membranes that offer new opportunities to design more robust materials with high free-volume elements via condensation polymerization mechanisms that are also able to strongly interact with ILs.

7.2.5 Other Polymer Materials

This chapter is by no means an exhaustive account of all polymer types that have been studied for CO_2/N_2 separations, and focus has been devoted to trends and materials reported within the last 10 years as post-combustion CO_2 capture technologies have been developed [3]. There are a number of other polymer materials classes that have been studied as CO_2/N_2 separation membranes, but with less promising results for use in post-combustion CO_2 capture [103]. For example, polyacetylenes can exhibit CO_2 permeabilities in excess of 10 000 barrer, but with CO_2/N_2 selectivities ≤10 [103]. Poly(dimethylsiloxane) (PDMS) and other silicones are well-known for having high CO_2 permeabilities as large as 4500 barrers but also low CO_2/N_2 selectivities of ∼13 [103].

CO$_2$ and N$_2$ permeabilities have been measured in many other classes of polymers such as polyarylates, polyamides, polycarbonates, and polysulfones as well as conventional hydrocarbon polymers (e.g. polyethylene, polypropylene, polystyrene, etc.), but the observed performances have been unremarkable and tend to fall within the main cluster of the Robeson plot (Figure 7.4) [103].

Perfluoropolymers derived from the co-polymerization of tetrafluoroethylene (TFE) and cyclic perfluorinated monomers that have also been of recent interest for their excellent stability and chemical resistance imparted by the fully fluorinated structure [44]. Poly(tetrafluoroethylene) (PTFE), perhaps best known as DuPont's Teflon, has not typically been of interest as a material for polymer membranes due to its low gas permeability, which arises from its semicrystalline nature and creates challenges for processing PTFE into membranes due to its insolubility in many solvents [44]. Copolymerization of TFE with cyclic perfluorinated monomers does produce polymer materials that can be processed into membranes [104–109], although these materials have performance well below the upper bound of the Robeson plot for CO$_2$/N$_2$. However, perfluoropolymer membranes may find use in separations involving hydrocarbons, as the fluorinated structure resists plasticization due to the well-known, but not well-understood, weak interactions between fluorocarbons and hydrocarbons [110].

7.3 Facilitated Transport

Facilitated transport relies upon the use of a fixed site (i.e. covalently linked to polymer) or mobile carrier site to enhance the solubility and/or diffusivity of the species of interest in the membrane [1, 111]. Facilitated-transport mechanisms enhance the permeability of CO$_2$ (or other reactive species, e.g. O$_2$, C$_2$H$_4$, etc.) in dense polymer membranes through the use of reversible chemical reactions [112]. For CO$_2$, the typical mechanism is through the use of tertiary (3°) amines or other bases that are not capable of direct reaction with CO$_2$ carbamates, but instead promote formation of the bicarbonate (HCO$_3^-$) anion in the presence of water [7, 8, 113]. CO$_2$ and water combine to form carbonic acid (H$_2$CO$_3$, 1st pK$_a$ = 6.35), which is neutralized to HCO$_3^-$ salts in the presence of a relatively weak base [7, 8]. The reaction can be readily reversed at low to moderate temperatures and/or with mild pressure differentials or vacuum [114]. N$_2$ will not interact with the fixed-site carrier and will only experience S-D driven transport, and thus the membrane's selectivity for CO$_2$/N$_2$ can increase by an order of magnitude or more.

Polymer materials containing amines, *N*-heterocycles (e.g. imidazoles), or other Brønsted-basic sites in the appropriate pK$_a$ range (~7.0–9.5) are likely candidate materials that can be used for facilitated transport of CO$_2$ as HCO$_3^-$. A representative illustration of facilitated transport with imidazole groups as fixed-site carriers is shown in Figure 7.12.

In the laboratory under controlled conditions, many facilitated-transport membranes with fixed-site and mobile carriers may display very high permeabilities of CO$_2$ (>10 000 barrers) and CO$_2$/N$_2$ selectivities ≫100. Clearly, such materials far exceed the upper bound of the Robeson plot for CO$_2$/N$_2$ (Figure 7.4) (although facilitated-transport membranes are not included in Robeson plots) and appear to suggest that cost-efficient CO$_2$ capture (Figure 7.3) is already within reach.

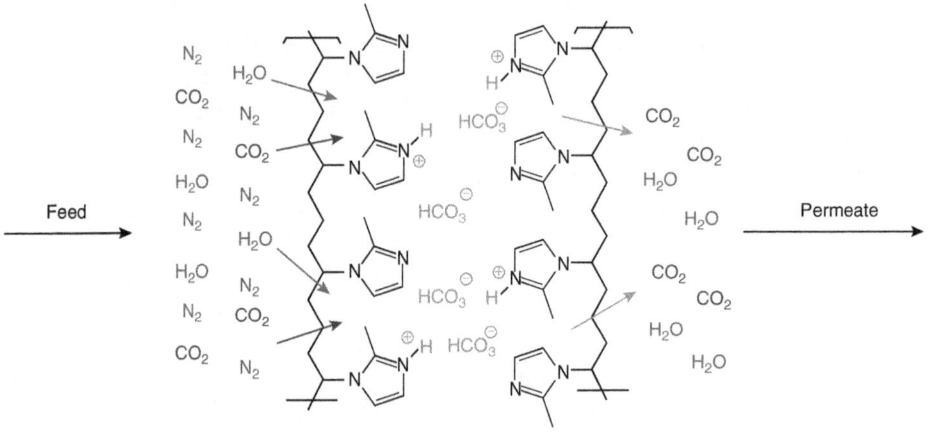

Facilitated CO_2 transport via bicarbonater (HCO_3^-) anion

Figure 7.12 Representative operation of facilitated transport of CO_2 as HCO_3^- anion in polymer membrane with imidazoles as fixed-site carriers.

However, despite the apparent breakthroughs relative to conventional polymer membranes operating via the S-D mechanism, facilitated-transport membranes have not yet been fully proven as viable for post-combustion CO_2 capture applications, as many types of carriers can experience significant degradation under operation with real flue gas via rapid and irreversible reactions with oxidizers such as O_2, SO_2, and NO_x. Such deactivation limits the applicable lifetimes of the carriers and causes membrane performance to rapidly diminish, wherein a polymer membrane operating via S-D remains, yet with inferior CO_2 permeability and selectivity [2]. Thus, the fixed site must be sufficiently stable and the overall material relatively inexpensive if replacement becomes necessary.

Facilitated transport via the bicarbonate mechanism in fixed-site carrier membranes (Figure 7.12) has been shown to be possible in post-combustion CO_2 capture, pre-combustion CO_2 capture, and natural gas sweetening. Hydrophilic membranes based on poly(vinyl amine) – poly(vinyl alcohol) (PVAm-PVA) copolymers with 1 amine groups promote facilitated transport of CO_2 as bicarbonate but only *after* the amine sites have been saturated as carbamates, using the relatively stable carbamate anion as a base [20–22]. These materials have been demonstrated to successfully exhibit facilitated transport of CO_2 under controlled conditions (CO_2, N_2, and H_2O) as well as in actual flue gas with SO_2, NO_x, and particulates [20–22, 115–117].

Other strategies incorporating amine groups as fixed and mobile carriers have included other types of fixed and mobile amines within various polymers [42, 118–126], ion-exchange [111, 127–131], as well as poly(ILs) [132, 133] and SILMs [134–140] containing amine functionalities. Polymer-inorganic hybrids, including materials containing carbon nanotubes, graphene oxide, and metal organic frameworks (MOFs), which promote facilitated transport of CO_2, have also been a recent trend in the literature [141–148].

Poly(ethylenimine) (PEI) materials produced from the ring opening of aziridine have also been utilized as a low-cost material for promoting facilitated transport of CO_2 [149]. PEI is typically highly branched and as a result contains mixtures of 1°, 2°, and 3° amines that are capable of promoting CO_2 transport as HCO_3^- in much the same way as PVAm. However, this branching also causes PEI to have poor mechanical properties, and it is typically a viscous liquid at ambient temperature, which may present challenges to casting ultrathin films. Linear PEI (i.e. $-[CH_2CH_2NH]_n-$), which contains only 2° amine groups and can be considered the nitrogen analogue of PEG (Figure 7.6), can be synthesized under controlled conditions but has not yet been studied as a material for gas-separation membranes. Poly(amidoamine) (PAMAM) dendrimers, which are also similar to branched PEI but contain only 1° and 3° amine groups based on a well-controlled synthesis, have also been studied for facilitated transport of CO_2 [150].

Other approaches to the facilitated transport of CO_2 have been inspired by nature. Materials containing active carbonic anhydrase (CA) enzyme or simplified mimics (Figure 7.13) have been studied for post-combustion CO_2 capture [17, 151–156]. CA facilitates the reaction $CO_2 + H_2O \rightarrow H_2CO_3$ at the central Zn^{2+} cation, which is coordinated by three histidine side chains and one molecule of water. The reaction rate of CA is extremely fast, and it is capable of catalyzing between 10^4 and 10^6 reactions per second; the reaction is considered to be mass-transfer limited based on the media surrounding the enzyme [157]. It is also clearly able to work in the presence of low concentrations of CO_2, as it is present in human and animal tissue where it serves to maintain acid–base balance and transport CO_2 out of cells. At least five families of CA are known to exist and are labeled α, β, γ, δ, and ε. There are 15 forms α-CA found in humans, 12 of which are catalytically active [158].

Yao and co-workers reported CO_2 permeances as high as 1100 gpu with CO_2/N_2 selectivities as large as 85 using membranes based on a simplified CA enzyme mimic incorporated into a poly(vinylimidazole) framework [151]. However, the authors noted that the material performance was highly dependent on the pH at which the Zn^{2+} cations (as $Zn[OAc]_2$) were introduced to, and subsequently coordinated with, poly(vinylimidazole).

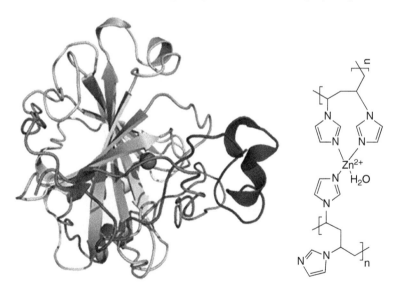

Figure 7.13 Depiction of CA as a ribbon diagram (left) and chemical structure of poly(vinylimidazole)-based mimic studied in Ref. [151] (right). Source: reprinted with permission from the Royal Society of Chemistry. (*See color plate section for color representation of this figure*).

Biological structures such as enzymes are often fragile outside of physiological conditions, and the stability of the CA enzyme in harsh flue conditions raises concerns about whether it can be feasibly and economically implemented within a membrane (or in combination with an amine solvent) for post-combustion CO_2 capture. Enzyme mimics may be poisoned through reaction with species such as SO_2 or HCl that are present in flue gas. However, recent efforts focused on the directed evolution of *Desulfovibrio vulgaris*, a sulfur-tolerant bacteria, have yielded a highly temperature- and alkaline-stable form of β-CA that has shown great promise when used in combination with aqueous amine solvents [155].

7.4 Polymer Membrane Contactors

Conventionally, gas absorption by liquid solvents has been performed in vertical packed or trayed columns, where efficient mass transfer depends on the interfacial area available within a given volume while maintaining a high void fraction so as to minimize pressure drop [159]. Such columns have been in use for over 85 years [160], and while they have been optimized to achieve sufficient gas–liquid contact, they also suffer from disadvantages in height/footprint and less than ideal surface area to volume (S/V) ratios ($<200 \, m^2 \, m^{-3}$). The presence of two contacting phases also leads to operational difficulties such as foaming and flooding.

Membrane contactors containing bundles of hollow fibers are viewed as capable of making significant impacts in minimizing the capital expenditures and footprint of post-combustion CO_2 capture processes, enabling greatly improved mass-transfer rates relative to conventional packed or trayed absorption columns. Major advantages of membrane contactors include high specific surface areas (up to $2000 \, m^2 \, m^{-3}$), increased mass transfer coefficients, reduced solvent inventory, and minimized process footprint. All of these improvements can make large contributions to achieving cost- and energy-efficient CO_2 capture.

Figure 7.14 illustrates the general operating principles of a hollow-fiber membrane contactor for gas separation, wherein the flue gas is fed on the tube side (e.g. "Fluid #1 in") and the solvent is fed on the shell side (e.g. "Fluid #2 in") [161]. The principles of

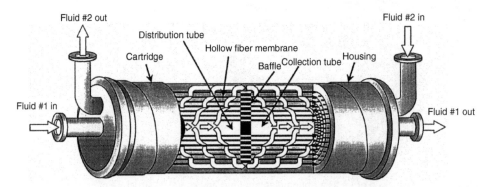

Figure 7.14 Schematic of membrane contactor operation for gas absorption into a liquid. Source: reprinted from Ref. [161] with permission from Elsevier.

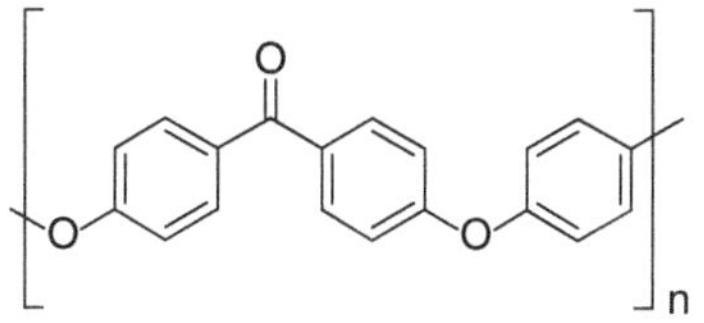

Figure 7.15 Chemical structure of a PEEK polymer material.

the S-D mechanism apply as CO_2 is transported through the active membrane layer in a non-dispersive manner. However, as CO_2 reaches the permeate side of the membrane (contactor shell side), it is immediately absorbed by the aqueous amine solvent on the permeate side. Absorbed CO_2 remains in the liquid (e.g. "Fluid #2 out"), which is sent to a regeneration stage (e.g. stripping column or another membrane contactor), while the cleaned flue gas is vented to the atmosphere (e.g. "Fluid #1 out"). This configuration keeps the CO_2 concentration on the permeate side at effectively zero and maintains a maximum driving force for transport across the hollow fiber membrane.

For post-combustion CO_2 capture, the liquid absorbent used in conjunction with a hollow fiber membrane contactor has typically been an aqueous amine solution [162], which is aggressive toward many types of polymers. As robust stability of the polymer is an obvious requirement to ensure long-term operation of the membrane contactor, the types of materials that can be used as the wetted interface within membrane contactors are quite limited. Here, polymers with outstanding chemical resistance are essential to enable membrane contactor operation, even if they have been previously observed to possess insufficiently promising performance in the context of the CO_2/N_2 Robeson plot (Figure 7.4). PTFE [163, 164], polypropylene [165–167], poly(vinylidene difluoride) [168, 169], polyamide-imides [170, 171], and polyether ether ketone (PEEK) [172] (Figure 7.15) materials are all promising candidates for using membrane contactors for post-combustion CO_2 capture applications. Further modification of polymer surfaces with fluorinated groups can also enhance material stability and wettability [170, 171, 173]. Specialized process conditions can also yield microporous polymers that further enhance CO_2 mass transfer [167, 172].

Notable projects that have utilized hollow-fiber membrane contactors for post-combustion CO_2 capture in the field include work by Aker–Kvaerner that employed PTFE hollow fibers [163, 164] and an ongoing project between Gas Technology Institute (GTI) and Porogen that makes use of PEEK fibers [174]. Challenges facing hollow-fiber membrane contactors include demonstration at larger scales and greater efficiencies and cost reductions associated with their manufacture.

7.5 Summary and Perspectives

Clearly, a great number of polymer materials have been examined for CO_2/N_2 separation as part of the effort to make post-combustion CO_2 capture a viable process that can be deployed worldwide as a means of reducing GHG emissions. Significant progress has been made using membranes made from both relatively common and inexpensive materials such as PEG as well as more complex polymers such as PIMs, polyimides, poly(ILs), and hybrids. We foresee that the trend to develop polymer materials with increasingly larger FFV will continue, but greater focus must also be given to methods by which to prevent or altogether stop aging in glassy materials. Facilitated transport will also continue to be of interest, but greater concern must be placed on developing highly stable carriers that will not be rapidly deactivated in the presence of multiple reactive components found within flue gas. This is a challenge that can be addressed by both chemical synthesis and well as biological methods. Finally, membrane contactors hold unique promise for process intensification and minimization of process footprints. However, for membranes and membrane contactor systems for post-combustion CO_2 capture to be competitive with traditional aqueous amine absorber-stripper systems, not only are improved materials needed but also advanced manufacturing techniques that streamline production of membrane-containing modules.

References

1 U.S. Department of Energy Office of Scientific and Technical Information. (2010). Basic Research Needs for Carbon Capture: Beyond 2020. https://www.osti.gov/servlets/purl/1291240.

2 U.S. Department of Energy National Energy Technology Laboratory. (2011). Advanced Carbon Dioxide Capture R&D Program: Technology Update May 2011.

3 Boot-Handford, M.E., Abanades, J.C., Anthony, E.J. et al. (2014). Carbon capture and storage update. *Energy & Environmental Science* 7 (1): 130–189.

4 Jones, C.W. (2011). CO2 capture from dilute gases as a component of modern global carbon management. *Annual Review of Chemical and Biomolecular Engineering* 2: 31–52.

5 International Energy Agency. (2012). Key World Energy Statistics 2012.

6 Rochelle, G.T. (2009). Amine scrubbing for CO2 capture. *Science* 325 (5948): 1652–1654.

7 Kidnay, A.J. and Parrish, W.R. (2006). *Fundamentals of Natural Gas Processing*. Boca Raton, FL: CRC Press and Taylor & Francis Group.

8 Astarita, G., Savage, D.W., and Bisio, A. (1983). *Gas Treating with Chemical Solvents*. New York: Wiley.

9 da Silva, E.F., Lepaumier, H., Grimstvedt, A. et al. (2012). Understanding 2-ethanolamine degradation in Postcombustion CO2 capture. *Industrial and Engineering Chemistry Research* 51 (41): 13329–13338.

10 Lepaumier, H., da Silva, E.F., Einbu, A. et al. (2011). Comparison of MEA degradation in pilot-scale with lab-scale experiments. *Energy Procedia* 4: 1652–1659.

11 Lepaumier, H., Grimstvedt, A., Vernstad, K. et al. (2011). Degradation of MMEA at absorber and stripper conditions. *Chemical Engineering Science* 66 (15): 3491–3498.

12 Merkel, T.C., Lin, H., Wei, X., and Baker, R. (2010). Power plant post-combustion carbon dioxide capture: an opportunity for membranes. *Journal of Membrane Science* 359 (1–2): 126–139.

13 Baker, R.W. and Lokhandwala, K. (2008). Natural gas processing with membranes: an overview. *Industrial & Engineering Chemistry Research* 47 (7): 2109–2121.

14 Baker, R.W. (2002). Future directions of membrane gas separation technology. *Industrial & Engineering Chemistry Research* 41 (6): 1393–1411.

15 Favre, E. (2011). Membrane processes and postcombustion carbon dioxide capture: challenges and prospects. *Chemical Engineering Journal* 171 (3): 782–793.

16 Favre, E. (2007). Carbon dioxide recovery from post-combustion processes: can gas permeation membranes compete with absorption? *Journal of Membrane Science* 294 (1–2): 50–59.

17 Shelley, S. (2009). Capturing CO2: membrane systems move forward. *Chemical Engineering Progress* 105 (4): 42–47.

18 Luis, P. and Van der Bruggen, B. (2013). The role of membranes in post-combustion CO2 capture. *Greenhouse Gases: Science and Technology* 3 (5): 318–337.

19 Bhown, A.S. and Freeman, B.C. (2011). Analysis and status of post-combustion carbon dioxide capture technologies. *Environmental Science & Technology*.

20 Deng, L., Kim, T.-J., and Hägg, M.-B. (2009). Facilitated transport of CO2 in novel PVAm/PVA blend membrane. *Journal of Membrane Science* 340 (1–2): 154–163.

21 Sandru, M., Kim, T.-J., and Hägg, M.-B. (2009). High molecular fixed-site-carrier PVAm membrane for CO2 capture. *Desalination* 240 (1–3): 298–300.

22 Grainger, D. and Hägg, M.-B. (2008). Techno-economic evaluation of a PVAm CO2-selective membrane in an IGCC power plant with CO2 capture. *Fuel* 87 (1): 14–24.

23 Wijmans, J.G. and Baker, R.W. (1995). The solution-diffusion model: a review. *Journal of Membrane Science* 107 (1–2): 1–21.

24 Graham, T. (1886). On the absorption and dialytic separation of gases by colloid septa. *Philoshophical Magazine* 32: 401–420.

25 Ravishankar, R., Martens, J.A., and Jacobs, P.A. (1997). The scientific legacy of the late Richard M. Barrer, FRS. *Microporous Materials* 8 (5–6): 283–284.

26 Bernardo, P., Drioli, E., and Golemme, G. (2009). Membrane gas separation: a review/state of the art. *Industrial & Engineering Chemistry Research* 48 (10): 4638–4663.

27 Robeson, L.M., Freeman, B.D., Paul, D.R., and Rowe, B.W. (2009). An empirical correlation of gas permeability and permselectivity in polymers and its theoretical basis. *Journal of Membrane Science* 341 (1–2): 178–185.

28 Robeson, L.M. (2008). The upper bound revisited. *Journal of Membrane Science* 320 (1–2): 390–400.

29 Freeman, B.D. (1999). Basis of permeability/selectivity tradeoff relations in polymeric gas separation membranes. *Macromolecules* 32 (2): 375–380.

30 Robeson, L.M. (1991). Correlation of separation factor versus permeability for polymeric membranes. *Journal of Membrane Science* 62 (2): 165–185.

31 Lin, H.Q. and Freeman, B.D. (2005). Materials selection guidelines for membranes that remove CO2 from gas mixtures. *Journal of Molecular Structure* 739 (1–3): 57–74.

32 Lee, W.M. (1980). Selection of barrier materials from molecular structure. *Polymer Engineering and Science* 20 (1): 65–69.

33 Lin, H., Van Wagner, E., Swinnea, J.S. et al. (2006). Transport and structural characteristics of crosslinked poly(ethylene oxide) rubbers. *Journal of Membrane Science* 276 (1–2): 145–161.

34 Raharjo, R.D., Lin, H., Sanders, D.F. et al. (2006). Relation between network structure and gas transport in crosslinked poly(propylene glycol) diacrylate. *Journal of Membrane Science* 283 (1+2): 253–265.

35 Kalakkunnath, S., Kalika, D.S., Lin, H., and Freeman, B.D. (2005). Segmental relaxation characteristics of cross-linked poly(ethylene oxide) copolymer networks. *Macromolecules* 38 (23): 9679–9687.

36 Lin, H. and Freeman, B.D. (2005). Gas and vapor solubility in cross-linked poly(ethylene glycol Diacrylate). *Macromolecules* 38 (20): 8394–8407.

37 Lin, H., Kai, T., Freeman, B.D. et al. (2005). The effect of cross-linking on gas permeability in cross-linked poly(ethylene glycol Diacrylate). *Macromolecules* 38 (20): 8381–8393.

38 Liu, S.L., Shao, L., Chua, M.L. et al. (2013). Recent progress in the design of advanced PEO-containing membranes for CO2 removal. *Progress in Polymer Science* 38 (7): 1089–1120.

39 Reijerkerk, S.R., Wessling, M., and Nijmeijer, K. (2011). Pushing the limits of block copolymer membranes for CO2 separation. *Journal of Membrane Science* 378 (1–2): 479–484.

40 Gu, Y., Cussler, E.L., and Lodge, T.P. (2012). ABA-triblock copolymer ion gels for CO2 separation applications. *Journal of Membrane Science* 423-424: 20–26.

41 Hu, X., Tang, J., Blasig, A. et al. (2006). CO2 permeability, diffusivity and solubility in polyethylene glycol-grafted polyionic membranes and their CO2 selectivity relative to methane and nitrogen. *Journal of Membrane Science* 281 (1+2): 130–138.

42 Yi, C., Wang, Z., Li, M. et al. (2006). Facilitated transport of CO2 through polyvinylamine/polyethylene glycol blend membranes. *Desalination* 193 (1–3): 90–96.

43 Xin, M.W., Qiu, G.Z., Peng, J.L. et al. (2015). Towards enhanced CO2 selectivity of the PIM-1 membrane by blending with polyethylene glycol. *Journal of Membrane Science* 493: 147–155.

44 Sanders, D.F., Smith, Z.P., Guo, R. et al. (2013). Energy-efficient polymeric gas separation membranes for a sustainable future: a review. *Polymer* 54 (18): 4729–4761.

45 Xiao, Y., Low, B.T., Hosseini, S.S. et al. (2009). The strategies of molecular architecture and modification of polyimide-based membranes for CO2 removal from natural gas – a review. *Progress in Polymer Science* 34 (6): 561–580.

46 Koros, W.J. and Mahajan, R. (2000). Pushing the limits on possibilities for large scale gas separation: which strategies? *Journal of Membrane Science* 175 (2): 181–196.

47 Swaidan, R., Al-Saeedi, M., Ghanem, B. et al. (2014). Rational design of intrinsically ultramicroporous polyimides containing bridgehead-substituted triptycene for highly selective and permeable gas separation membranes. *Macromolecules* 47 (15): 5104–5114.

48 Swaidan, R., Ghanem, B., Al-Saeedi, M. et al. (2014). Role of intrachain rigidity in the plasticization of intrinsically microporous triptycene-based polyimide membranes in mixed-gas CO2/CH4 separations. *Macromolecules* 47 (21): 7453–7462.

49 Wiegand, J.R., Smith, Z.P., Liu, Q. et al. (2014). Synthesis and characterization of triptycene-based polyimides with tunable high fractional free volume for gas separation membranes. *Journal of Materials Chemistry A* 2 (33): 13309–13320.

50 Mason, C.R., Maynard-Atem, L., Al-Harbi, N.M. et al. (2011). Polymer of intrinsic microporosity incorporating Thioamide functionality: preparation and gas transport properties. *Macromolecules* 44 (16): 6471–6479.

51 Emmler, T., Heinrich, K., Fritsch, D. et al. (2010). Free volume investigation of polymers of intrinsic microporosity (PIMs): PIM-1 and PIM1 copolymers incorporating Ethanoanthracene units. *Macromolecules* 43 (14): 6075–6084.

52 Ghanem, B.S., McKeown, N.B., Budd, P.M. et al. (2009). Synthesis, characterization, and gas permeation properties of a novel group of polymers with intrinsic microporosity: PIM-polyimides. *Macromolecules* 42 (20): 7881–7888.

53 Budd, P.M., McKeown, N.B., Ghanem, B.S. et al. (2008). Gas permeation parameters and other physicochemical properties of a polymer of intrinsic microporosity: Polybenzodioxane PIM-1. *Journal of Membrane Science* 325 (2): 851–860.

54 Budd, P.M., Msayib, K.J., Tattershall, C.E. et al. (2005). Gas separation membranes from polymers of intrinsic microporosity. *Journal of Membrane Science* 251 (1–2): 263–269.

55 Brunauer, S., Emmett, P.H., and Teller, E. (1938). Adsorption of gases in multimolecular layers. *Journal of the American Chemical Society* 60 (2): 309–319.

56 Hao, L., Liao, K.-S., and Chung, T.-S. (2015). Photo-oxidative PIM-1 based mixed matrix membranes with superior gas separation performance. *Journal of Materials Chemistry A* 3 (33): 17273–17281.

57 Rose, I., Carta, M., Malpass-Evans, R. et al. (2015). Highly permeable Benzotriptycene-based polymer of intrinsic microporosity. *ACS Macro Letters* 4 (9): 912–915.

58 Du, N., Park, H.B., Dal-Cin, M.M., and Guiver, M.D. (2012). Advances in high permeability polymeric membrane materials for CO2 separations. *Energy & Environmental Science* 5 (6): 7306–7322.

59 Du, N., Robertson, G.P., Song, J. et al. (2009). High-performance Carboxylated polymers of intrinsic microporosity (PIMs) with tunable gas transport properties†. *Macromolecules* 42 (16): 6038–6043.

60 Shamsipur, H., Dawood, B.A., Budd, P.M. et al. (2014). Thermally rearrangeable PIM-polyimides for gas separation membranes. *Macromolecules* 47 (16): 5595–5606.

61 Rogan, Y., Malpass-Evans, R., Carta, M. et al. (2014). A highly permeable polyimide with enhanced selectivity for membrane gas separations. *Journal of Materials Chemistry A* 2 (14): 4874–4877.

62 Bara, J.E., Camper, D.E., Gin, D.L., and Noble, R.D. (2010). Room-temperature ionic liquids and composite materials: platform technologies for CO2 capture. *Accounts of Chemical Research* 43 (1): 152–159.

63 Gin, D.L. and Noble, R.D. (2011). Designing the next generation of chemical separation membranes. *Science* 332: 674–676.

64 Noble, R.D. and Gin, D.L. (2011). Perspective on ionic liquids and ionic liquid membranes. *Journal of Membrane Science* 369 (1–2): 1–4.

65 Bara, J.E. (2013). Ionic liquids in gas separation membranes. In: *Encyclopedia of Membrane Science and Technology* (eds. E.M.V. Hoek and V.V. Tarabara), 1–23. Wiley.

66 Anderson, E.B. and Long, T.E. (2010). Imidazole- and imidazolium-containing polymers for biology and material science applications. *Polymer* 51 (12): 2447–2454.

67 Lodge, T.P. (2008). Materials science – a unique platform for materials design. *Science* 321 (5885): 50–51.

68 Bara, J.E. and Shannon, M.S. (2012). Beyond 1,3-difunctionalized imidazolium cations. *Nanomaterials and Energy* 1 (4): 237–242.

69 Horne, W.J., Andrews, M.A., Terrill, K.L. et al. (2015). Poly(ionic liquid) superabsorbent for polar organic solvents. *ACS Applied Materials & Interfaces* 7 (17): 8979–8983.

70 Ohno, H. (2007). Design of ion conductive polymers based on ionic liquids. *Macro-molecular Symposia* 249/250: 551–556.

71 Ohno, H., Yoshizawa, M., and Ogihara, W. (2004). Development of new class of ion conductive polymers based on ionic liquids. *Electrochimica Acta* 50 (2–3): 255–261.

72 Washiro, S., Yoshizawa, M., Nakajima, H., and Ohno, H. (2004). Highly ion conductive flexible films composed of network polymers based on polymerizable ionic liquids. *Polymer* 45 (5): 1577–1582.

73 Bakker, M.G., Frazier, R.M., Burkett, S. et al. (2012). Perspectives on supercapacitors, pseudocapacitors and batteries. *Nanomaterials and Energy* 1 (3): 136–158.

74 Armand, M., Endres, F., MacFarlane, D.R. et al. (2009). Ionic-liquid materials for the electrochemical challenges of the future. *Nature Materials* 8 (8): 621–629.

75 Scovazzo, P., Visser Ann, E., Davis James, H. et al. (2002). Supported ionic liquid membranes and facilitated ionic liquid membranes. In: *Ionic Liquids*, vol. 818 (eds. R.D. Rogers and K.R. Seddon), 69–87. American Chemical Society.

76 Scovazzo, P., Kieft, J., Finan, D.A. et al. (2004). Gas separations using non-hexafluorophosphate PF6 (−) anion supported ionic liquid membranes. *Journal of Membrane Science* 238 (1–2): 57–63.

77 Laciak, D.V., Pez, G.P., and Burban, P.M. (1992). Molten salt facilitated transport membranes. Part 2. Separation of ammonia from nitrogen and hydrogen at high temperatures. *Journal of Membrane Science* 65 (1–2): 31–38.

78 Pez, G.P. and Carlin, R.T. (1992). Molten salt facilitated transport membranes. Part 1. Separation of oxygen from air at high temperatures. *Journal of Membrane Science* 65 (1–2): 21–30.

79 Quinn, R., Pez, G.P., and Appleby, J.B. Molten salt hydrate membranes for the separation of gases. European patent EP0311903A2. 1988.

80 Scovazzo, P., Camper, D., Kieft, J. et al. (2004). Regular solution theory and CO2 gas solubility in room-temperature ionic liquids. *Industrial & Engineering Chemistry Research* 43 (21): 6855–6860.

81 Alexander Stern, S. (1994). Polymers for gas separations: the next decade. *Journal of Membrane Science* 94 (1): 1–65.

82 Gan, Q., Rooney, D., Xue, M. et al. (2006). An experimental study of gas transport and separation properties of ionic liquids supported on nanofiltration membranes. *Journal of Membrane Science* 280 (1+2): 948–956.

83 Bara, J.E., Hatakeyama, E.S., Gabriel, C.J. et al. (2008). Synthesis and light gas separations in cross-linked gemini room temperature ionic liquid polymer membranes. *Journal of Membrane Science* 316 (1–2): 186–191.

84 Bara, J.E., Gabriel, C.J., Hatakeyama, E.S. et al. (2008). Improving CO2 selectivity in polymerized room-temperature ionic liquid gas separation membranes through incorporation of polar substituents. *Journal of Membrane Science* 321 (1): 3–7.

85 Bara, J.E., Lessmann, S., Gabriel, C.J. et al. (2007). Synthesis and performance of polymerizable room-temperature ionic liquids as gas separation membranes. *Industrial & Engineering Chemistry Research* 46 (16): 5397–5404.

86 Bara, J.E., Noble, R.D., and Gin, D.L. (2009). Effect of "free" cation substituent on gas separation performance of polymer-room-temperature ionic liquid composite membranes. *Industrial & Engineering Chemistry Research* 48 (9): 4607–4610.

87 Bara, J.E., Gin, D.L., and Noble, R.D. (2008). Effect of anion on gas separation performance of polymer-room-temperature ionic liquid composite membranes. *Industrial & Engineering Chemistry Research* 47 (24): 9919–9924.

88 Bara, J.E., Hatakeyama, E.S., Gin, D.L., and Noble, R.D. (2008). Improving CO2 permeability in polymerized room-temperature ionic liquid gas separation membranes through the formation of a solid composite with a room-temperature ionic liquid. *Polymers for Advanced Technologies* 19 (10): 1415–1420.

89 Carlisle, T.K., McDanel, W.M., Cowan, M.G. et al. (2014). Vinyl-functionalized poly(imidazolium)s: a curable polymer platform for cross-linked ionic liquid gel synthesis. *Chemistry of Materials* 26 (3): 1294–1296.

90 Carlisle, T.K., Wiesenauer, E.F., Nicodemus, G.D. et al. (2013). Ideal CO2/light gas separation performance of poly(vinylimidazolium) membranes and poly(vinylimidazolium)-ionic liquid composite films. *Industrial and Engineering Chemistry Research* 52 (3): 1023–1032.

91 Carlisle, T.K., Nicodemus, G.D., Gin, D.L., and Noble, R.D. (2012). CO2/light gas separation performance of cross-linked poly(vinylimidazolium) gel membranes as a function of ionic liquid loading and cross-linker content. *Journal of Membrane Science* 397-398: 24–37.

92 Miller, A.L., Carlisle, T.K., LaFrate, A.L. et al. (2012). Design of functionalized room-temperature ionic liquid-based materials for CO2 separations and selective blocking of hazardous chemical vapors. *Separation Science and Technology* 47 (2): 169–177.

93 Hudiono, Y.C., Carlisle, T.K., LaFrate, A.L. et al. (2011). Novel mixed matrix membranes based on polymerizable room-temperature ionic liquids and SAPO-34 particles to improve CO2 separation. *Journal of Membrane Science* 370 (1–2): 141–148.

94 Li, P., Paul, D.R., and Chung, T.-S. (2012). High performance membranes based on ionic liquid polymers for CO2 separation from the flue gas. *Green Chemistry* 14: 1052–1063.

95 Li, P., Pramoda, K.P., and Chung, T.S. (2011). CO(2) separation from flue gas using polyvinyl-(room temperature ionic liquid)-room temperature ionic liquid composite membranes. *Industrial & Engineering Chemistry Research* 50 (15): 9344–9353.

96 Zhou, J., Mok, M.M., Cowan, M.G. et al. (2014). High-Permeance room-temperature ionic-liquid-based membranes for CO2/N2 separation. *Industrial & Engineering Chemistry Research* 53 (51): 20064–20067.

97 Gu, Y.Y. and Lodge, T.P. (2011). Synthesis and gas separation performance of triblock copolymer ion gels with a polymerized ionic liquid mid-block. *Macromolecules* 44 (7): 1732–1736.

98 Wiesenauer, E.F., Nguyen, P.T., Newell, B.S. et al. (2013). Imidazolium-containing, hydrophobic-ionic-hydrophilic ABC triblock copolymers: synthesis, ordered phase-separation, and supported membrane fabrication. *Soft Matter* 9 (33): 7923–7927.

99 Green, M.D., Allen, M.H. Jr., Dennis, J.M. et al. (2011). Tailoring macromolecular architecture with imidazole functionality: a perspective for controlled polymerization processes. *European Polymer Journal* 47 (4): 486–496.

100 Kammakakam, I., O'Harra, K.E., Bara, J.E., and Jackson, E.M. (2019). Design and synthesis of imidazolium-mediated Tröger's base-containing Ionene polymers for advanced CO2 separation membranes. *ACS Omega* 4 (2): 3439–3448.

101 Mittenthal, M.S., Flowers, B.S., Bara, J.E. et al. (2017). Ionic polyimides: hybrid polymer architectures and composites with ionic liquids for advanced gas separation membranes. *Industrial & Engineering Chemistry Research* 56 (17): 5055–5069.

102 Shaplov, A.S., Morozova, S.M., Lozinskaya, E.I. et al. (2016). Turning into poly(ionic liquid)s as a tool for polyimide modification: synthesis, characterization and CO2 separation properties. *Polymer Chemistry* 7 (3): 580–591.

103 Powell, C.E. and Qiao, G.G. (2006). Polymeric CO2/N2 gas separation membranes for the capture of carbon dioxide from power plant flue gases. *Journal of Membrane Science* 279 (1–2): 1–49.

104 Okamoto, Y., Du, Q., Koike, K. et al. (2016). New amorphous perfluoro polymers: perfluorodioxolane polymers for use as plastic optical fibers and gas separation membranes. *Polymers for Advanced Technologies* 27 (1): 33–41.

105 Okamoto, Y., Zhang, H., Mikes, F. et al. (2014). New perfluoro-dioxolane-based membranes for gas separations. *Journal of Membrane Science* 471: 412–419.

106 Tiwari, R.R., Smith, Z.P., Lin, H. et al. (2015). Gas permeation in thin films of "high free-volume" glassy perfluoropolymers: part II. CO2 plasticization and sorption. *Polymer* 61: 1–14.

107 Tiwari, R.R., Smith, Z.P., Lin, H. et al. (2014). Gas permeation in thin films of "high free-volume" glassy perfluoropolymers: part I. physical aging. *Polymer* 55 (22): 5788–5800.

108 Alentiev, A.Y., Shantarovich, V.P., Merkel, T.C. et al. (2002). Gas and vapor sorption, permeation, and diffusion in glassy amorphous Teflon AF1600. *Macromolecules* 35 (25): 9513–9522.

109 Bondar, V.I., Freeman, B.D., and Yampolskii, Y.P. (1999). Sorption of gases and vapors in an amorphous glassy Perfluorodioxole copolymer. *Macromolecules* 32 (19): 6163–6171.

110 Scott, R.L. (1958). The anomalous behavior of fluorocarbon solutions. *The Journal of Physical Chemistry* 62 (2): 136–145.

111 Noble, R.D. (1991). Facilitated transport mechanism in fixed site carrier membranes. *Journal of Membrane Science* 60 (2–3): 297–306.

112 Way, J.D. and Noble, R. (1992). Facilitated transport. In: *Membrane Handbook* (eds. W.S.W. Ho and K. Sirkar), 833–866. Springer US.

113 Meldon, J.H., Stroeve, P., and Gregoire, C.E. (1982). Facilitated transport of carbon dioxide: a review. *Chemical Engineering Communications* 16 (1–6): 263–300.

114 Tomizaki, K.-y., Shimizu, S., Onoda, M., and Fujioka, Y. (2009). Heats of reaction and vapor–liquid equilibria of novel chemical absorbents for absorption/recovery of pressurized carbon dioxide in integrated coal gasification combined cycle–carbon capture and storage process. *Industrial & Engineering Chemistry Research* 49 (3): 1214–1221.

115 He, X. and Hägg, M.-B. (2014). Energy efficient process for CO2 capture from flue gas with novel fixed-site-carrier membranes. *Energy Procedia* 63: 174–185.

116 He, X., Fu, C., and Hagg, M.-B. (2015). Membrane system design and process feasibility analysis for CO2 capture from flue gas with a fixed-site-carrier membrane. *Chemical Engineering Journal (Amsterdam, Netherlands)* 268: 1–9.

117 Sandru, M., Haukeboe, S.H., and Haegg, M.-B. (2010). Composite hollow fiber membranes for CO2 capture. *Journal of Membrane Science* 346 (1): 172–186.

118 Huang, J., Zou, J., and Ho, W.S.W. (2008). Carbon dioxide capture using a CO2-selective facilitated transport membrane. *Industrial and Engineering Chemistry Research* 47 (4): 1261–1267.

119 Guha, A.K., Majumdar, S., and Sirkar, K.K. (1990). Facilitated transport of carbon dioxide through an immobilized liquid membrane of aqueous diethanolamine. *Industrial and Engineering Chemistry Research* 29 (10): 2093–2100.

120 Yegani, R., Hirozawa, H., Teramoto, M. et al. (2007). Selective separation of CO2 by using novel facilitated transport membrane at elevated temperatures and pressures. *Journal of Membrane Science* 291 (1+2): 157–164.

121 Zhang, Y., Wang, Z., and Wang, S.C. (2002). Selective permeation of CO2 through new facilitated transport membranes. *Desalination* 145 (1–3): 385–388.

122 Matsuyama, H., Teramoto, M., Matsui, K., and Kitamura, Y. (2001). Preparation of poly(acrylic acid)/poly(vinyl alcohol) membrane for the facilitated transport of CO2. *Journal of Applied Polymer Science* 81 (4): 936–942.

123 El-Azzami, L.A. and Grulke, E.A. (2008). Carbon dioxide separation from hydrogen and nitrogen by fixed facilitated transport in swollen chitosan membranes. *Journal of Membrane Science* 323 (2): 225–234.

124 Saedi, S., Nikravesh, B., Seidi, F. et al. (2015). Facilitated transport of CO2 through novel imidazole-containing chitosan derivative/PES membranes. *RSC Advances* 5 (82): 67299–67307.

125 Zhao, Y. and Ho, W.S.W. (2013). CO2-selective membranes containing sterically hindered amines for CO2/H2 separation. *Industrial and Engineering Chemistry Research* 52 (26): 8774–8782.

126 Zhao, Y. and Ho, W.S.W. (2012). Steric hindrance effect on amine demonstrated in solid polymer membranes for CO2 transport. *Journal of Membrane Science* 415-416: 132–138.

127 Yamaguchi, T., Boetje, L.M., Koval, C.A. et al. (1995). Transport properties of carbon dioxide through amine functionalized carrier membranes. *Industrial and Engineering Chemistry Research* 34 (11): 4071–4077.

128 Way, J.D., Noble, R.D., Reed, D.L. et al. (1987). Facilitated transport of CO2 in ion exchange membranes. *AIChE Journal* 33 (3): 480–487.

129 LeBlanc, O.H. Jr., Ward, W.J., Matson, S.L., and Kimura, S.G. (1980). Facilitated transport in ion-exchange membranes. *Journal of Membrane Science* 6 (3): 339–343.

130 Langevin, D., Pinoche, M., Selegny, E. et al. (1993). Carbon dioxide facilitated transport through functionalized cation-exchange membranes. *Journal of Membrane Science* 82 (1–2): 51–63.

131 Yamaguchi, T., Koval, C.A., Nobel, R.D., and Bowman, C. (1996). Transport mechanism of carbon dioxide through perfluorosulfonate ionomer membranes containing an amine carrier. *Chemical Engineering Science* 51 (21): 4781–4789.

132 Kasahara, S., Kamio, E., Yoshizumi, A., and Matsuyama, H. (2014). Polymeric ion-gels containing an amino acid ionic liquid for facilitated CO_2 transport media. *Chemical Communications (Cambridge, United Kingdom)* 50 (23): 2996–2999.

133 McDanel, W.M., Cowan, M.G., Chisholm, N.O. et al. (2015). Fixed-site-carrier facilitated transport of carbon dioxide through ionic-liquid-based epoxy-amine ion gel membranes. *Journal of Membrane Science* 492: 303–311.

134 Scovazzo, P., Visser, A.E., Davis, J.H. Jr., et al. (2002). Supported ionic liquid membranes and facilitated ionic liquid membranes. *ACS Symposium Series* 818 (Ionic liquids): 69–87.

135 Myers, C., Pennline, H., Luebke, D. et al. (2008). High temperature separation of carbon dioxide/hydrogen mixtures using facilitated supported ionic liquid membranes. *Journal of Membrane Science* 322 (1): 28–31.

136 Luis, P., Neves, L.A., Afonso, C.A.M. et al. (2009). Facilitated transport of CO_2 and SO_2 through supported ionic liquid membranes (SILMs). *Desalination* 245 (1–3): 485–493.

137 Hanioka, S., Maruyama, T., Sotani, T. et al. (2008). CO_2 separation facilitated by task-specific ionic liquids using a supported liquid membrane. *Journal of Membrane Science* 314 (1+2): 1–4.

138 Kasahara, S., Kamio, E., Ishigami, T., and Matsuyama, H. (2012). Amino acid ionic liquid-based facilitated transport membranes for CO_2 separation. *Chemical Communications (Cambridge, United Kingdom)* 48 (55): 6903–6905.

139 Kasahara, S., Kamio, E., Ishigami, T., and Matsuyama, H. (2012). Effect of water in ionic liquids on CO_2 permeability in amino acid ionic liquid-based facilitated transport membranes. *Journal of Membrane Science* 415–416 (0): 168–175.

140 Kasahara, S., Kamio, E., Otani, A., and Matsuyama, H. (2014). Fundamental investigation of the factors controlling the CO_2 permeability of facilitated transport membranes containing amine-functionalized task-specific ionic liquids. *Industrial and Engineering Chemistry Research* 53 (6): 2422–2431.

141 Chang, J., Hong, G.H., and Kang, S.W. (2015). Highly permeable ionic liquid membrane by both facilitated transport and the increase of diffusivity through porous materials. *RSC Advances* 5 (85): 69698–69701.

142 Zhou, H., Xie, J., and Ban, S. (2015). Insights into the ultrahigh gas separation efficiency of lithium doped carbon nanotube membrane using carrier-facilitated transport mechanism. *Journal of Membrane Science* 493: 599–604.

143 Liao, J., Wang, Z., Gao, C. et al. (2015). A high performance PVAm-HT membrane containing high-speed facilitated transport channels for CO_2 separation. *Journal of Materials Chemistry A* 3 (32): 16746–16761.

144 Shen, Y., Wang, H., Liu, J., and Zhang, Y. (2015). Enhanced performance of a novel polyvinyl amine/chitosan/graphene oxide mixed matrix membrane for CO_2 capture. *ACS Sustainable Chemistry & Engineering* 3 (8): 1819–1829.

145 Li, P., Wang, Z., Li, W. et al. (2015). High-performance multilayer composite membranes with mussel-inspired Polydopamine as a versatile molecular bridge for CO_2 separation. *ACS Applied Materials & Interfaces* 7 (28): 15481–15493.

146 Ansaloni, L., Zhao, Y., Jung, B.T. et al. (2015). Facilitated transport membranes containing amino-functionalized multi-walled carbon nanotubes for high-pressure CO2 separations. *Journal of Membrane Science* 490: 18–28.

147 Zhao, S., Cao, X., Ma, Z. et al. (2015). Mixed-matrix membranes for CO2/N2 separation comprising a poly(vinylamine) matrix and metal-organic frameworks. *Industrial and Engineering Chemistry Research* 54 (18): 5139–5148.

148 Xin, Q., Ouyang, J., Liu, T. et al. (2015). Enhanced interfacial interaction and CO2 separation performance of mixed matrix membrane by incorporating Polyethylenimine-decorated metal-organic frameworks. *ACS Applied Materials & Interfaces* 7 (2): 1065–1077.

149 Matsuyama, H., Terada, A., Nakagawara, T. et al. (1999). Facilitated transport of CO2 through polyethylenimine/poly(vinyl alcohol) blend membrane. *Journal of Membrane Science* 163 (2): 221–227.

150 Kovvali, A.S., Chen, H., and Sirkar, K.K. (2000). Dendrimer membranes: a CO2-selective molecular gate. *Journal of the American Chemical Society* 122 (31): 7594–7595.

151 Yao, K., Wang, Z., Wang, J., and Wang, S. (2012). Biomimetic material-poly(N-vinylimidazole)-zinc complex for CO2 separation. *Chemical Communications* 48 (12): 1766–1768.

152 Bao, L. and Trachtenberg, M.C. (2006). Facilitated transport of CO2 across a liquid membrane: comparing enzyme, amine, and alkaline. *Journal of Membrane Science* 280 (1+2): 330–334.

153 Zhang, Y.-T., Dai, X.-G., Xu, G.-H. et al. (2012). Modeling of CO2 mass transport across a hollow fiber membrane reactor filled with immobilized enzyme. *AIChE Journal* 58 (7): 2069–2077.

154 Cowan, R.M., Ge, J.J., Qin, Y.J. et al. (2003). CO2 capture by means of an enzyme-based reactor. *Annals of the New York Academy of Sciences* 984 (Advanced Membrane Technology): 453–469.

155 Alvizo, O., Nguyen, L.J., Savile, C.K. et al. (2014). Directed evolution of an ultrastable carbonic anhydrase for highly efficient carbon capture from flue gas. *Proceedings of the National Academy of Sciences* 111 (46): 16436–16441.

156 Favre, N. and Pierre, A.C. (2011). Synthesis and behavior of hybrid polymer-silica membranes made by sol gel process with adsorbed carbonic anhydrase enzyme, in the capture of carbon dioxide. *Journal of Sol-Gel Science and Technology* 60 (2): 177–188.

157 Lindskog, S. (1997). Structure and mechanism of carbonic anhydrase. *Pharmacology & Therapeutics* 74 (1): 1–20.

158 Lovejoy, D.A., Hewett-Emmett, D., Porter, C.A. et al. (1998). Evolutionarily conserved, "Acatalytic" carbonic anhydrase-related protein XI contains a sequence motif present in the neuropeptide Sauvagine: the human CA-RP XI gene (CA11) is embedded between the secretor gene cluster and the DBP gene at 19q13.3. *Genomics* 54 (3): 484–493.

159 McCabe, W.L., Smith, J.C., and Harriott, P. (2001). *Unit Operations of Chemical Engineering*, 6e. New York: McGraw-Hill.

160 Bottoms, R.R. (1930). Process for separating acidic gases. US patent 1,783,901, filed 7 October 1930 and issued 2 December 1930.

161 Gabelman, A. and Hwang, S.-T. (1999). Hollow fiber membrane contactors. *Journal of Membrane Science* 159 (1–2): 61–106.

162 Li, J.-L. and Chen, B.-H. (2005). Review of CO2 absorption using chemical solvents in hollow fiber membrane contactors. *Separation and Purification Technology* 41 (2): 109–122.

163 Groenvold, M.S., Falk-Pedersen, O., Imai, N., and Ishida, K. (2005). KPS membrane contactor module combined with Kansai/MHI advanced solvent, KS-1 for CO2 separation from combustion flue gases. In: *Carbon Dioxide Capture for Storage in Deep Geologic Formations* (ed. D.C. Thomas), 133–155. Elsevier.

164 Herzog, H. and Falk-Pedersen, O. (2001). The Kvaerner membrane contactor: lessons from a case study in how to reduce capture costs. In: *Proceedings of the 5th International Conference on Greenhouse Gas Control Technologies* (eds. D.J. Williams et al.), 121–132. CSIRO.

165 Wang, R., Zhang, H.Y., Feron, P.H.M., and Liang, D.T. (2005). Influence of membrane wetting on CO2 capture in microporous hollow fiber membrane contactors. *Separation and Purification Technology* 46 (1–2): 33–40.

166 Feron, P.H.M. and Jansen, A.E. (2002). CO2 separation with polyolefin membrane contactors and dedicated absorption liquids: performances and prospects. *Separation and Purification Technology* 27 (3): 231–242.

167 De Montigny, D., Tontiwachwuthikul, P., and Chakma, A. (2006). Using polypropylene and polytetrafluoroethylene membranes in a membrane contactor for CO2 absorption. *Journal of Membrane Science* 277 (1–2): 99–107.

168 Zhang, H.-Y., Wang, R., Liang, D.T., and Tay, J.H. (2008). Theoretical and experimental studies of membrane wetting in the membrane gas-liquid contacting process for CO2 absorption. *Journal of Membrane Science* 308 (1+2): 162–170.

169 Yeon, S.-H., Lee, K.-S., Sea, B. et al. (2005). Application of pilot-scale membrane contactor hybrid system for removal of carbon dioxide from flue gas. *Journal of Membrane Science* 257 (1–2): 156–160.

170 Zhang, Y., Wang, R., Zhang, L., and Fane, A.G. (2012). Novel single-step hydrophobic modification of polymeric hollow fiber membranes containing imide groups: its potential for membrane contactor application. *Separation and Purification Technology* 101: 76–84.

171 Zhang, Y., Wang, R., Yi, S. et al. (2011). Novel chemical surface modification to enhance hydrophobicity of polyamide-imide (PAI) hollow fiber membranes. *Journal of Membrane Science* 380 (1–2): 241–250.

172 Li, S., Rocha, D.J., Zhou, S.J. et al. (2013). Post-combustion CO2 capture using super-hydrophobic, polyether ether ketone, hollow fiber membrane contactors. *Journal of Membrane Science* 430: 79–86.

173 Rahbari-Sisakht, M., Ismail, A.F., Rana, D., and Matsuura, T. (2012). A novel surface modified polyvinylidene fluoride hollow fiber membrane contactor for CO2 absorption. *Journal of Membrane Science* 415-416: 221–228.

174 U.S. Department of Energy National Energy Technology Laboratory. (2013). Advanced Carbon Dioxide Capture R&D Program: May 2013 Update.

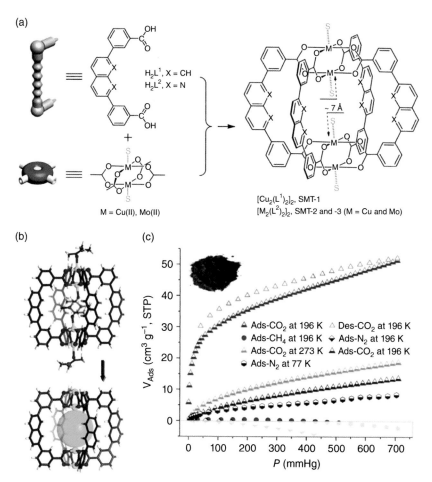

Figure 2.3 (a) Schematic representation of the construction of SMT-1–3: S is the coordinated solvent molecule, which can be removed through sample activation to yield an empty molecular cage for gas adsorption. (b) The molecular structure of SMT-1 with (top) and without (bottom) coordinated solvent molecules. The color schemes: Cu, cyan; O, red; N, blue; C, brown; and H, light gray. The green sphere represents the free space inside the molecular cage. (c) Gas adsorption isotherms of SMT-1, showing selective adsorption of CO_2 over CH_4 and N_2 (inset: a picture of an activated SMT-1 sample). Source: reproduced from Ref. [32] with permission from *Nature Communications*.

Materials for Carbon Capture, First Edition. Edited by De-en Jiang, Shannon M. Mahurin and Sheng Dai.
© 2020 John Wiley & Sons Ltd. Published 2020 by John Wiley & Sons Ltd.

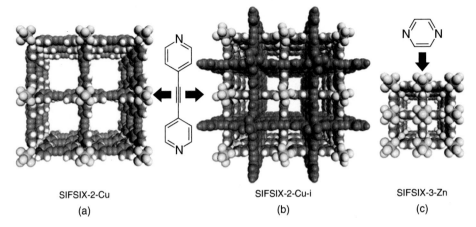

SIFSIX-2-Cu

(a)

SIFSIX-2-Cu-i

(b)

SIFSIX-3-Zn

(c)

Figure 2.4 (a) SIFSIX-2-Cu; BET apparent surface area (N_2 adsorption) 3140 m^2 g^{-1}. (b) SIFSIX-2-Cu-i: BET apparent surface area (N_2 adsorption) 735 m^2 g^{-1}. (c) SIFSIX-3-Zn; apparent surface area (determined from CO_2 adsorption isotherm) 250 m^2 g^{-1}. Color code: C (gray), N (blue), Si (yellow), F (light blue), H (white). All guest molecules are omitted for clarity. Note that the green net represents the interpenetrated net in SIFSIX-2-Cu-i. The nitrogen-containing linker present in SIFSIX-2-Cu and SIFSIX-2-Cu-i is 4,4′-dipyridylacetylene (dpa), whereas that in SIFSIX-3-Zn is pyrazine (pyr). Source: reproduced from Ref. [33] with permission from *Nature*.

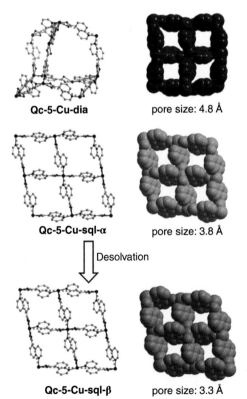

Qc-5-Cu-dia　　pore size: 4.8 Å

Qc-5-Cu-sql-α　　pore size: 3.8 Å

Desolvation

Qc-5-Cu-sql-β　　pore size: 3.3 Å

Figure 2.5 Pore size for Qc-5-Cu-dia, Qc-5-Cu-sql-α, and Qc-5-Cu-sql-β polymorphs: C (gray), Cu (maroon), O (red), N (blue), H (white). Source: reproduced from Ref. [34] with permission from *Angewandte Chemie International Edition*.

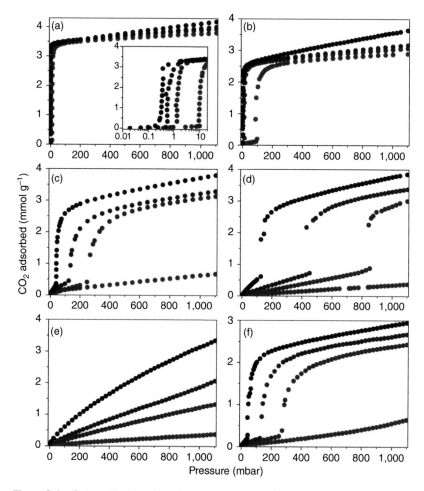

Figure 2.6 Carbon dioxide adsorption isotherms at 25 °C (blue), 40 °C (blue-violet), 50 °C (red-violet), and 75 °C (red) for (a) mmen-Mg$_2$(dobpdc); (b) mmen-Mn$_2$(dobpdc); (c) mmen-Fe$_2$(dobpdc); (d) mmen-Co$_2$(dobpdc); (e) mmen-Ni$_2$(dobpdc); and (f) mmen-Zn$_2$(dobpdc). Source: reproduced from Ref. [43] with permission from *Nature*.

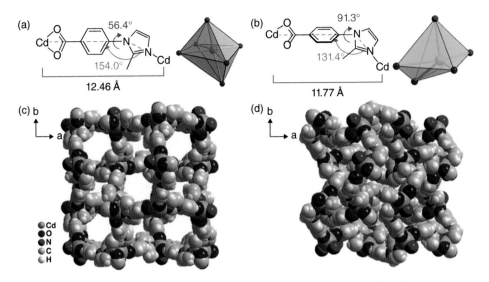

Figure 2.7 Cd···Cd distances, bond angles, and dihedral angles of the miba ligand and the coordination environment of the Cd(II) atoms of (a) 1a and (b) 1b. 3D framework of 1a (c) and 1b (d) [66]. Source: adapted by permission of the American Chemical Society.

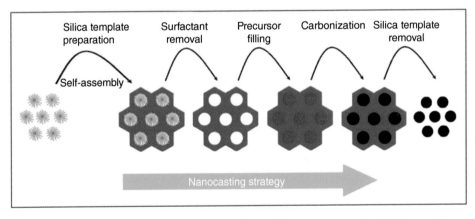

Figure 3.1 Typical method for the preparation of ordered mesoporous carbon (OMC) materials: the nanocasting strategy from mesoporous silica hard templates. Source: reproduced from Ref. [14] with permission from the Royal Society of Chemistry.

Figure 3.9 Schematic of the rapid synthesis of mesoporous carbons with lysine as the catalyst. Source: reproduced from Ref. [63] with permission from the Royal Society of Chemistry.

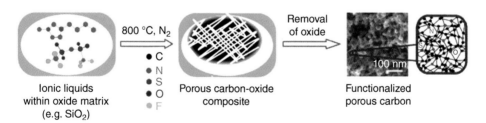

Figure 3.19 Schematic illustration of synthesis of ionic liquids (IL)-derived carbon materials. Source: reproduced from Ref. [109] with permission from John Wiley & Sons.

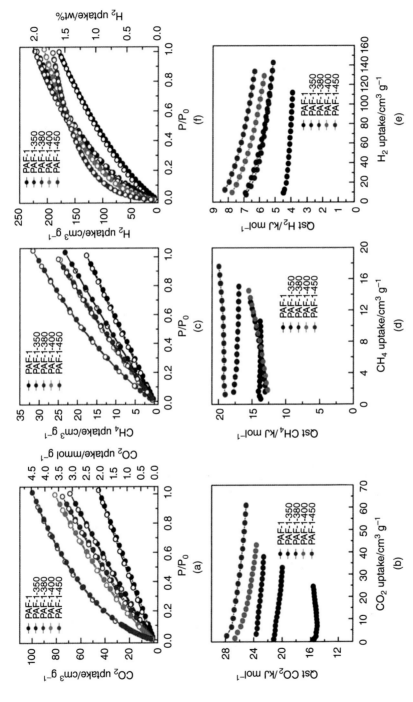

Figure 3.24 (a) CO_2 adsorption (solid symbols) and desorption (open symbols) isotherms of PAF-1 and carbonized samples at 273 K; (b) Q_{st} CO_2 of PAF-1 and carbonized samples as a function of the amount of CO_2 adsorbed. (c) CH_4 adsorption (solid symbols) and desorption (open symbols) isotherms of PAF-1 and carbonized samples at 273 K; (d) Q_{st} CH_4 of PAF-1 and carbonized samples as a function of the amount of CH_4 adsorbed. (e) H_2 adsorption (solid symbols) and desorption (open symbols) isotherms of PAF-1 and carbonized samples at 77 K; (f) Q_{st} H_2 of PAF-1 and carbonized samples as a function of the amount of H_2 adsorbed. Source: reproduced from Ref. [140] with permission from the Royal Society of Chemistry.

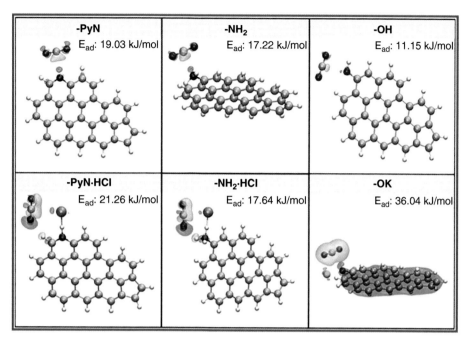

Figure 3.29 Optimized configurations of CO_2 adsorption on carbon clusters with different polar groups (cyan, C; white, H; blue, N; red, O; light green, K) and corresponding contour plots of the differential charge density. The contour value is ± 0.001 au. The purple and lime regions represent the charge accumulation and charge depletion regions, respectively. Source: reproduced from Ref. [185] with permission from the American Chemical Society.

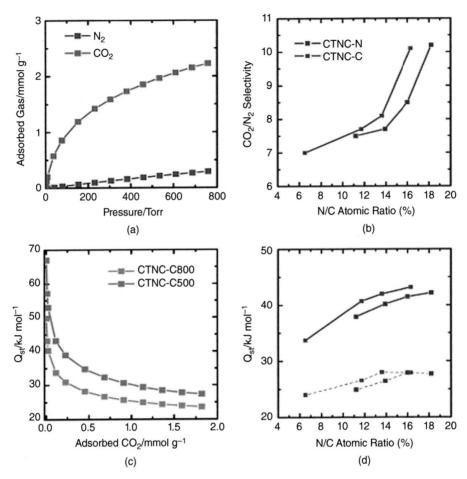

Figure 3.30 (a) Comparison of 25 °C adsorption isotherms of CO_2 and N_2 for CTNC-N700; (b) correlation between CO_2/N_2 selectivity and N/C atomic ratio; (c) isosteric heats of CO_2 adsorption (Q_{st}) by CTNC-C500 and CTNC-C800; (d) correlation between Q_{st} and N/C atomic ratio at different CO_2 coverages (red: CTNC-N; blue: CTNC-C; solid line: 0.1 mmol g^{-1} of CO_2 adsorbed; dash line: 1.8 mmol g^{-1} of CO_2 adsorbed). Source: reproduced from Ref. [186] with permission from the Royal Society of Chemistry.

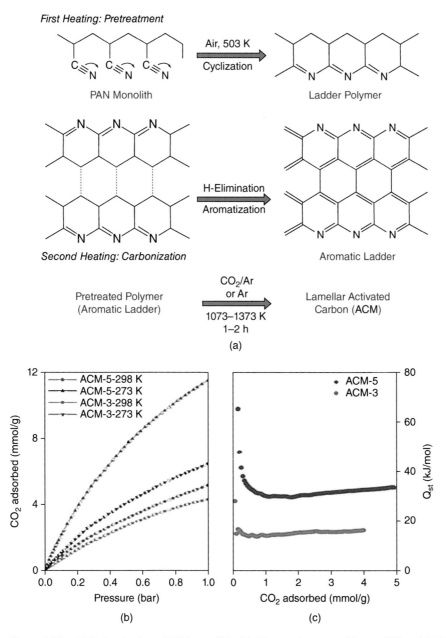

Figure 3.31 (a) Carbonization of PAN monolith; (b) CO_2 sorption up to 1 bar, at 273 and 25 °C: adsorption (filled symbols) and desorption (empty symbols); (c) isosteric heat of CO_2 adsorption (Q_{st}) as a function of CO_2 adsorbed. Source: reproduced from Ref. [182] with permission from the Royal Society of Chemistry.

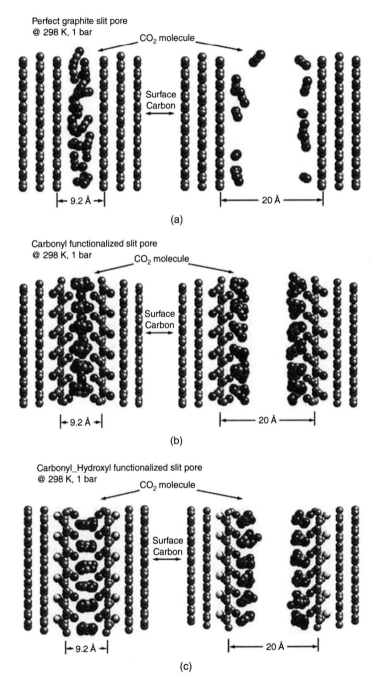

Figure 3.32 Comparisons of CO_2 adsorbed in functionalized micropores with that in the perfect graphite slit pore: left, side views of adsorbed CO_2 in various functionalized graphite slit pores with pore width of 9.2 Å; right, side views of adsorbed CO_2 in various functionalized graphite slit pores with pore width of 20 Å. Source: reproduced from Ref. [196] with permission from the American Chemical Society.

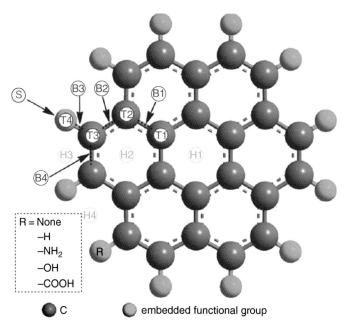

Figure 3.33 Initial configurations of CO_2–CH_4 adsorption on the edge-functionalized basis unit. Nomenclature: H, position above the center of a benzene ring in the basis unit; T, position at the top of the C atom or the atom connected to the functional group; B, position above the bond center; and S, side position in the plane of the functional group. Source: reproduced from Ref. [202] with permission from the Royal Society of Chemistry.

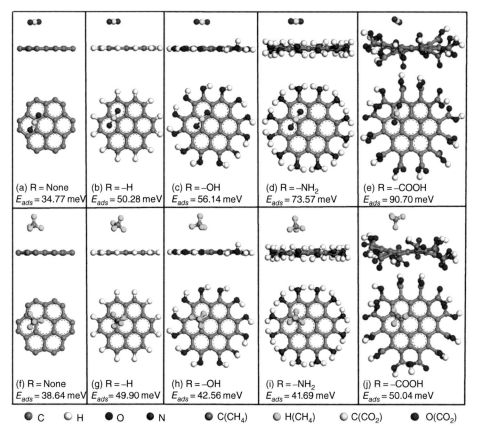

Figure 3.34 Stable adsorption configurations (side view [up] and top view [down]) of (a–e) CO_2 and (f–j) CH_4 on the edge-functionalized basis unit at the B1 site. Source: reproduced from Ref. [202] with permission from the Royal Society of Chemistry.

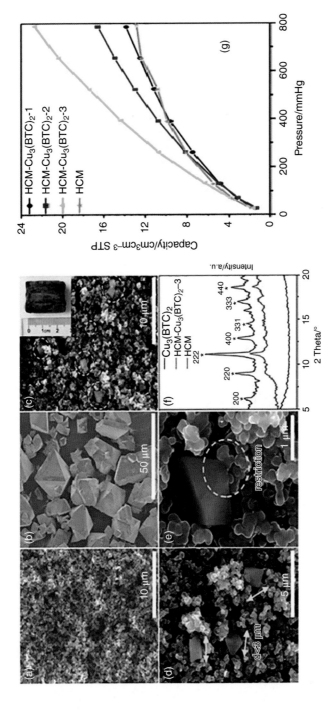

Figure 3.39 SEM micrographs of (a) HCM, (b) $Cu_3(BTC)_2$, and (c–e) HCM-$Cu_3(BTC)_2$-3. (f) XRD patterns of HCM, $Cu_3(BTC)_2$, and HCM-$Cu_3(BTC)_2$-3. (g) CO_2 adsorption isotherms on a volumetric basis. Source: reproduced from Ref. [233] with permission from the American Chemical Society.

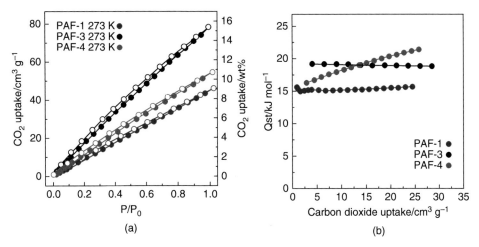

Figure 4.1 (a) CO_2 adsorption (solid symbols) and desorption (open symbols) isotherms of PAF-1, PAF-3, and PAF-4 at 273 K; (b) QstCO$_2$ of PAF-1, PAF-3, and PAF-4 as a function of the amount of CO_2 adsorbed. Source: Reproduced from Ref. [38] with permission from the Royal Society of Chemistry.

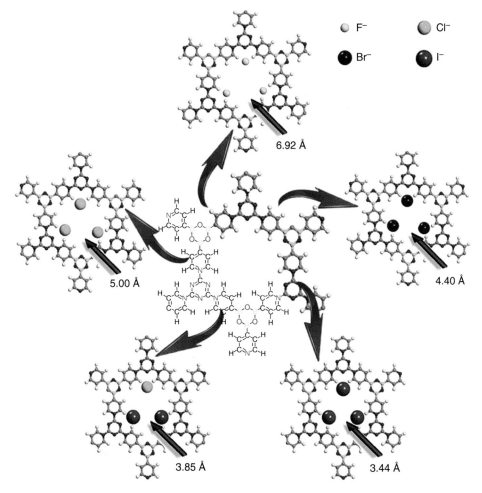

Figure 4.5 Scheme of preparation of F-PAF-50, Br-PAF-50, 2I-PAF-50, and 3I-PAF-50 from Cl-PAF-50. Source: Reproduced from Ref. [43] with permission from Springer Nature.

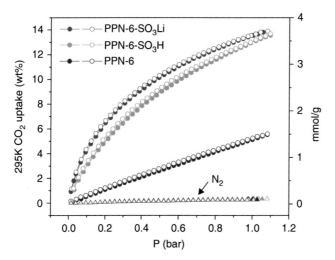

Figure 4.7 Gravimetric CO_2 and N_2 adsorption/desorption isotherms at 295 K. Source: Reproduced from Ref. [44] with permission from the American Chemical Society.

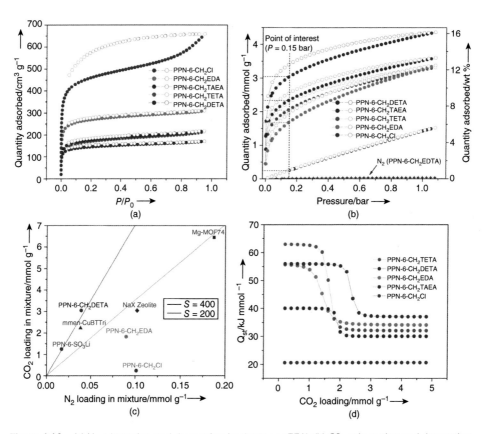

Figure 4.10 (a) N_2 adsorption and desorption isotherms at 77 K; (b) CO_2 adsorption and desorption isotherms, as well as PPN-6-CH$_2$DETA N_2 adsorption, at 295 K; (c) the component loadings of N_2 and CO_2 calculated by IAST with bulk gas-phase partial pressures of 85 kPa and 15 kPa for N_2 and CO_2, respectively, with PPN-6-CH$_2$DETA, PPN-6-CH$_2$EDA, PPN-6-CH$_2$Cl, PPN-6-SO$_3$Li, NaX zeolite, MgMOF-74, mmen-CuBTTri; loadings for calculated selectivities of 200 and 400 are shown as a guide; (d) isosteric heats of adsorption Qst for the adsorption of CO_2, calculated using the dual-site Langmuir isotherm fits. Source: Reproduced from Ref. [46] with permission from John Wiley & Sons.

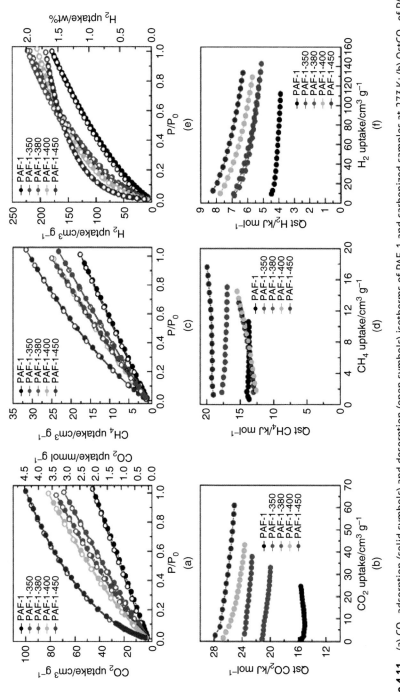

Figure 4.11 (a) CO_2 adsorption (solid symbols) and desorption (open symbols) isotherms of PAF-1 and carbonized samples at 273 K; (b) $QstCO_2$ of PAF-1 and carbonized samples as a function of the amount of CO_2 adsorbed. (c) CH_4 adsorption (solid symbols) and desorption (open symbols) isotherms of PAF-1 and carbonized samples at 273 K; (d) $QstCH_4$ of PAF-1 and carbonized samples as a function of the amount of CH_4 adsorbed. (e) H_2 adsorption (solid symbols) and desorption (open symbols) isotherms of PAF-1 and carbonized samples at 77 K; (f) $QstH_2$ of PAF-1 and carbonized samples as a function of the amount of H_2 adsorbed. Source: Reproduced from Ref. [42] with permission from the Royal Society of Chemistry.

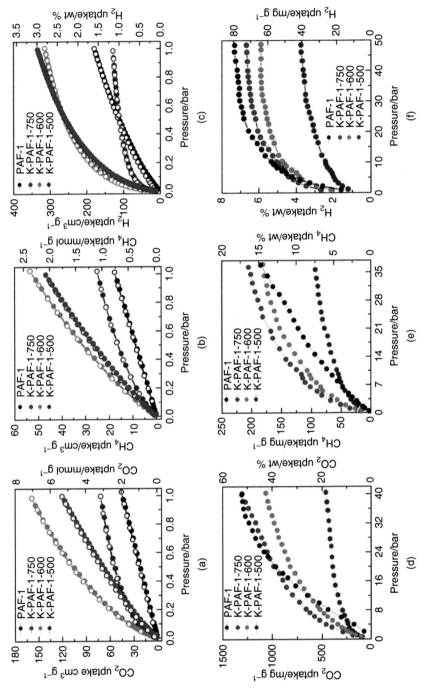

Figure 4.13 (a) CO_2 uptake of PAF-1 and KOH-activated samples at 273 K and 1 bar; (b) CH_4 uptake of PAF-1 and KOH-activated samples at 273 K and 1 bar; (c) H_2 uptake of PAF-1 and KOH-activated samples at 77 K and 1 bar; (d) high-pressure CO_2 uptake of KOH-activated samples at 298 K; (e) high-pressure CH_4 uptake of KOH-activated samples at 298 K; (f) high-pressure H_2 uptake of KOH-activated samples at 77 K. Source: Reproduced from Ref. [47] with permission from Springer Nature.

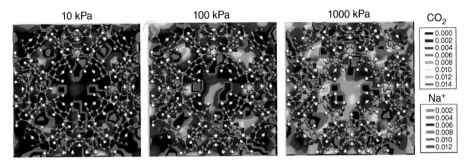

Figure 5.4 Density distribution contours of CO_2 molecules and Na + ions for a 15 : 85 CO_2/H_2 mixture at (left) 10, (center) 100, and (right) 1000 kPa using GCMC simulations. Source: reprinted from Ref. [13] with the permission of the American Chemical Society.

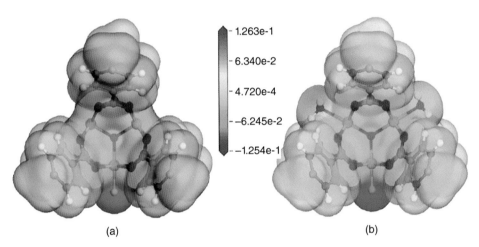

(a)   (b)

Figure 5.5 Electrostatic potential maps by density functional theory (DFT) calculations around the Cr3O trimer in (a) dehydrated and (b) hydrated MIL-101 [19]. These maps were drawn to investigate the adsorption of CO_2 and CH_4 in dehydrated and hydrated MIL-101 and the effect of terminal water molecules on adsorption Reprinted with the permission of the American Chemical Society.

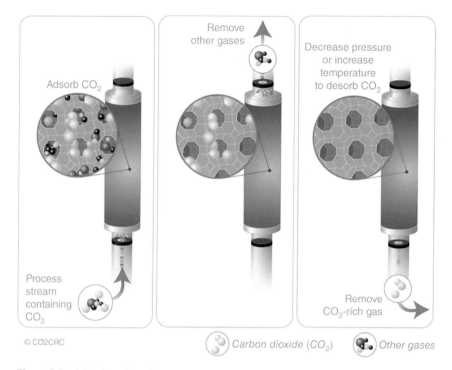

Figure 5.7 Adsorbent-based separation. CO_2 molecules are attracted to the surface of an adsorbent (www.co2crc.com.au/aboutccs/cap_adsorption.html). Source: http://hdl.handle.net/2027.42/90951. Licensed under CCBY 4.0.

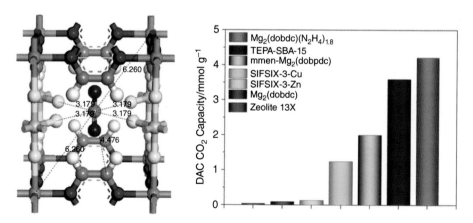

Figure 5.8 (Left) Depiction of a single CO_2 molecule within the pore structure of SIFSIX-3-Cu [38]; (right) comparison of selected MOF materials with zeolite 13X in direct CO_2 capture [39]. Sources: (left) reproduced with the permission of the American Chemical Society; (right) reproduced with the permission of John Wiley & Sons.

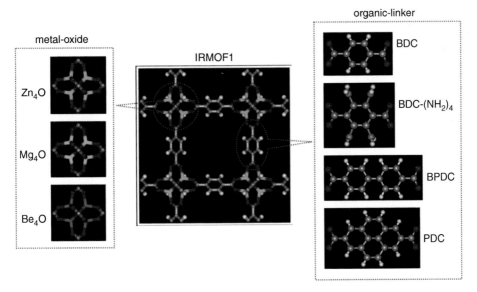

Figure 5.9 Schematic tailoring of the metal oxide and organic linker in IRMOF1. Color code: Zn, green; Mg, cyan; Be, purple; O, red; N, blue; C, gray; H, white. Source: reprinted from Ref. [45] with the permission of the American Chemical Society.

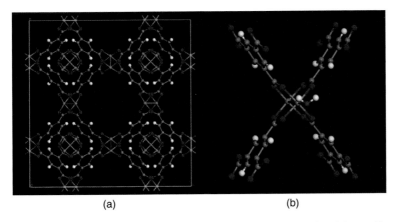

Figure 5.14 (a) Dry Cu-BTC unit cell; (b) hydrated Cu-BTC (4 wt %) with coordinated water molecule from density functional theory (DFT). Color code: Cu, orange; O, red; C, gray; H, white. The oxygen atom of the coordinated water molecule is shown in blue. Source: reprinted from Ref. [72] with the permission of the American Chemical Society.

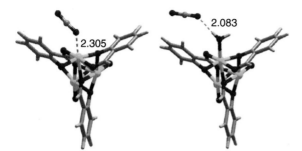

Figure 5.15 Optimized structures of the Mg-MOF-74 cluster with (right) and without (left) water coordination interacting with CO_2. Color code: Mg, green; O, red; C, gray; H, lavender. Source: reprinted from Ref. [74] with the permission of the American Chemical Society.

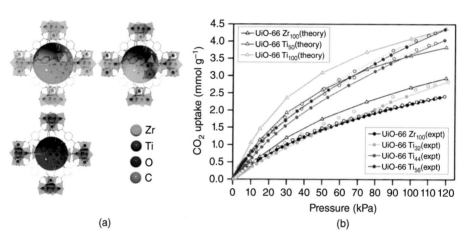

(a)   (b)

Figure 5.17 (a) The Zr-based metal organic framework (MOF) UiO-66 with postsynthetic exchange with Ti(IV) to deliver heterometallic MOFs, with decreased octahedral cages sizes; (b) CO_2 uptake as a function of pressure and Ti concentration. Source: reprinted from Ref. [80] with the permission of the Royal Society of Chemistry.

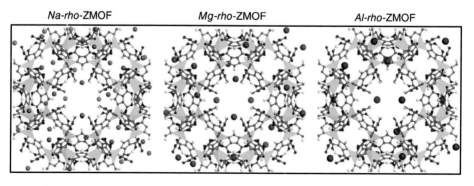

Figure 5.18 Atomic structures of rho-ZMOFs. Color code: In, cyan; N, blue; C, gray; O, red; H, white; Na^+, purple; Mg^{2+}, pink; Al^{3+}, dark blue. Source: reprinted from Ref. [81] with the permission of Elsevier.

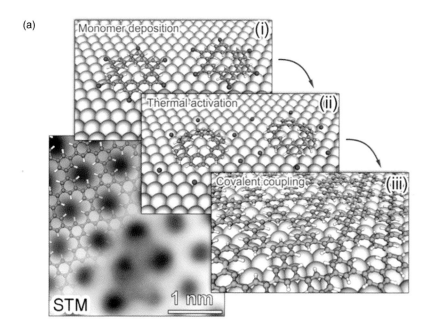

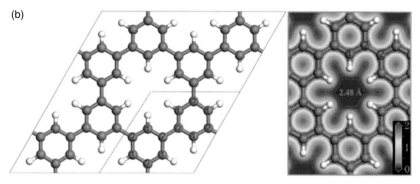

Figure 6.8 (a) Synthetic route of the surface-promoted aryl-aryl coupling to fabricate poly-phenylene and its STM image supported on a Ag(111) surface. Source: reproduced from Ref. [62] with permission from 2010, John Wiley and Sons. (b) The 2*2 supercell of poly-phenylene and the electron density of a nanopore. A unit cell was indicated by dash line in the supercell. Color code: C, green; H, white. Source: reproduced from Ref. [63] with permission from the Royal Society of Chemistry.

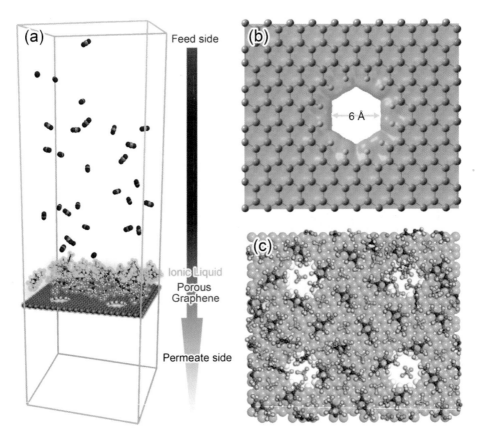

Figure 6.15 (a) A bichamber system for simulation of gas permeance through the composite membrane; top and bottom sides of the box are impermeable. (b) Structure of a hydrogen-terminated, 6 Å pore inside the graphene layer with an electronic density contour. (c) Top view of the composite membrane from the feed side: a monolayer of the [emim][BF₄] IL on the porous graphene. Color code in (c): C, gray; H, white; N, blue; B, red; F, green. Cations are highlighted with a blue edge, anions with red. Source: reproduced from Ref. [113] with permission from 2017, American Chemical Society.

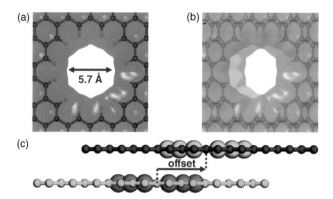

Figure 6.17 Construction of a porous bilayer graphene membrane: (a) the single-layer graphene with a larger pore than the sizes of small gas molecules, such as CO_2; (b) the bilayer nanoporous graphene membrane from stacking two single-layer membranes from (a); (c) side view of the bilayer nanoporous graphene membrane that shows the offset of the pore centers. Isosurfaces of electron density are shown in (a) and (b) to define the pore shape, and larger spheres in (c) indicate the pore rims. Source: reproduced from Ref. [114] with permission from the American Chemical Society.

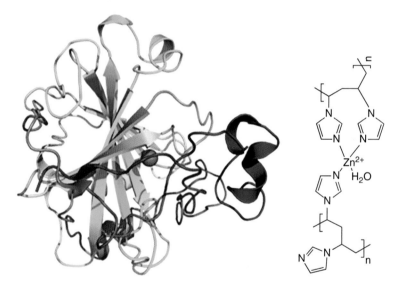

Figure 7.13 Depiction of CA as a ribbon diagram (left) and chemical structure of poly(vinylimidazole)-based mimic studied in Ref. [151] (right). Source: reprinted with permission from the Royal Society of Chemistry.

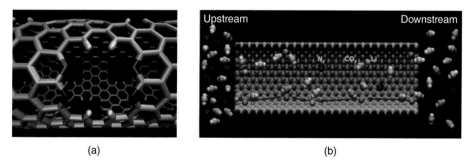

Figure 8.6 (a) Nanopores on a (19,0) zigzag single-walled CNT. Source: reprinted from Ref. [39] with permission from the American Chemical Society. (b) Schematic illustration for facilitated separation of CO_2 from N_2 by CNTs incorporated with Li + inside cylindrical nanopores. Source: reprinted from Ref. [40] with permission from Elsevier.

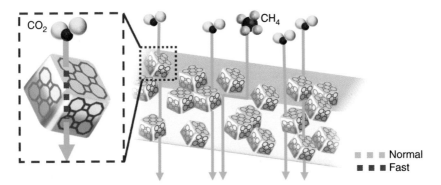

Figure 9.5 Schematic illustration of a mixed-matrix membrane with zeolite fillers.

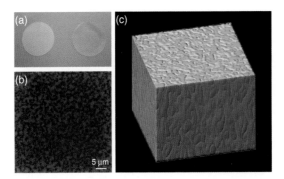

Figure 10.6 Structural analysis of poly(amidoamine)/poly(ethylene glycol) (PAMAM/PEG) membranes prepared by photopolymerization. Average PEG unit is 14. (a) Left: with PAMAM dendrimer; right: without the dendrimer; $\varphi = 1.0$ cm. (b) A fluorescent image by LSCM of PAMAM/PEG membrane (50/50 by wt.) at 50 μm depth from the surface. (c) A reconstructed 3D image of PAMAM/PEG membrane (50/50 by wt.) in $35 \times 35 \times 30$ μm. PEG-rich, and the dendrimer-rich phases are colored in green and yellow, respectively.

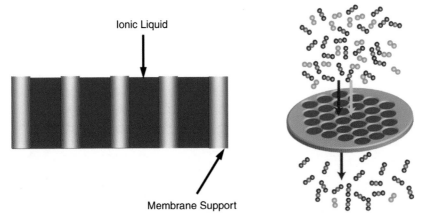

Figure 12.4 (Left) Graphic depicting the integrations of an ionic liquid (IL) in a supported ionic liquid membrane (SILM); (right) the selective transport of gas through the SILM.

8

Carbon Membranes for CO_2 Separation

Kuan Huang[1] and Sheng Dai[2,3]

[1]Key Laboratory of Poyang Lake Environmental and Resources Utilization of Ministry of Education, School of Resources Environmental and Chemical Engineering, Nanchang University, Nanchang, Jiangxi, China
[2]Chemical Sciences Division, Oak Ridge National Laboratory, Oak Ridge, TN, USA
[3]Department of Chemistry, University of Tennessee, Knoxville, TN, USA

CHAPTER MENU

8.1 Introduction

Generally, CO_2 capture technologies can be classified into three types: absorption in liquids [1], adsorption on solids [2], and membrane separation [3]. Among the three types, membrane separation is particularly attractive because it is much lower in energy consumption, operation cost, and capital investment than the other two. In membrane separation, CO_2 is permitted to pass through the thin barrier faster than other gas components, resulting in the selective elimination of CO_2 from industrial streams. Polymeric membranes have been the most extensively investigated membranes for CO_2 separation in past decades, because they exhibit excellent mechanical strength and industrial processability [4]. However, polymeric membranes suffer from low chemical and thermal stability, and undesirable plasticization in a CO_2 atmosphere. Moreover, the CO_2 separation performance of polymeric membranes is limited by a trade-off relationship between permeability and selectivity [5]. That is, polymeric membranes with high permeability normally have low selectivity, and vice versa. To this end, the development of new membranes with not only desirable physiochemical properties but also high performance for CO_2 separation is very intriguing.

Carbon membranes with intrinsic porosity have been regarded as a class of promising alternatives to polymeric membranes for CO_2 separation application [6]. They have

Materials for Carbon Capture, First Edition. Edited by De-en Jiang, Shannon M. Mahurin and Sheng Dai.
© 2020 John Wiley & Sons Ltd. Published 2020 by John Wiley & Sons Ltd.

many unique features such as high chemical and thermal stability, and are applicable for long-term use under harsh conditions (e.g. high temperature and pressure). Most importantly, the pore size and chemical functionality of carbon membranes can be easily tuned by making use of the diverse precursors and routes for carbon synthesis. To date, many state-of-the-art carbon membranes have been reported, including graphene membranes [7], carbon nanotube (CNT) membranes [8], and carbon molecular sieve (CMS) membranes [6]. In this chapter, we will discuss the fundamentals of, and recent progress, in CO₂ separation by carbon membranes. Carbon materials have also been extensively used as the fillers for polymeric membranes to fabricate mixed-matrix membranes (MMMs), in which the advantages of carbons and polymers are simultaneously attained [9, 10]. Since the major frameworks of MMMs are polymers, the use of carbon materials as the fillers for polymers is not included in the present chapter.

8.2 Theory

To start with, it is necessary to provide some basic knowledge about membrane-based gas separation. Typically, there are three mechanism for the separation of gases in membranes, as depicted in Figure 8.1. In dense membranes (e.g. polymeric membranes), the transportation of gases in membranes is subject to a solution-diffusion mechanism: gases dissolve in membranes at the feed side, then diffuse across the dense layer, and finally are released at the permeate side. Therefore, the selective separation of gases by dense membranes is a combining result of solubility difference and diffusivity difference. In porous membranes (e.g. carbon membranes), the selective separation of gases is a result of a size-sieving or adsorption-diffusion effect, depending of the pore size of the membranes. When the pore size of the membranes is larger than the kinetic diameter of gas i, but smaller than the kinetic diameter of gas j, gas j will be prevented from passing through the pore channels. When the pore size of the membranes is larger than the kinetic diameters of both gases, the gases are adsorbed on the pore walls at the feed side, then diffuse across the open channels, and finally are released at the permeate side.

Permeability and selectivity are two of most important parameters for the evaluation of gas-separation performance of membranes. The permeability of gas component i (P_i) is calculated by Eq. (8.1):

$$P_i = \frac{Q_i l}{A \Delta p} \tag{8.1}$$

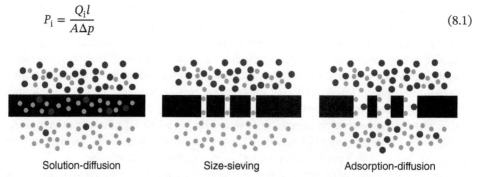

| Solution-diffusion | Size-sieving | Adsorption-diffusion |

Figure 8.1 Three typical mechanisms for separation of gases in membranes.

where Q_i is the gas flux in cm^3 (STP)·s^{-1}, l is the membrane thickness in cm, A is the membrane area in cm^2, and Δp is the transmembrane pressure difference in cm Hg. The gas permeability is often reported in barrer (1 barrer = 10^{-10} cm^3 (STP) cm cm^{-2} s^{-1} cm Hg^{-1}). The gas permeance ($\frac{P_i}{l}$) is also practically used, and often reported in gas permeation units (GPUs, 1 GPU = 10^{-6} cm^3 (STP)·cm^{-2} s^{-1} cm Hg^{-1}). The selectivity of gas pairs i and j (α_{ij}) is calculated by Eq. (8.2):

$$\alpha_{ij} = \frac{P_i}{P_j} \tag{8.2}$$

In solution-diffusion process, the gas permeability equals the product of solubility coefficient (S_i) and diffusion coefficient (D_i), as expressed by Eq. (8.3):

$$P_i = S_i D_i \tag{8.3}$$

Therefore, the selectivity of gas pairs i and j (α_{ij}) can also be calculated by Eq. (8.4):

$$\alpha_{ij} = \frac{S_i}{S_j} \cdot \frac{D_i}{D_j} \tag{8.4}$$

8.3 Graphene Membranes

Graphene is a kind of two-dimensional material with carbon atoms uniformly assembled in aromatic rings through sp^2 hybridization to form a planar sheet [11]. Owing to its distinctive structure and properties, graphene has attracted tremendous attention and displays potential applications in many fields such as energy [12], catalysis [13], and separation [14]. The emergence of graphene provides a great opportunity for the fabrication of membranes with atomic-scale thickness [7]. However, perfect graphene is impermeable to gases, because the electron density of the planar sheet is substantial, and the stacking of the aromatic rings is very tight. Therefore, it is necessary to create defective nanopores on the planar sheet or expand the interlayer distance of the graphene. Figure 8.2 shows the two possible pathways for gas transportation in graphene membranes.

Early in 2009, Jiang et al. first envisioned the potential use of nanoporous graphene as the ultimate membranes for gas separation [15]. They theoretically designed two single-layered nanoporous graphene membranes: one has a nanopore with dimensions 3.0 × 3.8 Å and

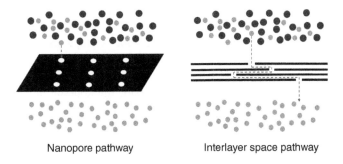

Nanopore pathway   Interlayer space pathway

Figure 8.2 Two possible pathways for gas transportation in graphene membranes.

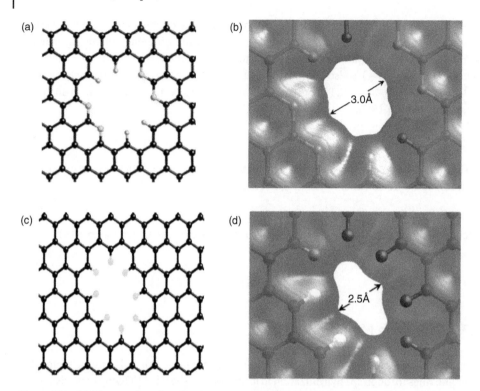

Figure 8.3 (a) Nitrogen-functionalized pore on a graphene nanosheet (black: C, green: N, cyan: H). (b) Pore electron density isosurface of nitrogen-functionalized porous graphene (isovalue: 0.02 e Å^{-3}). (c) All-hydrogen passivated pore on a graphene nanosheet (black: C, cyan: H). (d) Pore electron density isosurface of hydrogen-passivated porous graphene (isovalue: 0.02 e Å^{-3}). Source: reprinted from Ref. [15] with permission from the American Chemical Society.

nitrogen functionality, and another has a nanopore with dimensions 2.5 × 3.8 Å and hydrogen passivation (see Figure 8.3). Through first-principle density functional calculations, it was demonstrated that the nanopores with angstrom size impose different barriers for the permeance of H$_2$ and CH$_4$. As a result, the designed graphene membranes display exceptionally high H$_2$/CH$_4$ selectivity on the order of 10^8~10^{23}. Inspired by this work, various nanoporous graphene membranes were theoretically investigated for CO$_2$ separation [16–21]. It was found that the CO$_2$ separation performance of nanoporous graphene membranes is greatly affected by the pore size, pore density, and pore chemistry. By manipulating these pore parameters, graphene membranes with high performance for CO$_2$ separation can be rationally designed. For example, Bai et al. [18] anticipated that graphene membranes containing 12-member-ring nanopores functionalized with nitrogen have CO$_2$ permeance of 10^5~10^6 GPU and CO$_2$/CH$_4$ selectivity of up to 100, far surpassing the Robeson upper bound for polymeric membranes.

Defective nanopores can be in situ formed during the synthesis of graphene [22] or post-synthetically drilled using various etching technique [23]. Although the theoretical basis for CO$_2$ separation by nanoporous graphene membranes has been well established, it is still challenging to create nanopores with precisely controlled pore parameters, resulting in the unsatisfying performance of fabricated graphene membranes. For example, Strano et al. [24] prepared a series of single-layered graphene membranes with nanopores with a mean size of 2.5 nm spontaneously formed on the planar sheet by chemical vapor deposition (CVD). The permeance of CO_2 was found to be even lower than that of CH_4 in these graphene membranes, because such nanopores have no size-sieving function for gases, and the transportation of gases is subject to the Knudsen diffusion mechanism. According to the Knudsen diffusion mechanism, gas molecules tend to collide with the pore walls more frequently than with each other when the size of the nanopores is smaller than the mean free path of the gas movements. In this case, the separation of gases is based on the difference in velocity for the gas movements, and the permeance of the gases is inversely proportional to the molecular weight of the gases.

In comparison with the creation of nanopores on planar sheet, the expansion of the interlayer distance of graphene is more easily realized. Many methods have been developed to expand the interlayer distance of graphene so far: the most classic one was reported by Hummers et al. [25], through which graphene is oxidized by $H_2SO_4/NaNO_3/KMnO_4$ to result in graphene oxide (GO). GO contains abundant oxygen species (e.g. hydroxyl, carboxyl, and epoxy groups) in the interlayer space, and has the interlayer distance at sub-nanometer scale, which is very suitable for the selective separation of CO_2 from N_2 and CH_4. Moreover, the oxygen species can interact with CO_2 through enhanced hydrogen-bonding or van der Waals interaction, thereby facilitating the transportation of CO_2 in the interlayer space.

The transformation of GO into ordered laminar membranes is a prerequisite for the application of GO in CO_2 separation. In this respect, spin-casting and vacuum-filtration techniques have been widely adopted for the fabrication of GO membranes (see Figure 8.4). Choi et al. [26] prepared few-layered GO membranes with a highly interlocked layer structure by spin casting a GO solution on a polyethersulfone (PES) membrane. The GO membranes exhibit CO_2 permeability of ~8500 barrer and CO_2/N_2 selectivity of ~20 in a

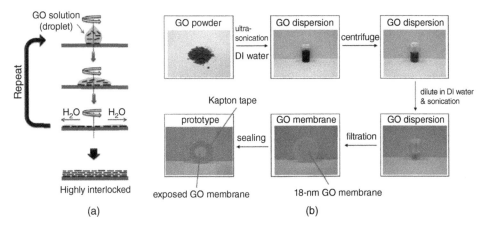

Figure 8.4 (a) Fabrication of GO membranes by spin casting. Source: reprinted with permission from Ref. [26]. Copyright 2013 American Association for the Advancement of Science. (b) Fabrication of GO membranes by vacuum filtration. Source: reprinted from Ref. [28] with permission from the American Association for the Advancement of Science.

dry state, which is better than most reported polymeric membranes. If used in a hydrated state, the GO membranes show much improved CO_2/N_2 selectivity, which is very useful for the capture of CO_2 from flue gas. Park et al. [27] prepared ultrathin GO membranes with a thickness below 5 nm also by spin casting a GO solution on a PES membrane. The GO membranes exhibit CO_2 permeability of ~9800 barrer and CO_2/CH_4 selectivity of ~50 in the presence of water vapor, which is very useful for the capture of CO_2 from natural gas. Instead, Yu et al. [28] fabricated ultrathin GO membranes with a thickness approaching 1.8 nm by vacuum filtration of a GO solution on an anodic aluminum oxide (AAO) support. The GO membranes exhibit high flux for small gases such as H_2 but show poor permeance for CO_2, which is probably a result of their highly stacked structure.

Although pristine GO membranes have shown encouraging performance for CO_2 separation, the oxygen species in the interlayer space of GO provides a great opportunity to adjust the chemical property of GO. This fact enables the rational design of functionalized GO membranes with further improved CO_2 separation performance. Guiver et al. [29] designed and fabricated ultrathin GO membranes using borate as the crosslinker of GO nanosheets and facilitated transportation carrier for CO_2. The GO membranes exhibit CO_2 permeance of up to 650 GPU and CO_2/CH_4 selectivity of 75 in the presence of water vapor. The same group further designed and fabricated GO membranes intercalated with poly(ethylene glycol) diamines (PEGDA) [30]. As demonstrated in Figure 8.5, PEGDA reacts with the epoxy groups in the interlayer space of GO to form CO_2-philic nanodomains, which renders a high capacity for CO_2 adsorption, while the non-CO_2-philic nanodomains render low-friction diffusion for gases. The GO membranes with heterogeneous nanodomains exhibit CO_2 permeance of 175.5 GPU and CO_2/CH_4 selectivity of 69.5 in a dry state. In a similar manner, GO membranes were functionalized with other CO_2-philic compounds such as piperazine [31], ionic liquids [32], polydopamine (PDA) [33], and ethylenediamine [34] by other researchers to achieve high CO_2 separation performance.

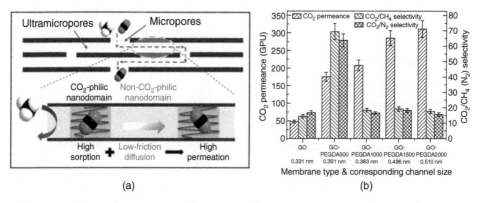

(a) (b)

Figure 8.5 (a) Gas transportation in PEGDA-functionalized GO membranes containing CO_2-philic and non-CO_2-philic nanodomains. (b) Mixed-gas permeation performance of PEGDA-functionalized GO membranes in a dry state (50/50 vol%, total feed pressure 2 bar, 30 °C). Source: reprinted from Ref. [30] with permission from John Wiley & Sons.

8.4 Carbon Nanotube Membranes

A CNT can be regarded as the allotrope of graphene with the planar sheet curled to a cylindrical structure [35]. CNTs have unusual properties compared to graphene, which are valuable for its application in electronics, optics, and other fields of materials science and technology [36]. Theoretical calculations have demonstrated that the diffusion of gases in the cylindrical nanopores of CNTs is much faster than that in polymers, zeolites, and other porous materials, because the interior wall of CNTs are very smooth, resulting in nearly specular reflection when gas molecules collide with the smooth wall [37, 38]. As a consequence, CNTs are a somewhat promising material for the fabrication of high-flux membranes. However, the cylindrical nanopores of CNTs are normally large in diameter and do not have size-sieving functions for gas molecules. To address this issue, Johnson et al. [39] proposed to drill nanopores with angstrom size on the wall of CNTs, while Ban et al. [40] proposed to introduce Li^+ in the cylindrical nanopores of CNTs (see Figure 8.6). Through theoretical calculations, it was anticipated that CNT membranes with such structural design exhibit both high permeability and high selectivity for gases. The (19,0) zigzag single-walled CNTs with two adjacent six-member carbon rings removed, four dangling bonds saturated by hydrogen, and four unsaturated carbon atoms substituted by nitrogen display high performance for the selective separation of CO_2 from CH_4 through the size-sieving effect. Alternatively, mobile Li^+ in the cylindrical nanopores of CNTs can bind with CO_2 strongly and act as the facilitated transportation carrier for CO_2, resulting in dramatically improved selectivity of CO_2/N_2.

However, it is also difficult to prepare CNT membranes with precisely controlled pore parameters and satisfying CO_2 separation performance, just as is the case for nanoporous graphene membranes. Labropoulos et al. [41] prepared CNT membranes with vertically aligned pores at sub-nanometer scale by growing CNTs in the interior of one-dimensional and monodispersed pores of c-oriented aluminophosphate (AlPO) molecular sieve films. Raj et al. [42] prepared CNT membranes by templated thermal cracking of asphaltene inside the pores of anodized alumina. It was demonstrated that the permeance of CO_2 in the CNT

(a)   (b)

Figure 8.6 (a) Nanopores on a (19,0) zigzag single-walled CNT. Source: reprinted from Ref. [39] with permission from the American Chemical Society. (b) Schematic illustration for facilitated separation of CO_2 from N_2 by CNTs incorporated with Li + inside cylindrical nanopores. Source: reprinted from Ref. [40] with permission from Elsevier. (*See color plate section for color representation of this figure*).

membranes prepared by both groups is even lower than the permeance of N_2 and CH_4, and the transportation of gases is subject to the Knudsen diffusion mechanism.

8.5 Carbon Molecular Sieve Membranes

The definition of *molecular sieve* is very broad, and any porous materials with size-sieving function can be classified into this category. In the literature, CMS membranes mainly refer to those with amorphous structure [6], to differentiate them from graphene and CNTs, which have well-defined structure. CMS membranes are normally prepared by the carbonization of polymer precursors at high temperature, which is a rather complex process involving multiple reactions. Although research on CMS membranes dates back to as early as the 1990s [43], the underlying mechanism of the formation of CMS membranes and their chemical and porous structures are still not fully understood. According to the model proposed by Koros et al. [44] (see Figure 8.7), the preparation of CMS membranes comprises four steps: (i) aromatization and fragmentation take place to create carbon strands during the ramp process; (ii) meanwhile, entropy drives the arrangement of carbon strands to form carbon plates due to *excluded volume* effects; (iii) then, carbon plates organize during the final ramp and thermal soak process; (iv) finally, thermal soak and cooling enable the

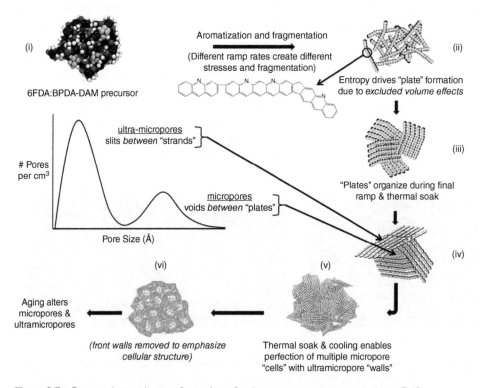

Figure 8.7 Proposed steps in transformation of polymer precursors to amorphous CMS membranes with a bimodal structure in pore-size distribution. Source: reprinted from Ref. [44] with permission from Elsevier.

Table 8.1 Polymer precursors and CO_2 separation performance of CMS membranes.

Categories	Precursors	CO_2 permeability or permeance	CO_2/N_2 selectivity	CO_2/CH_4 selectivity	Refs.
Polyimides	6FDA-TMPDA/azide	280 barrer	31.7	164	[48]
	HBPIs	1085 barrer	—	52	[49]
	PIM-6FDA-OH	512 barrer	—	88	[50]
	6FDA-mPDA/DABA	2610 barrer	—	118	[51]
	Matrimid	~360 barrer	~30	~150	[52]
	Kapton	36.7 barrer	41	—	[53]
	PMDA-ODA	314.8 barrer	34.6	—	[54]
	TB-PI	1406 barrer	37	112	[55]
	ODPA-FDA	1085.69 barrer	143.6	—	[56]
Celluloses	Cellophane paper	17.0 barrer	40	100	[57]
	Ionic liquid-regenerated celluloses	13.4 barrer	83.8	—	[58]
Phenolic resins	Phloroglucinol-formaldehyde	4418 GPU	1.86	—	[59]
	Resorcinol-formaldehyde	33.24 barrer	2.3	—	[60]
	Phenol-formaldehyde	~1380 GPU	~0.6	—	[61]
Others	Sulfonated poly(phenylene oxide)	14.4 GPU	—	173	[62]
	Trimerized acetyl compounds	1149.3 barrer	43.2	—	[63]
	Triazine frameworks	716 barrer	47.5	—	[64]

perfection of multiple micropore cells with ultramicropore walls. The resultant CMS membranes normally feature a bimodal structure for the pore size distribution: the slits between carbon strands constitute ultramicropores (<7 Å) for the size-sieving function, and the voids between carbon plates constitute micropores (7~20 Å) for the adsorption-diffusion function. The CMS membranes can be prepared as flat sheet membranes [45] or hollow fiber membranes [46], and as free-standing membranes [45] or supported membranes [47].

The CO_2 separation performance of CMS membranes is greatly affected by the polymer precursors, and CMS membranes prepared from different polymer precursors exhibit different CO_2 separation performance. Table 8.1 lists the polymer precursors and CO_2 separation performance of some representative CMS membranes reported in the last 10 years. Generally, the polymer precursors can be classified into four categories: polyimides [48–56], celluloses [57, 58], phenolic resins [59–61], and others [62–64]. Among the four categories, polyimides are the most widely used because they have high chain rigidity, melting points, and thermal stability. The polyimides for preparation of CMS membranes are normally synthesized by the polycondensation of aromatic acid dianhydrides and diamines. By varying the monomers and/or adjusting the synthetic conditions for polycondensation, the structure of polyimides can be tuned, enabling the optimization of CO_2 separation performance for the resultant CMS membranes.

Chung et al. [49] designed and synthesized a series of hyperbranched polyimides (HBPIs) by using tris-1,3-bis(3-aminophenoxy) benzene (TrisAPB) as the monomer. The polycondensation of dianhydrides and TrisAPB results in the growth of polyimide chains in three directions. It was found that HBPIs are very useful for the preparation of CMS membranes with high CO_2 separation performance, because they have multiple end groups, high solubility, reduced viscosity, and many open and accessible cavities. Koros et al. [51] compared the CO_2 separation performance of CMS membranes prepared from un-crosslinked and crosslinked 6FDA-mPDA/DABA. It was found that the CMS membranes prepared from crosslinked 6FDA-mPDA/DABA show very high CO_2 permeability and CO_2/CH_4 selectivity, because the decarboxylation reaction taking place during the crosslinking process creates microvoids in the space originally occupied by carboxylic groups, and the microvoids are retained during the subsequent carbonization process. In addition, the CO_2 separation performance of CMS membranes is also influenced by the content of DABA in 6FDA-mPDA/DABA. Specifically, 6FDA-mPDA/DABA with higher DABA content results in CMS membranes with higher CO_2 permeability, owing to the presence of more decarboxylation and crosslinking sites. Recently, Jin et al. [55] synthesized a Tröger base (TB)-derived polyimide and used it as the precursor for CMS membranes. The TB unit constructed from bridged bicyclic amine is highly contorted and shape-persistent, endowing the polyimide with improved chain rigidity and abundant intrinsic microporosity. The prepared CMS membranes were found to exhibit outstanding CO_2 separation performance. In other studies, it was found that PMDA-ODA with lower intrinsic viscosity results in CMS membranes with lower CO_2 permeability but higher CO_2/N_2 selectivity [54], and ODPA-FDA with higher free volume than commercially available Matrimid results in CMS membranes with higher CO_2 permeability and CO_2/N_2 selectivity [56]. The intrinsic viscosity and free volume of polyimides are ultimately determined by their structure and can also be manipulated by changing the monomers and polycondensation conditions.

As another class of polymer precursors that is widely used for the preparation of CMS membranes, celluloses have good biodegradability, low cost, and high carbon yield. However, it is difficult to dissolve them in traditional organic solvents due to the strong inter- and intra-molecular hydrogen bonds existing in celluloses. Regenerated celluloses are solution-processible and can be casted into membranes for carbonization. They are normally produced by a viscose process [65]. Cellophane paper is a typical kind of regenerated cellulose and has been used to prepare CMS membranes with high CO_2/N_2 and CO_2/CH_4 selectivity by one-step carbonization [57]. Unfortunately, the traditional viscose process involves the use of metastable cellulose xanthogenate solution and produces large amounts of hazardous byproducts and waste solutions. To develop a greener process for the preparation of regenerated celluloses and corresponding CMS membranes, Mendes et al. [58] employed ionic liquids to dissolve celluloses, and then casted the solutions into membranes by spin coating and washing with solvents to remove ionic liquids. Ionic liquids are organic salts with melting points below or near room temperature, and they have unique properties such as a wide liquid range, high thermal stability, and negligible volatility [66]. It has been previously demonstrated that ionic liquids exhibit high solubility for celluloses because of the special electron environment within them [67]. Although the CO_2 selectivity of CMS membranes prepared from celluloses is quite impressive, the CO_2 permeability is somewhat poor relative to that of CMS membranes prepared from polyimides. This is probably a result of the underdeveloped porosity in cellulose-derived CMS membranes.

Phenolic resins are often used for the preparation of CMS membranes with ordered mesopores via a soft-template route [59, 60]. However, mesoporous CMS membranes are normally poor in CO_2 selectivity due to the large diameters of mesopores, which have no size-sieving function for gases. Furthermore, phenolic resins suffer from significantly structural shrinkage and reduced functionality during the carbonization process. Dai et al. [68] developed a hyper-crosslinking strategy to construct mesoporous phenolic resin membranes with robust frameworks and enriched microporosity. In this strategy, 1,4-bis(chloromethyl)benzene reacts with the neighboring aromatic rings to form molecular bridges in a highly crosslinked state. The hyper-crosslinked mesoporous phenolic resin membranes were found to exhibit a higher adsorption capacity for CO_2 and selectivity for CO_2/N_2 than their un-crosslinked counterparts, thereby showing promising application in the preparation of CMS membranes with improved CO_2 separation performance.

It is worth noting that the development of advanced synthetic methods in recent years enables the preparation of CMS membranes with novel architectures. For example, Mahurin et al. [63, 64] prepared a series of polymeric membranes by superacid-catalyzed trimerization of acetyl compounds or letrozole. Subsequent carbonization of these polymeric membranes results in CMS membranes with rich micropores, a narrow distribution of small mesopores, and abundant nitrogen species. The CMS membranes show superior performance for CO_2 separation owing to the enhanced selective adsorption and diffusion of CO_2 in pore channels.

In principle, there are unlimited variations in the structure of polymer precursors for CMS membranes. However, the synthesis of new polymers with preferred structure is quite costly and time-consuming. As an alternative, it is more feasible to fabricate mixtures of polymers that are cheap and readily available for carbonization. The properties of mixed polymer precursors can be finely tuned by changing the mixture compositions. As a result, researchers can prepare CMS membranes according to specific separation requirements. Table 8.2 summarizes the mixed polymer precursors reported for the preparation of CMS membranes and their performance for CO_2 separation. The mixed polymer precursors can be formulated by a thermally stable polymer and a thermally labile polymer, such as cellulose+polyvinylpyrrolidone [69], polyetherimide+polyvinylpyrrolidone [70] and

Table 8.2 Mixed polymer precursors and CO_2 separation performance of CMS membranes.

Precursor 1	Precursor 2	CO_2 permeability or permeance	CO_2/N_2 selectivity	CO_2/CH_4 selectivity	Refs.
Cellulose	Polyvinylpyrrolidone	~281 barrer	~25	—	[69]
Polyetherimide	Polyvinylpyrrolidone	1.66 barrer	41.50	55.33	[70]
Polyetherimide	Polyethylene glycol	211 barrer	64.72	38.02	[71]
Polyimide	Polyetherimide	~420 barrer	~40	—	[72]
Polyimide	Polybenzimidazole	78.3 barrer	229.72	69.29	[73]
Polyimide	Cellulose	213.6 barrer	—	68.2	[74]
Poly(4-styrenesulfonic acid)	Poly(N-methylpyrrole)	7.19 barrer	—	167	[75]
Polyimide	Polysilsesquioxane	2465 barrer	—	56	[76]

polyetherimide+polyethylene glycol [71]. In this case, the decomposition of thermally labile polymer during the carbonization process creates additional pores in the resultant CMS membranes, which helps improve the permeability of gases. For example, the CO_2 permeability of CMS membranes prepared from mixtures of deacetylated cellulose acetate and polyvinylpyrrolidone [69] is much higher than that of CMS membranes prepared from bare regenerated celluloses [57, 58]. The mixed polymer precursors can also be formulated by two thermally stable polymers, such as polyimide+polyetherimide [72], polyimide+polybenzimidazole [73], and polyimide+cellulose [74]. In this case, the CMS membranes prepared from mixed polymer precursors normally show compromised performance for CO_2 separation in comparison with those prepared from an individual polymer precursor. For instance, the CO_2 permeability and CO_2/N_2 selectivity of CMS membranes prepared from mixtures of Matrimid and Ultem are lower than those of CMS membranes prepared from bare Matrimid, but higher than those of CMS membranes prepared from bare Ultem 1010 [72].

During the carbonization of polymer precursors, it is inevitable to form defective pores due to the high rigidity of the resultant carbon strands and carbon plates. On the other hand, the thermal relaxation of polymer chains during the carbonization process may lead to the collapse of porous microstructure in the resultant CMS membranes. Both phenomena are unwanted because the former may cause a significant sacrifice in gas selectivity, while the latter may lead to a significant sacrifice in gas permeability of the resultant CMS membranes. It is normally difficult to address these issues if a single polymer precursor is used for the preparation of CMS membranes. However, the fabrication of mixed polymer precursors for carbonization offers a chance in which the molecular interaction between different polymers can be utilized. In an early work, Hong et al. [75] fabricated a multilayered matrix consisting of poly(4-styrenesulfonic acid) and poly(N-methylpyrrole) on a porous ceramic support, as shown in Figure 8.8. The complexation of sulfonic acid groups in poly(4-styrenesulfonic acid) with tertiary amine groups in poly(N-methylpyrrole) forms a robust electrostatic interaction between different polymer layers, which restrains the size of carbon nanoparticles after carbonization. The formation

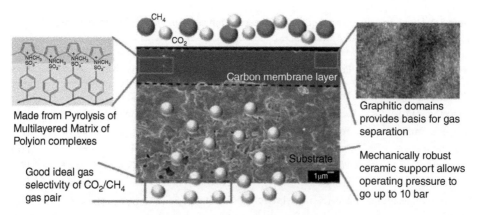

Figure 8.8 CO_2-permselective CMSs membranes prepared from a multilayered matrix of polyion complexes. Source: reprinted from Ref. [75] with permission from the American Chemical Society.

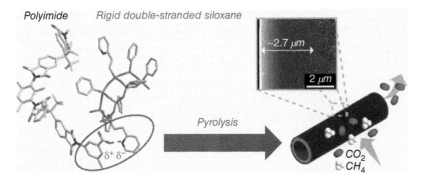

Figure 8.9 High-flux CMS membranes prepared from mixtures of polyimide and ladder-structured polysilsesquioxane. Source: reprinted from Ref. [76] with permission from Elsevier.

of defective pores is thus significantly avoided, resulting in an extremely high CO_2/CH_4 selectivity of 167 for prepared CMS membranes. Very recently, Lee et al. [76] reported high-flux CMS membranes from mixtures of 6FDA-DAM:DABA and ladder-structured polysilsesquioxane, as shown in Figure 8.9. The hydrogen-bonding interaction formed between 6FDA-DAM:DABA and polysilsesquioxane can effectively delay the thermal relaxation of polyimide chains during the carbonization process, thus maintaining the open porous microstructure of prepared CMS membranes. As a result, the CMS membranes display the high CO_2 permeability of 2465 barrer while still achieving a competitive CO_2/CH_4 selectivity of 56.

Besides the polymer precursors, carbonization conditions such as heating rate, carbonization temperature, carbonization atmosphere, and cooling time also have a significant impact on the performance of CMS membranes for CO_2 separation. Ismail et al. [77] found that CMS membranes prepared by carbonizing mixtures of Ultem and polyvinylpyrrolidone at a lower heating rate have better CO_2 separation performance than those prepared at a higher heating rate. Koros et al. [78] found that increasing the carbonization temperature makes the nanopores in CMS membranes prepared from 6FDA-DETDA:DABA tighter, leading to a sacrifice in CO_2 permeability but an improvement in CO_2/CH_4 selectivity. By carbonizing Matrimid at a temperature >800 °C, new CMSs with unprecedentedly high CO_2/CH_4 selectivity of ~150 were obtained by the same group [52], owing to the formation of ultraselective micropores that can exclude CH_4 molecules. Salleh et al. [79] found that CMS membranes prepared by carbonizing Matrimid under Ar atmosphere have higher CO_2/N_2 and CO_2/CH_4 selectivity than those prepared under N_2 atmosphere. This is probably due to the favorable release of volatile compounds under Ar atmosphere during the carbonization process, which can sweep away by-products to avoid blocking the nanopores in the CMS membranes. Hägg et al. [80] systematically examined the effects of carbonization conditions on the CO_2 separation performance of CMS membranes prepared from mixtures of cellulose acetate and polyvinylpyrrolidone. It was found that the importance of carbonization parameters follows the sequence of carbonization atmosphere > carbonization temperature > heating rate > cooling time. However, with these findings, it must be pointed out that it is difficult to figure out a universal rule for the dependence of the CO_2 separation performance of CMS membranes on the carbonization conditions, because the reactions involved in the carbonization process are rather complex, and the characterization techniques currently available are far from being capable of fully understanding such a process.

Usually, the adjustment of carbonization conditions is not enough to produce CMS membranes with satisfying performance for CO$_2$ separation, and additional post-carbonization treatments such as oxygen doping [81], amine doping [82], and steam activation [83] are needed. The major objective of these modifications is to optimize the porous structure of CMS membranes. Figure 8.10 shows the evolution of the pore-size distribution of CMS membranes after oxygen doping, and a comparison of oxygen doping and amine doping in CMS membranes. Oxygen doping makes the ultramicropores tighter by inserting oxygen species into the slits between carbon strands, but it has little effect on the micropores, resulting in a significant improvement in CO$_2$ selectivity. However, there is a minor change in CO$_2$ permeability after oxygen doping. The effect of amine doping on CMS membranes is basically similar to that of oxygen doping. The difference is that paraphenylenediamine

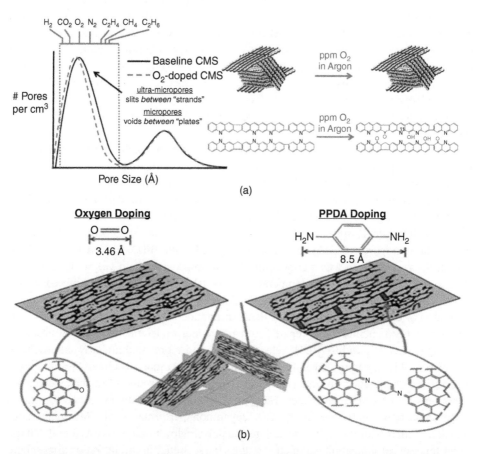

Figure 8.10 (a) Evolution of the pore-size distribution of CMS membranes after oxygen doping. Source: reprinted from Ref. [44] with permission from Elsevier. (b) Comparison of oxygen doping and amine doping in CMS membranes. Source: reprinted from Ref. [82] with permission from John Wiley & Sons.

Table 8.3 Functionalized CMS membranes and CO_2 separation performance.

Precursors	Functionality	CO_2 permeability or permeance	CO_2/N_2 selectivity	CO_2/CH_4 selectivity	Refs.
Poly(aryl ether ketone)	Silver nanoparticles	95.5 barrer	34	67	[84]
Polyimide	Cs_2CO_3	~27 GPU	~42	—	[85]
Phenolic resin	Boehmite	6.74 GPU	15.4	—	[86]
Polyimide	ZIF-108	24.75 barrer	43.0	130	[87]
Polyimide	γ-Fe_2O_3	0.0075 GPU	12.5	8.3	[88]
Polyfurfuryl alcohol	Zeolite L	171 GPU	20.43	35.75	[89]
Polyetherimide	SAPO-34	687 GPU	—	41	[90]
Polyimide	SiO_2	138.56	36.61	—	[91]
Polyimide	Diatomaceous earth	262.3 barrer	5.4	—	[92]
Phenolic resin	Ionic liquids	181 barrer	~36	—	[93]
Polyimide	CNT	6661 barrer	26.33	41.37	[94]

has large kinetic diameter oxygen species and is restricted to the occupation of larger ultramicropores. Steam activation was demonstrated to be more effective for the modification of porous structure of CMS membranes. The activation temperature and activation time are two important parameters that can be finely adjusted to achieve the preferred pore size distribution. Therefore, simultaneous improvements in CO_2 permeability and CO_2 selectivity can be realized by steam activation.

CMS membranes with improved performance for CO_2 separation can also be obtained by incorporating other materials such as metal complexes [84–88], zeolites [89, 90], silica oxides [91, 92], ionic liquids [93], and CNT [94] into the carbon matrix. Table 8.3 lists some representative functionalized CMS membranes reported in the last 10 years and their CO_2 separation performance. For example, Ban et al. [87] fabricated CMS membranes by carbonizing the MMMs composed of P48 and ZIF-108. The decomposition of ZIF-108 generates nanoporous carbon that is quite compatible with the carbon matrix formed by the decomposition of P48, making the pore-size distribution of prepared CMS membranes rather narrow. Most interestingly, the prepared CMS membranes display a unimodal structure for the pore-size distribution, with the pore size centered at 5.5 Å. As a result, the functionalized CMS membranes exhibit the remarkable CO_2/CH_4 selectivity of 130. In a similar manner, CMS membranes integrated with zeolites, silica oxides, and CNT were successfully fabricated by carbonizing the corresponding MMMs. Obviously, the effect of these functionalities on the CO_2 separation performance of CMS membranes is mainly a result of the change in porous structure. In another example, Mahurin et al. [93] physically impregnated ionic liquids into the mesopores of CMS membranes derived from phenolic resin. Ionic liquids have been previously demonstrated to have the ability to selectively absorb CO_2 from N_2 and CH_4 [95]. The occupation of non-selective mesopores by ionic liquids leads to significantly improved CO_2/N_2 selectivity for prepared CMS membranes.

8.6 Conclusions and Outlook

In summary, basic knowledge about membrane-based gas separation was briefly introduced in this chapter, followed by a detailed discussion of recent progress in the design and synthesis of carbon membranes for CO$_2$ separation. Based on the carbon structure, carbon membranes are divided into three categories: graphene membranes with well-defined planar structure, CNT membranes with well-defined cylindrical structure, and CMS membranes with amorphous structure. Through the tuning of the pore-size distribution and chemical functionality, various carbon membranes with high CO$_2$ separation performance have been fabricated. Given their high chemical and thermal stability, and excellent resistance to CO$_2$ plasticization, carbon membranes are believed to have great potential application in the selective capture of CO$_2$ from industrial streams.

However, there are some important issues remaining to be addressed in this field, and considerable future work is still needed. The first is how to precisely control the pore size of carbon membranes. Although size-sieving is very effective for the separation of gases, the discrimination of gas molecules with minor differences in kinetic diameters requires tuning the pore size at a sub-angstrom scale. For example, the kinetic diameters of CO$_2$, N$_2$, and CH$_4$ are 3.3, 3.64, and 3.8 Å, respectively. The difference in kinetic diameters of CO$_2$ and N$_2$ is 0.34 Å, and that for CO$_2$ and CH$_4$ is 0.5 Å. The methods for tuning the pore size of carbon membranes at a sub-nanometer scale have been well established. Nonetheless, it is very difficult to tune the pore size of carbon membranes at a sub-angstrom scale in the current state.

The second issue is how to disclose the underlying mechanism for the formation of carbon membranes and related CO$_2$ separation process. Due to the complex structure of carbon membranes, it is difficult to fully understand such mechanisms, which should be very instructive for the rational design of carbon membranes and optimization of CO$_2$ separation performance.

The third issue is how to investigate the CO$_2$ separation performance of carbon membranes under industrial conditions. To date, most research on carbon membranes for CO$_2$ separation were conducted under laboratory conditions. The performance of carbon membranes for CO$_2$ separation under industrial conditions (e.g. the presence of O$_2$ and moisture, elevated temperature and pressure, and high gas velocity), which is very important for their practical applications, remains to be systematically examined.

The fourth issue is how to prepare carbon membranes with large area, high mechanical strength, and excellent long-term stability. Overall, more future work should be dedicated to the development of advanced synthetic methods and instrumental technologies to promote advancement in carbon membranes for CO$_2$ separation.

Acknowledgments

This work was supported by the US Department of Energy, Office of Basic Energy Sciences, Chemical Sciences, Geosciences and Biosciences Division. K.H. also appreciates the sponsorship of the Natural Science Foundation of Jiangxi Province (20171BAB203019) and Nanchang University.

References

1 Rochelle, G.T. (2009). Amine scrubbing for CO_2 capture. *Science* 325: 1652–1654.

2 Choi, S., Drese, J.H., and Jones, C.W. (2009). Adsorbent materials for carbon dioxide capture from large anthropogenic point sources. *ChemSusChem* 2: 796–854.

3 Brunetti, A., Scura, F., and Barbieri, G. (2010). Membrane technologies for CO_2 separation. *J. Membr. Sci.* 359: 115–125.

4 Liu, J.Y., Hou, X.D., Park, H.B., and Lin, H.Q. (2016). High-performance polymers for membrane CO_2/N_2 separation. *Chem. Eur. J.* 22: 15980–15990.

5 Robreson, L.M. (2008). The upper bound revisited. *J. Membr. Sci.* 320: 390–400.

6 Sanyal, O., Zhang, C., Wenz, G.B. et al. (2018). Next generation membranes – using tailored carbon. *Carbon* 127: 688–698.

7 Zhou, F.L., Fathizadeh, M., and Yu, M. (2018). Single-to few-layered, graphene-based separation membranes. *Annu. Rev. Chem. Biomol.* 9: 17–39.

8 Song, C.S., Kwon, T., Han, J.H. et al. (2009). Controllable synthesis of single-walled carbon nanotube framework membranes and capsules. *Nano Lett.* 9: 4279–4284.

9 Chuah, C.Y., Goh, K., Yang, Y. et al. (2018). Harnessing filler materials for enhancing biogas separation members. *Chem. Rev.* 118: 8655–8769.

10 Cheng, Y.D., Wang, Z.H., and Zhao, D. (2018). Mixed matrix membranes for natural gas upgrading: current status and opportunities. *Int. Eng. Chem. Res.* 57: 4139–4169.

11 Geim, A.K. and novoselov, K.S. (2007). The rise of graphene. *Nat. Mater.* 6: 183–191.

12 Sun, Y.Q., Wun, Q., and Shi, G.Q. (2011). Graphene based new energy materials. *Energy Environ. Sci.* 4: 1113–1132.

13 Machado, B.F. and Serp, P. (2012). Graphene-based materials for catalysis. *Catal. Sci. Technol.* 2: 54–75.

14 Hu, M. and Mi, B.X. (2013). Enabling graphene oxide nanosheets as separation membranes. *Environ. Sci. Technol.* 47: 3715–3723.

15 Jiang, D.E., Cooper, V.R., and Dai, S. (2009). Porous graphene as the ultimate membrane for gas separation. *Nano Lett.* 9: 4019–4024.

16 Lee, J. and Aluru, N.R. (2013). Water-solubility-driven separation of gases using graphene membrane. *J. Membr. Sci.* 428: 546–553.

17 Wu, T.T., Xue, Q.Z., Ling, C.C. et al. (2014). Fluorine-modified porous graphene as membrane for CO_2/N_2 separation: molecular dynamic and fist-principles simulations. *J. Phys. Chem. C* 118: 7369–7376.

18 Sun, C.Z., Wen, B.Y., and Bai, B.F. (2015). Application of nanoporous graphene membranes in natural gas processing: molecular simulations of CH_4/CO_2, CH_4/H_2S and CH_4/N_2 separation. *Chem. Eng. Sci.* 138: 616–621.

19 Wang, Y., Yang, Q.Y., Zhong, C.L., and Li, J.P. (2017). Theoretical investigation of gas separation in functionalized nanoporous graphene membranes. *Appl. Surf. Sci.* 407: 532–539.

20 Sun, C.Z. and Bai, B.F. (2018). Improved CO_2/CH_4 separation performance in negatively charged nanoporous graphene membranes. *J. Phys. Chem. C* 122: 6178–6185.

21 Wang, S., Tian, Z., Dai, S., and Jiang, D. (2018). Effect of pore density on gas permeation through nanoporous graphene membranes. *Nanoscale* 10: 14660–14666.

22 Banhart, F., Kotakoski, J., and Krasheninnikov, A.V. (2011). Structural defects in graphene. *ACS Nano* 5: 26–41.

23 Fischbein, M.D. and Drndic, M. (2008). Electron beam nanosculpting of suspended graphene sheets. *Appl. Phys. Lett.* 93: 113107.

24 Yuan, Z., Benck, J.D., Eatmon, Y. et al. (2018). Stable, temperature-dependent gas mixture permeation and separation through suspended nanoporous single-layer graphene membranes. *Nano Lett.* 18: 5057–5069.

25 Hummers, W.S. and Offeman, R.E. (1958). Preparation of graphitic oxide. *J. Am. Chem. Soc.* 80: 1339.

26 Kim, H.W., Yoon, H.W., Yoon, S. et al. (2013). Selective gas transport through few-layered graphene and graphene oxide membranes. *Science* 342: 91–95.

27 Kim, H.W., Yoon, H.W., Yoo, B.M. et al. (2014). High-performance CO_2-philic graphene oxide membranes under wet-conditions. *Chem. Commun.* 50: 13563–13566.

28 Li, H., Song, Z., Zhang, X. et al. (2013). Ultrathin, molecular-sieving graphene oxide membranes for selective hydrogen separation. *Science* 342: 95–98.

29 Wang, S., Wu, Y., Zhang, N. et al. (2016). A highly permeable graphene oxide membrane with fast and selective transport nanochannels for efficient carbon capture. *Energy Environ. Sci.* 9: 3107–3112.

30 Wang, S., Xie, Y., He, G. et al. (2017). Graphene oxide membranes with heterogeneous nanodomains for efficient CO_2 separations. *Angew. Chem. Int. Ed.* 56: 14246–14251.

31 Zhou, F., Tien, H.N., Xu, W.L. et al. (2017). Ultrathin graphene oxide-based hollow fiber membranes with brush-like CO_2-philic agent for highly efficient CO_2 capture. *Nat. Commun.* 8: 2107.

32 Ying, W., Cai, J., Zhou, K. et al. (2018). Ionic liquid selectively facilitates CO_2 transport through graphene oxide membrane. *ACS Nano* 12: 5385–5393.

33 Ren, Y., Peng, D., Wu, H. et al. (2019). Enhanced carbon dioxide flux by catechol-Zn^{2+} synergistic manipulation of graphene oxide membranes. *Chem. Eng. Sci.* 195: 230–238.

34 Zhou, F., Huynh, N.T., Dong, Q. et al. (2019). Ultrathin, ethylenediamine-functionalized graphene oxide membranes on hollow fibers for CO_2 capture. *J. Membr. Sci.* 573: 184–191.

35 Tasis, D., Tagmatarchis, N., Bianco, A., and Prato, M. (2006). Chemistry of carbon nanotubes. *Chem. Rev.* 106: 1105–1136.

36 Baughman, R.H., Zakhidov, A.A., and de Heer, W.A. (2002). Carbon nanotubes – the route toward applications. *Science* 297: 787–792.

37 Skoulidas, A.I., Sholl, D.S., and Johnson, J.K. (2006). Adsorption and diffusion of carbon dioxide and nitrogen through single-walled carbon nanotube membranes. *J. Chem. Phys.* 124: 54708.

38 Ban, S. and Huang, C. (2012). Molecular simulation of CO_2/N_2 separation using vertically-aligned carbon nanotube membranes. *J. Membr. Sci.* 417: 113–118.

39 Bucior, B.J., Chen, D., Liu, J., and Johnson, J.K. (2012). Porous carbon nanotube membranes for separation of H_2/CH_4 and CO_2/CH_4 mixtures. *J. Phys. Chem. C* 116: 25904–25910.

40 Zhou, H., Xie, J., and Ban, S. (2015). Insights into the ultrahigh gas separation efficiency of lithium doped carbon nanotube membrane using carrier-facilitated transport mechanism. *J. Membr. Sci.* 493: 599–604.

41 Labropoulos, A., Veziri, C., Kapsi, M. et al. (2015). Carbon nanotube selective membranes with subnanometer, vertically aligned pores, and enhanced gas transport properties. *Chem. Mater.* 27: 8198–8210.

42 Kueh, B., Kapsi, M., Veziri, C.M. et al. (2018). Asphaltene-derived activated carbon and carbon nanotube membranes for CO_2 separation. *Energy Fuel* 32: 11718–11730.

43 Jones, C.W. and Koros, W.J. (1994). Carbon molecular-sieve gas separation membranes. I. Preparation and characterization based on polyimide precursors. *Carbon* 32: 1419–1425.

44 Rungta, M., Wenz, G.B., Zhang, C. et al. (2017). Carbon molecular sieve structure development and membrane performance relationships. *Carbon* 115: 237–248.

45 Fu, S.L., Sanders, E.S., Kulkarni, S.S., and Koros, W.J. (2015). Carbon molecular sieve membrane structure-property relationships for four novel 6FDA based polyimide precursors. *J. Membr. Sci.* 487: 60–73.

46 Joglekar, M., Itta, A.K., Kumar, R. et al. (2019). Carbon molecular sieve membranes for CO_2/N_2 separations: evaluating subambient temperature performance. *J. Membr. Sci.* 569: 1–6.

47 Richter, H., Voss, H., Kaltenborn, N. et al. (2017). High-flux carbon molecular sieve membranes for gas separation. *Angew. Chem. Int. Ed.* 56: 7760–7763.

48 Low, B.T. and Chung, T.S. (2011). Carbon molecular sieve membranes derived from pseudo-interpenetrating polymer networks for gas separation and carbon capture. *Carbon* 49: 2104–2112.

49 Sim, Y.H., Wang, H., Li, F.Y. et al. (2013). High performance carbon molecular sieve membranes derived from hyperbranched polyimide precursors for improved gas separation applications. *Carbon* 53: 101–111.

50 Swaidan, R., Ma, X.H., Litwiller, E., and Pinnau, I. (2013). High pressure pure- and mixed-gas separation of CO_2/CH_4 by thermally-rearranged and carbon molecular sieve membranes derived from a polyimide of intrinsic microporosity. *J. Membr. Sci.* 447: 387–394.

51 Qiu, W.L., Zhang, K., Li, F.S. et al. (2014). Gas separation performance of carbon molecular sieve membranes based on 6FDA-mPDA/DABA(3:2) polyimide. *ChemSusChem* 7: 1186–1194.

52 Zhang, C. and Koros, W.J. (2017). Ultraselective carbon molecular sieve membranes with tailored synergistic sorption selective properties. *Adv. Mater.* 29: 1701631.

53 Mondal, S., Elkamel, A., Reinalda, D., and Wang, K.A. (2017). Preparation and morphology study of carbon molecular sieve membrane derived from polyimide. *Can. J. Chem. Eng.* 95: 1993–1998.

54 Qin, G.T., Cao, X.F., Wen, H. et al. (2017). Fine ultra-micropore control using the intrinsic viscosity of precursors for high performance carbon molecular sieve membranes. *Sep. Purif. Technol.* 177: 129–134.

55 Wang, Z.G., Ren, H.T., Zhang, S.X. et al. (2018). Carbon molecular sieve membranes derived from Tröger's base-based microporous polyimide for gas separation. *ChemSusChem* 11: 916–923.

56 Hu, C.P., Polintan, C.K., Tayo, L.L. et al. (2019). The gas separation performance adjustment of carbon molecular sieve membrane depending on the chain rigidity and free volume characteristic of the polymeric precursor. *Carbon* 143: 343–351.

57 Campo, M.C., Magalhaes, F.D., and Mendes, A. (2010). Carbon molecular sieve membranes from cellophane paper. *J. Membr. Sci.* 350: 180–188.

58 Rodrigues, S.C., Andrade, M., Moffat, J. et al. (2019). Preparation of carbon molecular sieve membranes from an optimized ionic liquid-regenerated cellulose precursor. *J. Membr. Sci.* 572: 390–400.

59 Mahurin, S.M., Lee, J.S., Wang, X., and Dai, S. (2011). Ammonia-activated mesoporous carbon membranes for gas separations. *J. Membr. Sci.* 368: 41–47.

60 Zhang, B., Shi, Y., Wu, Y.H. et al. (2014). Preparation and characterization of supported ordered nanoporous carbon membranes for gas separation. *J. Appl. Polym. Sci.* 131: 39925.

61 Qin, G.T., Zhang, Y.P., and Wei, W. (2017). Facile synthesis of a nitrogen-doped carbon membrane for CO₂ capture. *Mater. Lett.* 209: 75–77.

62 Yoshimune, M. and Haraya, K.J. (2013). CO₂/CH₄ mixed gas separation using carbon hollow fiber membranes. *Energy Procedia* 37: 1109–1116.

63 Zhu, X., Tian, C.C., Chai, S.H. et al. (2013). New tricks for old molecules: development and application of porous N-doped, carbonaceous membranes for CO₂ separation. *Adv. Mater.* 25: 4152–4158.

64 Zhu, X., Chai, S.H., Tian, C.C. et al. (2013). Synthesis of porous, nitrogen-doped adsorption/diffusion carbonaceous membranes for efficient CO₂ separation. *Macromol. Rapid Commun.* 34: 452–459.

65 Livazovic, S., Li, Z., Behzad, A.R. et al. (2015). Cellulose multilayer membranes manufacture with ionic liquid. *J. Membr. Sci.* 490: 282–293.

66 Giernoth, R. (2010). Task-specific ionic liquids. *Angew. Chem. Int. Ed.* 49: 2843–2849.

67 Pinker, A., Marsh, K.N., Pang, S., and Staiger, M.P. (2009). Ionic liquids and their interaction with cellulose. *Chem. Rev.* 109: 6712–6728.

68 Zhang, J.S., Qiao, Z.A., Mahurin, S.M. et al. (2015). Hypercrosslinked phenolic polymers with well-developed mesoporous frameworks. *Angew. Chem. Int. Ed.* 54: 4582–4586.

69 He, X.Z. and Hagg, M.B. (2012). Structural, kinetic and performance characterization of hollow fiber carbon membranes. *J. Membr. Sci.* 23: 390–391.

70 Salleh, W.N.W. and Ismail, A.F. (2013). Effect of stabilization temperature on gas properties of carbon hollow fiber membrane. *J. Appl. Polym. Sci.* 127: 2840–2846.

71 Zainal, W.N.H.W., Tan, S.H., and Ahmad, M.A. (2017). Carbon membranes prepared from a polymer blend of polyethylene glycol and polyetherimide. *Chem. Eng. Technol.* 40: 94–102.

72 Fu, Y.J., Hu, C.C., Lin, D.W. et al. (2017). Adjustable microstructure carbon molecular sieve membranes derived from thermally stable polyetherimide/polyimide blends for gas separation. *Carbon* 113: 10–17.

73 Behnia, N. and Pirouzfar, V. (2018). Effect of operating pressure and pyrolysis conditions on the performance of carbon membranes for CO₂/CH₄ and O₂/N₂ separation derived from polybenzimidazole/Matrimid and UIP-S precursor blends. *Polym. Bull.* 75: 4341–4358.

74 Sazali, N., Salleh, W.N.W., Ismail, A.F. et al. (2019). Exploiting pyrolysis protocols on BTDA-TDI/MDI (P84) polyimide/nanocrystalline cellulose carbon membrane for gas separations. *J. Appl. Polym. Sci.* 361: 46901.

75 Chen, X.W., Khoo, K.G., Kim, M.W., and Hong, L. (2014). Deriving a CO_2-permselective carbon membranes from a multilayered matrix of polyion complexes. *ACS Appl. Mater. Interfaces* 6: 10220–10230.

76 Shin, J.H., Yu, H.J., An, H. et al. (2019). Rigid double-stranded siloxane-induced high-flux carbon molecular sieve hollow fiber membranes for CO_2/CH_4 separation. *J. Membr. Sci.* 570: 504–512.

77 Salleh, W.N.W. and Ismail, A.F. (2012). Effects of carbonization heating rate on CO_2 separation of derived carbon membranes. *Sep. Purif. Technol.* 88: 174–183.

78 Fu, S., Wenz, G.B., Sanders, E.S. et al. (2016). Effects of pyrolysis condition on gas separation properties of 6FDA/DETDA:DABA(3:2) derived carbon molecular sieve membranes. *J. Membr. Sci.* 520: 699–711.

79 Sazali, N., Salleh, W.N.W., Nordin, N.A.H.M., and Ismail, A.F. (2015). Matrimid-based carbon tubular membrane: effect of carbonization environment. *J. Ind. Eng. Chem.* 32: 167–171.

80 He, X.Z. and Hägg, M.B. (2011). Optimization of carbonization process for preparation of high performance hollow fiber carbon membranes. *Ind. Eng. Chem. Res.* 50: 8065–8072.

81 Kiyono, M., Williams, P.J., and Koros, W.J. (2010). Generalization of effect of oxygen exposure on formation and performance of carbon molecular sieve membranes. *Carbon* 48: 4442–4449.

82 Wenz, G.B. and Koros, W.J. (2016). Tuning carbon molecular sieves for natural gas separations: a diamine molecular approach. *AICHE J.* 63: 751–760.

83 Lee, H.C., Monji, M., Parsley, D. et al. (2013). Use of steam activation as a post-treatment technique in the preparation of carbon molecular sieve membranes. *Ind. Eng. Chem. Res.* 52: 1122–1132.

84 Xiao, Y., Mei, L.C., Chung, T.S. et al. (2010). Asymmetric structure and enhanced gas separation performance induced by in situ growth of silver nanoparticles in carbon membranes. *Carbon* 48: 408–416.

85 Kai, T., Kazama, S., and Fujioka, Y. (2009). Development of cesium-incorporated carbon membranes for CO_2 separation under humid conditions. *J. Membr. Sci.* 342: 14–21.

86 Rodrigues, S.C., Whitley, R., and Mendes, A. (2014). Preparation and characterization of carbon molecular sieve membranes based on resorcinol–formaldehyde resin. *J. Membr. Sci.* 459: 207–216.

87 Jiao, W., Ban, Y., Shi, Z. et al. (2016). High performance carbon molecular sieving membranes derived from pyrolysis of metal-organic framework ZIF-108 doped polyimide matrices. *Chem. Commun.* 52: 13779–13782.

88 Favvas, E.P., Heliopoulos, N.S., Karousos, D.S. et al. (2019). Mixed matrix polymeric and carbon hollow fiber membranes with magnetic iron-based nanoparticles and their application in gas mixture separation. *Mater. Chem. Phys.* 223: 220–229.

89 Yin, X., Wang, J., Chu, N. et al. (2010). Zeolite L/carbon nanocomposite membranes on the porous alumina tubes and their gas separation properties. *J. Membr. Sci.* 348: 181–189.

90 Sen, M. and Das, N. (2017). In situ carbon deposition in polyetherimide/SAPO mixed matrix membrane for efficient CO_2/CH_4 separation. *J. Appl. Polym. Sci.* 134: 45508.

91 Lua, A.C. and Shen, Y. (2013). Preparation and characterization of polyimide–silica composite membranes and their derived carbon–silica composite membranes for gas separation. *Chem. Eng. J.* 220: 441–451.

92 Liu, S.S., Zhang, B., Wu, Y.H. et al. (2018). Effects of diatomaceous earth addition on the microstructure and gas permeation of carbon molecular sieving membranes. *ChemistrySelect* 3: 8428–8435.

93 Chai, S.H., Fulvio, P.F., Hillesheim, P.C. et al. (2014). Synthesis of free-standing mesoporous carbon-carbon black nanocomposite membranes as supports of room temperature ionic liquids for CO$_2$ separation. *J. Membr. Sci.* 468: 73–80.

94 Lin, L., Song, C., Jiang, D., and Wang, T. (2017). Preparation and enhanced gas separation performance of carbon/carbon nanotubes (C/CNTs) hybrid membranes. *Sep. Purif. Technol.* 188: 73–80.

95 Chen, F.F., Huang, K., Fan, J.P., and Tao, D.J. (2018). Chemical solvent in chemical solvent: a class of hybrid materials for effective capture of CO$_2$. *AICHE J.* 64: 632–639.

9

Composite Materials for Carbon Capture

Sunee Wongchitphimon[1], Siew Siang Lee[1], Chong Yang Chuah[2], Rong Wang[1,3] and Tae-Hyun Bae[1,2]

[1]*Singapore Membrane Technology Centre, Nanyang Technological University, Singapore, Singapore*
[2]*School of Chemical and Biomedical Engineering, Nanyang Technological University, Singapore, Singapore*
[3]*School of Civil and Environmental Engineering, Nanyang Technological University, Singapore, Singapore*

9.1 Introduction

Because of rising atmospheric CO_2 concentrations that have caused climate change, the implementation of carbon capture has been suggested to reduce CO_2 emissions from fossil fuel–burning resources such as coal-fired power plants. Carbon capture has also been used in energy production processes such as natural gas treatment, hydrogen purification, and biogas upgrading [1]. However, conventional capture methods usually consume a large amount of energy and cause undesirable side effects. Efforts to improve the energy efficiency of the carbon capture process have led to the development of various materials.

Materials for Carbon Capture, First Edition. Edited by De-en Jiang, Shannon M. Mahurin and Sheng Dai.
© 2020 John Wiley & Sons Ltd. Published 2020 by John Wiley & Sons Ltd.

Among them, this chapter introduces composite materials for energy-efficient CO_2 capture processes exploiting adsorption and membrane technology.

9.1.1 Technologies for CO_2 Capture

Carbon dioxide can be selectively captured from gas mixtures based on physicochemical properties of gases such as kinetic diameter, boiling point, polarizability, and quadrupole moment. Technologies used for carbon-capture processes include absorption, adsorption, membrane technology, as well as cryogenic process; and are typically chosen depending on the CO_2 concentration in feed gases, plant size, and separation efficiency required. The most common method for CO_2 removal has been absorption through packed and plate columns, which require a huge space and large investment cost. This absorption is based on differences in solubility in a solvent or the chemical reaction between CO_2 and a liquid sorbent. Absorption exhibits good results in terms of removal efficiency and is highly effective for low concentrations of CO_2. However, this method requires high energy to heat liquid sorbents, which typically have a high heat capacity with regeneration temperature reaching above 100 °C. Furthermore, the liquid absorbents used in CO_2 capture are typically harmful. For example, monoethalnolamine (MEA) has received much attention as a chemical absorbent especially for post-combustion carbon capture in conventional coal-fired power plant, due to its high rate of reaction, low cost, and easy regeneration. However, MEA has a high equipment corrosion rate and loses its capacity over absorption-desorption cycles owing to gradual degradation of the amine [2, 3].

Adsorption has been considered as an energy-efficient alternative to absorption-based CO_2 capture. Since solid adsorbents have a lower heat capacity than that of absorption media, the energy required for regeneration of the media can be lowered significantly [4, 5]. The product can be recovered, and the adsorbent is reactivated for the next adsorption cycle by various means. Depending on the methodology for desorption, the adsorption processes are typically classified as temperature swing adsorption (TSA), pressure swing adsorption (PSA), and vacuum swing adsorption (VSA). In general, the properties of the adsorbents required vary for the different processes, since the working capacity and separation efficiency are always a function of pressure and temperature. In the adsorption process, molecular separation is achieved based on the affinity or the interaction between an adsorbate (CO_2 for carbon-capture process) and adsorbent, such that the polarizability and the quadrupole moment of CO_2 play an important role in the adsorptive carbon-capture process. Microporous solids such as zeolites, porous carbons, porous silicas, and metal–organic frameworks (MOFs) are widely used as adsorbents owing to their large surface areas, which can ensure high CO_2 uptake capacity [6, 7]. The pore surfaces are often further tuned to improve the affinity toward CO_2 and thus increase the separation efficiency or selectivity. The adsorption system is typically operated via packed-bed columns that are filled with solid porous particulate adsorbents. To improve mass/heat transfer efficiency and minimize the pressure drop in the column, structured adsorbers such as a monolith of inorganic materials can be used instead of conventional packed-bed adsorbers.

In spite of various advantages of the adsorption process over the conventional absorption method, adsorbent media still need to be regenerated by pressure or temperature swings, which requires a periodic stoppage of the system. In contrast, membrane-based gas separation is a physical process that can continuously separate a desired component from a feed mixture using the partial pressure differences between the feed and the permeate side as the driving force. In general, this process is also highly energy-efficient, since it does not involve phase transitions as in distillation or periodic heating/cooling cycles. Membrane separation is based on differences in mass-transfer rates of gases in semi-permeable barriers. Thus, the affinity between transporting gas molecules and membrane materials, the kinetic diameter of gases, and the sizes of pores (or the free volume, in the case of polymeric materials) are important factors affecting separation efficiency. In spite of many advantages of membrane technology, industrial applications have been hampered by the limited performance of current membranes.

In addition to the aforementioned technologies, carbon capture can be carried out with other separation methods, which have been comprehensively reviewed elsewhere [8–10]. The main scope of this book is carbon-capture materials for adsorption and membrane technology in which the importance of materials is even higher than in other CO_2 capture processes. Most composite materials have also been developed for applications in adsorptive and membrane-based carbon capture. Thus, the contents in the following sections of this chapter will be narrowed down to these two separation methods.

9.1.2 Composite Materials for Adsorptive CO_2 Capture

For applications in adsorptive CO_2 capture, researchers have focused on developing microporous solid adsorbents such as zeolites, MOFs, and aminosilicas, which are mostly in particulate form. For industrial applications, the adsorbent particles need to be packed into a column after being fabricated into a bead where the solid adsorbents are bound with clays or other polymeric materials. When such conventional packed columns are operated in a TSA cycle, which is advantageous when a large quantity of low-pressure gas is treated, as in post-combustion CO_2 capture, the heat transfer to adsorbents can be the rate-limiting step. One way to resolve this issue is to design a structured adsorber. However, the majority of adsorbents are inorganic solids having poor processibility, such that the large-scale production of structured adsorbers is challenging and costly. Composite materials can be an effective way to fabricate structured adsorbers, since polymeric materials are typically inexpensive and provide good processibility and mechanical stability.

Recently, the concept of a hollow-fiber composite sorbent has been proposed for CO_2 capture with a rapid TSA cycle. As shown in Figure 9.1a, a large quantity of sorbent particles are embedded in a porous polymer matrix fabricated into a hollow fiber. The lumen side of the hollow fiber is coated with an impermeable layer to protect the sorbent particles from the heating medium (typically hot steam) and cooling water that flow through the lumen side. Figure 9.1b shows how the hollow-fiber adsorber works in sorption and desorption modes. Since heating and cooling media are supplied to the lumen side of the individual hollow fiber, the heat transfer is much faster than that for conventional packed-bed columns where the heat is supplied from external jackets or heating pipes. In addition to such high efficiency in the CO_2 capture process, hollow-fiber composite adsorbers can also provide many other advantages. Because of the processibility of polymers, hollow-fiber adsorbents can be fabricated in a large scale using inexpensive polymers. The composite materials are

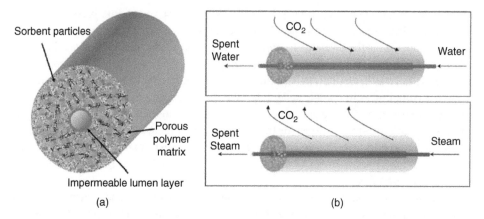

Figure 9.1 (a) The structure of a hollow-fiber composite adsorbent; (b) its adsorption–desorption cycle. Source: reprinted from Ref. [11] with permission from the American Chemical Society.

mechanically stable compared to pure inorganic monoliths or films, which typically suffer from intrinsic brittleness. Furthermore, a wide range of sorbent particles can be incorporated in the composites simply by adding the fillers into the dope solution [12]. Thus, the materials can be readily optimized for a given CO_2 capture condition.

9.1.3 Composite Materials for Membrane-Based CO_2 Capture

Advantages of membrane-based gas separation such as high energy efficiency and ease of operation have rendered it highly attractive as a carbon-capture method. Since the separation efficiency strongly depends on the properties of the membrane materials, significant efforts have been devoted to development of novel membrane materials. So far, polymers have been widely used for fabrication of gas-separation membranes due to their processibility and low cost [13–16]. Polymeric membranes can be fabricated on an industrial scale using the phase-inversion technique, and thus the advantages of polymers are still highly attractive for industrial application of membrane processes. However, most polymeric membranes in which the mass transport is governed by solution-diffusion showed limited separation performance, as indicated in the Robeson plot [17]. Therefore, permeation flux and selectivity, the key aspects to determining the efficiency of a given process, must be improved with new membrane materials in order for this process to be more widely applicable. Various nanoporous materials including zeolites, mesoporous silica, porous carbons, and MOFs are currently being studied as next-generation membrane materials, since the molecular-sieving property of these nanoporous materials gives rise to high separation efficiency. In fact, numerous lab-scale studies have successfully demonstrated high separation performance of such membranes. However, industrial-scale fabrication of nanoporous materials in the form of pure thin film is still challenging. An alternative technically viable option is to synthesize composite membranes or mixed-matrix membranes where nanoporous materials are incorporated in the polymer matrix, providing mechanical stability and excellent processibility [13].

As depicted in Figure 9.2, the nanoporous fillers (sometimes non-porous fillers) can be embedded in a dense polymeric film or an asymmetric hollow-fiber membrane, which is a

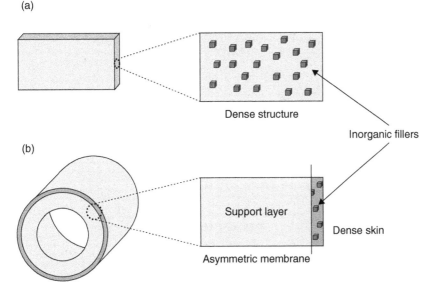

Figure 9.2 Mixed-matrix membranes in (a) dense film and (b) asymmetric hollow-fiber configurations.

popular membrane configuration for industrial applications. The former is self-supporting and thus has been fabricated in most lab-scale studies to investigate the intrinsic mass transport properties of composite materials. However, for industrial application, the membrane should be made into an asymmetric structure where the skin layer is supported by a porous support to minimize mass transfer resistance. The gas-transport property of a resultant mixed-matrix membrane can be predicted by mathematical models [18]. One of the most widely used models is the Maxwell model with the following equation:

$$P_{MMM} = P_p \left[\frac{P_d + 2P_p - 2\varphi_d(P_p - P_d)}{P_d + 2P_p + 2\varphi_d(P_p - P_d)} \right]$$

where, P_{MMM}, P_p, and P_d are the permeabilities of mixed-matrix, polymer, and dispersed (filler) phases, respectively, and φ_d is the volume fraction of the filler.

The Maxwell model predicts that judicious selection of polymer-filler pairs is highly important for a successful design of mixed-matrix membranes. This is also highlighted in a computational study on mixed-matrix membranes comprising MOFs [19]. As shown in Figure 9.3, when the gas permeabilities of both phases are well-matched, the enhancement in gas-separation performance can be maximized. Otherwise, the properties of the resulting membranes will be strongly governed by the continuous phase (polymer), such that only marginal or no improvement can be observed in the mixed-matrix membranes. A high degree of control over the filler/polymer interfacial morphology is also a prerequisite to attaining a successful mixed-matrix membrane fabrication, since non-ideal interfaces typically result in undesirable gas-separation performance. The non-idealities will be discussed in Section 9.3.

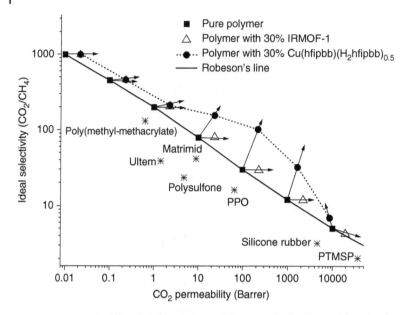

Figure 9.3 Predicted CO_2/CH_4 separation performance of hypothetical mixed-matrix membranes comprising two different metal–organic frameworks. Squares represent hypothetical pure polymer membranes on the upper bound limit established in 2008. Source: reprinted from Ref. [19] with permission from the Royal Society of Chemistry.

9.2 Fillers for Composite Materials

9.2.1 Zeolites

Zeolites are aluminosilicate crystals that are made up of interconnected AlO_4 and SiO_4 tetrahedra that can be present in various types of three-dimensional (3-D) frameworks of uniform channels and micropores. Zeolites are relatively low cost and robust with high chemical and thermal stability. Naturally occurring zeolites have a low level of purity and uniformity. Hence, zeolites for industrial applications including CO_2 capture have to be synthetically produced to ensure a certain level of purity and to attain a structure that is unavailable in nature, such as the zeolite LTA in Figure 9.4. Zeolites can be synthesized via hydrothermal, solvothermal, or even solvent-free methods [21–23]. Organic molecules known as *structure-directing agents* (SDAs) are often used to "direct" the structural formation of zeolites. Aluminosilicate zeolites can accommodate a wide variety of cations such as Na^+, K^+, Li^+, Ca^{2+}, Mg^{2+}, etc. to balance their negative charges and bear a chemical formula of $M_2/nO.Al_2O_3.xSiO_2.yH_2O$. Such extra-framework cations would endow zeolites with an adsorptive selectivity toward CO_2. In addition, their intra-crystalline multi-sized channels and pores facilitate size and shape selectivity, and thus good separation and transport of gas molecules with certain kinetic diameters (Figure 9.5). For example, for application in kinetic CO_2 separation, zeolites containing relatively small pore windows such as LTA, CHA, and MFI have been employed as fillers of mixed-matrix membranes.

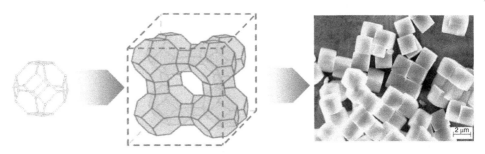

Figure 9.4 Basic crystal structure representation and scanning electron microscopic (SEM) image of zeolite A (LTA framework). Source: SEM image adapted from Ref. [20] with permission from Elsevier.

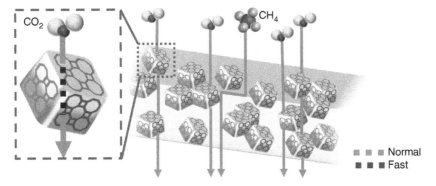

Figure 9.5 Schematic illustration of a mixed-matrix membrane with zeolite fillers. (*See color plate section for color representation of this figure*).

However, owing to their hydrophilic nature, aluminosilicate zeolites demonstrate a high affinity toward water vapor, which can reduce their CO_2 adsorption capacity in real applications such as post-combustion carbon capture. Furthermore, inorganic zeolites generally have a poor compatibility with organic polymer phases, resulting in the formation of non-selective voids and defects in composite membranes [24]. General strategies that have been proposed to enhance the viability of zeolites as filler are shown in Table 9.1.

9.2.2 Metal–Organic Frameworks

MOFs are highly porous crystals consisting of organic linkers-bridged metal (single ions or clusters) in one-, two-, or three-dimensional (1-D, 2-D, 3-D) coordination networks. Three commonly used organic linkers are carboxylate-, azolate-, and phosphonate-based ligands [4]. Figure 9.6 shows crystal structures and SEM images of two commonly studied MOFs, MOF-5 and HKUST-1. MOFs are more versatile than other fillers due their highly tunable pore structure and shape, modifiable chemical functionality, low density, as well as high porosity and surface areas [24]. Modifying the chemical functionality of MOFs, in particular, could control the CO_2 transport properties of a mixed-matrix membrane. For example, MOFs with coordinatively opened metal sites or amine functionalities can

Table 9.1 Comparison of significant porous fillers – zeolites, MOFs, CMSs, and mesoporous silica.

	Zeolites	MOFs	CMSs	Mesoporous silica
Physicochemical properties	• Surface area: 10–1100 m² g⁻¹ • Pore size: 0.3–1.0 nm • Pore volume: 0.07–0.5 cm³ g⁻¹ • Hydrophilic(aluminosilicate) or hydrophobic (all-silica) • Surface tunability • Good chemical and thermal stability (crystallinity not affected by moisture)	• Surface area: 1000–6000 m² g⁻¹ • Pore size: 0.6–1.2 nm • Pore volume: 0.6 to more than 2.0 cm³ g⁻¹ • Hydrophilic or hydrophobic • Excellent physicochemical tunability • Moderate chemical stability (metal–ligand bond susceptible to moisture)	• Surface area: 77–2000 m² g⁻¹ • Pore size: 0.3–0.5 nm • Pore volume: 0.06–0.73 cm³ g⁻¹ • Uniform crystal size • Hydrophobic • Best chemical and thermal stability (> zeolites and MOFs)	• Surface area: 25–2000 m² g⁻¹ • Pore size: 2–50 nm • Pore volume: 0.04–0.90 cm³ g⁻¹ • High coverage of silanol groups • Hydrophilic
Synthesis	• Sol-gel • Hydrothermal (with structure-directing agents) • Solvent-free method (mix, grind, and heat)	• Hydrothermal • Solvothermal • Solvent-free method (mechano-chemical, i.e. ball-milling)	• Pyrolysis of carbon-containing chemical compounds and polymers • Carbon vapor deposition (CVD)	• Hydrothermal • Sol–gel method
Values	• Selective uptake of CO_2 • Well-defined pores • Robust • Molecular-sieving property • Production in large-scales	• Highly porous with great versatility for physical and chemical manipulation for enhanced properties • Molecular-sieving property • Better filler-polymer interaction due to the organic ligand	• Good adsorption and transport properties • Molecular-sieving property • Interact well with polymer due to hydrophobicity • Less affected by moisture	• High permeability • Facilitate filler-polymer interaction via penetration of polymer chains through the mesopores • High selectivity toward CO_2 by the polar silanol groups • Further tuning functionality via silane chemistry
Limitation	• Water molecules may hinder adsorption and transport of CO_2 • Poor filler-polymer compatibility • Agglomeration tendency, dispersibility problem • Limited options for modification	• Weak metal-ligand bonds (susceptible to hydrolysis in the presence of H_2O)	• Rigid, thus could lead to formation of non-selective voids • Lower enthalpy of and capacity for CO_2 adsorption due to uniform electrical potential on their surface	• Filler-polymer hydrogen bonds (between silanol groups and polymer chains) can block pores and render them inaccessible by gas molecules • Low diffusion selectivity due to large pores • Poor chemical compatibility with polymer matrix

Enhancement strategies	Enhance zeolites-polymer interaction: • Surface hydrophobization via introduction (i.e. alcohol) to promote hydrogen-bond formation between zeolites and polymer • Use silane-coupling agent to enhance zeolite-polymer interaction • Synthesize hollow zeolite spheres to minimize agglomeration and enhance permeability • Prime zeolites with polymer via sonication to promote dispersion • Increase surface roughness via growth of inorganic nanostructures Enhance CO_2 adsorption or selectivity: • Impregnation of amine functional groups • Impregnation of exposed metal cations	Enhance chemical stability: • Use azolate-based linkers due to their basicity; thus can form stronger metal–ligand bonds • Use trivalent or tetravalent cations to yield robust MOFs • Incorporate hydrophobic functional groups Enhance CO_2 adsorption or selectivity: • Introduction of amine functional groups • Functionalization with other polarizing organic functional groups such as hydroxy, nitro, cyano, thio, and halide groups • Coordinately open metal cations such as Cu^{2+}, Zn^{2+}, Mg^{2+}, Ni^{2+}, Al^{3+}, Cr^{3+} ($\uparrow$ charge density, $\uparrow$ polarization to attract CO_2)	Enhance CMS-polymer interaction: • Surface treatment with strong inorganic acid to introduce OH^- or COO^- groups so as to increase dispersion and increase free volume of polymer • Polymer wrapping • Use surfactant to promote dispersion Enhance CO_2 adsorption or selectivity: • Surface functionalization with OH^- or amine groups • Impregnation of metal cations	Enhance silica-polymer interaction: • Grafting of organosilanes to enhance dispersion and silica-polymer interaction Enhance CO_2 adsorption or selectivity: • Use of silane agents with amine groups • Introduction of amine groups
References	[2, 13, 21, 23–29]	[2, 24, 28, 30–33]	[24, 34]	[25, 30, 35]

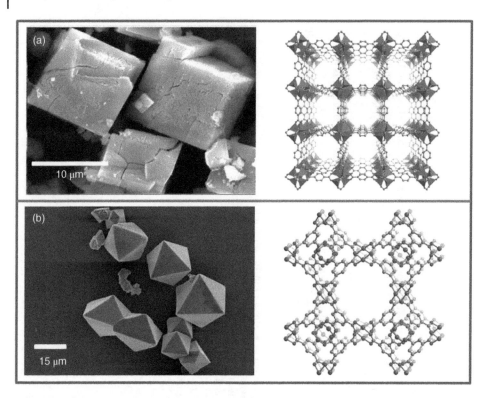

Figure 9.6 SEM images and crystal structure representation of (a) MOF-5; (b) HKUST-1. Sources: SEM images adapted from Ref. [36] for (a) with permission from Elsevier, and from Ref. [37] for (b) with permission from Elsevier; crystal structure images adapted from Ref. [4] with permission from the American Chemical Society.

promote selective adsorption and diffusion of CO_2 through their pores. As in a zeolite molecular sieve, MOFs possessing an optimum pore size can provide a molecular-sieving effect to resulting mixed-matrix membranes. Owing to organic groups in the framework, the interaction with polymers is relatively good, such that high-quality composite membranes with no filler-polymer interfacial voids are often successfully made without any compatibilization processes.

MOFs are usually synthesized via hydrothermal, solvothermal, or microwave heating techniques [30, 31]. Metal-binding solvents such as dimethyl formaldehyde often act as structure-directing templates for MOF formation. In such case, the solvent is eventually removed via heat treatment, producing exposed metal sites in a highly porous crystalline structure. Organic functional groups can be post-synthetically introduced to those metal sites via coordination bonding. Mechano-chemical synthesis such as ball milling of metal acetates and organic ligands is also an alternative path to making MOFs [32]. Nonetheless, the metal–ligand bonds in MOFs are relatively weak, as they are moisture-sensitive and thus easily hydrolyzed, resulting in degradation of the crystal structure as well as reduction of the surface area [24]. Refer to Table 9.1 for approaches adoptable to improve MOFs' efficiency.

9.2.3   Other Particulate Materials – Carbon Molecular Sieves and Mesoporous Silica

Carbon molecular sieves (CMSs) are carbonaceous spherical materials with a narrow pore-size distribution. CMSs possess high surface area to volume ratios, uniform size, and micropores; as such, they exhibit excellent adsorption and transport properties for discrimination and segregation of molecular gases. Microporous CMSs are obtainable from carbon-containing chemical compounds via pyrolysis or carbon vapor deposition techniques [34]. CMSs are hydrophobic with great chemical and thermal stability; thus, they are less affected by the presence of water and they interact relatively better with the polymer. Despite that, formation of non-specific voids is also common for composite materials bearing CMSs as fillers due to their rigidity.

Ordered mesoporous silica (Figure 9.7a) is amorphous silica particles with a regular arrangement of mesopores. It exists in various particle sizes, shapes, and pore diameters. Mesoporous silica is conventionally synthesized via a template-assisted hydrothermal or sol–gel process. It possesses high coverage of silanol groups that may interact well with polymer chains through hydrogen bonds; and it may attract polar CO_2, thus promoting an adsorptive selectivity toward CO_2 molecules. Furthermore, organic functional groups can be introduced via various means such as co-condensation and post-synthetic grafting. For example, to render a higher affinity to CO_2, alkyl amine groups have been introduced to various mesoporous silicas. Mesoporosity, besides facilitating high permeability of gas molecules, often promotes penetration of polymer chains to create a good filler-polymer interface.

9.2.4   1-D Materials – Carbon Nanotubes

Carbon nanotubes (CNTs) are 1-D cylindrical frameworks with covalently bonded carbon atoms and one or both closed ends (Figure 9.7b). They result from rolling up a single

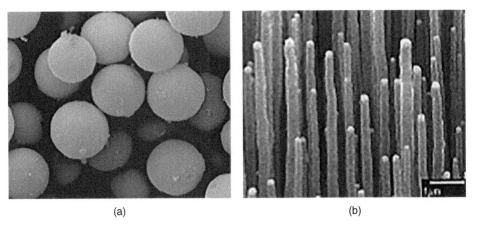

(a)   (b)

Figure 9.7   SEM images of (a) mesoporous silica MCM-41; (b) carbon nanotube. Sources: SEM images adapted from Ref. [38] for (a) with permission from Elsevier, and from Ref. [39] for (b) with permission from the American Association for the Advancement of Science (AAAS).

graphite sheet (single-walled CNT [SWCNT]) or multiple graphite sheets (multi-walled CNT [MWCNT]) [24, 40]. The basic synthesis of CNTs is similar to that of CMSs. They consist of large surface area and high aspect ratio with diameter ranging from a few to tens of nanometers and length reaching several millimeters. Their smooth surfaced-channel property can be exploited to enhance gas transport in a mixed-matrix membrane. Furthermore, CNTs offer CO_2 adsorption sites on both internal and external surfaces. Hence, they have much higher CO_2 volumetric adsorption capability than CMSs. Nonetheless, CNTs exhibit poor compatibility with the polymer matrix and inherently strong agglomeration tendency and thus low dispersibility, which necessitates modification of CNTs to promote their viability as a filler in a mixed-matrix membrane (Table 9.2). In addition, to maximize their mass-transport efficiency in composite membranes, CNTs should be aligned to the permeation direction, which is a big challenge for the fabrication of membranes on an industrial scale.

9.2.5 2-D Materials – Layered Silicate and Graphene

Layered silicate or clay is a naturally occurring aluminosilicate mineral whose crystal structure is determined by the ratios of silicon and oxygen. It is usually hydrated with water or attached to hydroxyl groups and is generally a few nanometers thick and hundreds to thousands of nanometers long. It is abundant and relatively low cost, and it possesses a high cationic exchange capacity, strong molecular barrier properties, and a very high aspect ratio. The silanol groups on the surface of the layered silicate interact with the polymer via —Si—O—Si— covalent bonds. Due to the weak interlayer force, this nanoporous layered material can be easily intercalated (or swollen) with an organic compound and exfoliated into individual sheets or layers for enhanced dispersibility and contact with polymer chains; and coupled with its large surface area, gas separation of membranes can be greatly enhanced. When nanoporous layered material is dispersed in a polymer membrane, the resulting tortuosity and reduced polymer-chain mobility would facilitate gas separation through the mixed matrix membrane by limiting the permeability of larger and slower gas molecules such as CH_4, while smaller CO_2 could easily diffuse through the nanopores (Figure 9.8). The ultimate structure of mixed-matrix membranes is determined by the filler-polymer interfacial interaction, materials, and preparation methods. The three most commonly adopted fabrication methods are exfoliation-adsorption, in situ intercalative polymerization, and polymer melt exfoliation or intercalation [25, 45]. In addition, surface modification of layered materials such as silylation using amine-functionalized silane, cation exchange, as well as esterification could enhance polymer-filler interaction as well as selectivity and permeability of CO_2. Further exfoliation using ultrasonication and calcination could be employed to further improve dispersion, polymer-filler interaction, and mass-transport properties of membrane [25]. Nonetheless, swelling should be optimized in order to prevent plasticization that could adversely affect the membrane selectivity.

Graphene is another type of 2-D carbon nanomaterial, with an emerging potential as CNTs' substitute for mixed-matrix membranes due to its economics and excellent mechanical, structural, thermal, and electrical properties [46]. It has a high aspect ratio and is impermeable, and therefore could be used as a filler to modify membrane permeability

Table 9.2 List of other inorganic fillers (porous and non-porous).

	Carbon nanotubes (CNTs)	Layered silicate	Graphene	Non-porous silica
Physiochemical properties	• Large surface area (>CMSs) • High aspect ratio (>30) • Smooth-surfaced channels • High mechanical strength • Hydrophobic	• Few nanometers thick • Hundreds to thousands nanometers long • High aspect ratio • Silanol groups on surface • Strong molecular barrier • Hydrophilic	• Large surface area to volume ratio • Large active sites (>CNTs) • High mechanical strength • Excellent thermal and electrical conductivity • Hydrophobic	• Hydrophilic • Crystal size of 30–100 nm • Low surface area (<40 $m^2\,g^{-1}$) • Low pore volume (<0.2 $cm^3\,g^{-1}$, pore diameter <2 nm) • Impermeable • Contains silanol groups
Synthesis	• Arc discharge • Laser ablation • Plasma torch • Chemical vapor deposition (CVD)	• Hydrothermal	• Chemical vapor decomposition on metal substrates • Thermal decomposition of carbon-based wafer • Mechanical exfoliation of pyrolytic graphite • Chemical and thermal reduction of graphene oxide	• Sol–gel using organosilane as precursor
Values	• High permeability through well-defined 1-D pore channels • Higher adsorption sites (internal and external) (>CMSs)	• Easily dispersed into individual layer • Exfoliation with polymer promotes polymer-filler interaction, enhances molecular-sieving and gas-transport properties • High cationic exchange capacity • Reduce diffusivity and solubility of molecular gas (↑ selectivity) • Molecular-sieve property	• Modify membrane permeability by providing tortuous paths	• Modify membrane permeability by modifying polymer chain packing and increasing polymer free volume (↑ permeability) • Silanol groups facilitate solubility and diffusivity of CO_2
Limitation	• Poor chemical compatibility with polymer matrix • Agglomeration tendency and thus low dispersibility	• Poor chemical compatibility with polymer matrix	• Low dispersibility	• Poor chemical compatibility with polymer matrix • High agglomeration tendency
Enhancement strategies	Similar methods for CMSs may be adoptable	Enhance layered silicate–polymer interaction: • Surfactant cation exchange to make clay hydrophobic by replacing hydrated cations on clay with cationic surfactants such as alkylammonium • Anion exchange – modify edge with a coupling agent to enhance silica-polymer adhesion resulting in less void formation • Silane functionalization with organosilane to enhance dispersion, intercalation, and silica-polymer interaction Enhance CO_2 adsorption or selectivity: • Silylation using silane agents with amine groups	Enhance CO_2 adsorption: Surface functionalization with amine, —COOH, —OH, —CH$_3$ Similar methods for CMSs may be adoptable	
References	[24, 41–43]–>	[24, 25]	[24, 42–44]–>	[2, 24]

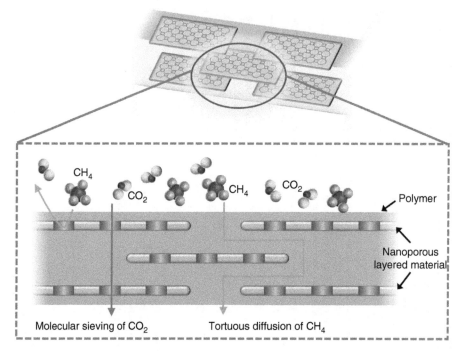

Figure 9.8 Schematic illustration of nanoporous layered materials in a mixed-matrix membrane.

properties by providing tortuous paths that restrict molecular transport of heavier gases and thus promoting gas separation. Nevertheless, feasibility of bulk processing and achieving a good dispersion of graphene in mixed matrix-membranes remains challenging.

Table 9.1 compares general physicochemical properties of zeolites with significant fillers such as MOFs, CMSs, and mesoporous silica; while Table 9.2 summarizes the general properties of other alternative porous as well as non-porous materials.

9.3 Non-Ideality of Filler/Polymer Interfaces

In an ideal mixed-matrix membrane, polymer chains are well-adhered onto the outer surface of well-dispersed fillers, as depicted in Figure 9.9. However, during membrane formation, shrinking of the polymer matrix followed by reduced polymer-chain diffusion (or mobility) as the solvent evaporates would exert stress in particular at polymer-filler interfaces, which could lead to non-ideal nanoscale morphologies. Consequently, the mass-transport properties of resulting mixed-matrix membrane would deviate from the theoretical prediction estimated from the properties of the pristine filler and neat polymer membrane. The degree of deformity varies with different polymer-filler combinations as well as preparation steps and conditions. Typically, there are three types of non-idealities: (i) sieve-in-a-cage, (ii) polymer matrix rigidification, and (iii) plugged filler pores.

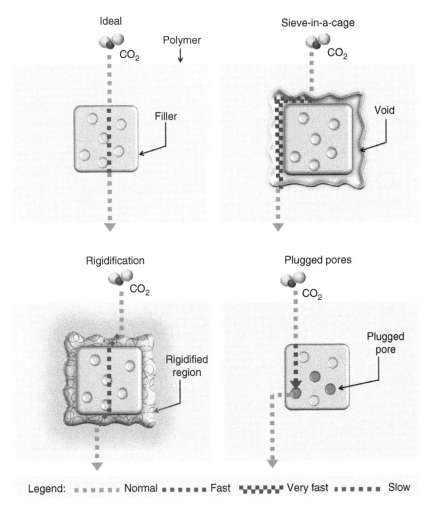

Figure 9.9 Schematic of common non-ideal polymer-filler interfacial morphologies.

9.3.1 Sieve-in-a-Cage

Sieve-in-a-cage describes the morphology where voids are formed at the polymer-filler interfaces. It is a result of polymer chains being detached from the filler surface during membrane formation primarily due to different thermal coefficients of expansion between the polymer and filler. This morphology is typically observed in composite membranes comprising inorganic filler, such as zeolites and glassy polymers. During membrane formation, volume loss in a pure dilating polymer membrane due to solvent evaporation could result in elliptical voids [47]. When fillers are suspended in this polymer phase, the rigidity of fillers and polymer-filler repulsion in addition to the stress during solvent evaporation would restrict a synchronized elongation between the two inorganic–organic phases, hence drawing the polymer chains away from the filler and yielding a delaminated or de-wetted filler surface, called the sieve-in-a-cage morphology. In a pervasive case, it looks like a continuous or non-continuous drain in the polymer phase surrounding the filler. Its formation would increase the overall free volume of a resulting mixed-matrix membrane.

As shown in Figure 9.10a, the voids are generally much larger than the size of pores in fillers, thus promoting a rapid diffusion of CO_2, bypassing the pores of the fillers. However, since non-selective Knudsen diffusion is dominant in such large voids, the selectivity of the membrane cannot be improved even with a high loading of a CO_2-selective molecular sieve. The Maxwell model predicts that selectivity can remain unchanged for membranes with sieve-in-a-cage morphology, as shown in Figure 9.11. However, sieve-in-a-cage often accompanies other problems, such as the agglomeration of fillers, that lead to the formation of larger defects in mixed-matrix membranes. In such cases, decreased selectivity is observed. To avoid this problem, the surfaces of fillers have been modified using various methods as described in Table 9.1. Figure 9.10b shows a zeolites/polymer interface that is successfully controlled via a surface treatment of zeolites.

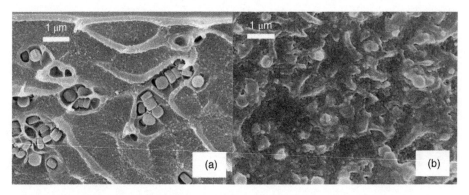

Figure 9.10 Zeolite MFI-polyimide interfaces (a) before and (b) after the surface treatment of zeolites. Source: reprinted from Ref. [29] with permission from the American Chemical Society.

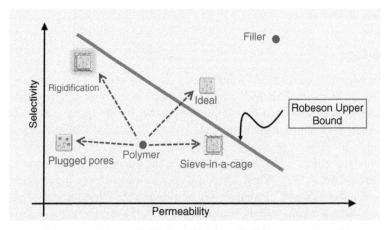

Figure 9.11 Probable effect of non-ideal polymer-filler interfaces on permeability and selectivity of membrane. Source: reprinted from Ref. [48] with permission from Elsevier.

9.3.2 Polymer Matrix Rigidification

Rigidification of a polymer region, induced by addition of fillers [49], is a result of compressive stress and adsorption of polymer chains around and onto the surface of dispersed fillers due to reduced polymer-chain mobility at the space-constraint polymer-filler interface, following reduction of free volume during membrane formation at temperatures above the membrane glass-transition temperature, T_g (vitrification) [47]. Consequently, adsorption and permeation of gas molecules, especially larger ones, would be restricted in this region, with possible diffusion-selectivity enhancement. The higher the filler loading, the lower the permeability, and the higher the selectivity as the degree of rigidification intensifies. Nevertheless, at loading approaching 100% filler content, the permselectivity is expected to approach near that of the pristine filler, as the polymer phase will be all rigidified [47]. Different fillers may exhibit different effective loading where this effect will take place. A rise in glass-transition temperature and effective gas permeation energy compared to that of the pristine polymer membrane is a good indication of the occurrence of a rigidified region [50]. It is reported that chemical coupling using silane chemistry, a method widely used for improving zeolite/polymer adhesion, often causes this non-ideality due to strong covalent bonding between rigid fillers and surrounding polymer chains [51].

9.3.3 Plugged Filler Pores

Plugged filler pores are a phenomenon where the pores on the filler surface are sealed partially or fully by water, solvent, rigidified polymer chains, gas molecules, or other chemicals used during the synthesis of a mixed-matrix membrane, thus resulting in impermeability or reduced permeability through the filler. For instance, hydrophilic fillers such as zeolites can have water molecules strongly bound in their pores. If all the pores of fillers are perfectly plugged, the decrease in permeability without a change in selectivity is predicted by the Maxwell model, as shown in Figure 9.11. However, partial blockage of pores can change the mass-transport properties of fillers, such that the prediction of permeation properties of mixed-matrix membranes is very difficult. Plugged filler pores may co-exist with rigidification of the polymer region at the polymer-filler interface. In such case, an enhancement in selectivity together with decreased permeability could be observed.

Figure 9.11 illustrates the possible deviation of membrane transport properties as a result of each non-ideal case. It is assumed that polymer-filler pairs are well-matched to increase both permeability and selectivity. As clearly shown in the figure, non-idealities always cause undesirable gas-separation properties compared to the performance predicted from ideal filler/polymer interfaces.

9.4 Composite Adsorbents

As mentioned in Section 9.1.2, composite materials are advantageous in designing structured adsorbers in which adsorption and desorption processes are facilitated. One representative example is hollow-fiber adsorbents for rapid TSA. The first hollow-fiber composite adsorbent has been fabricated by incorporating commercial zeolite Na-X into a cellulose acetate (CA) matrix, as shown in Figure 9.12 [11]. The spinning conditions

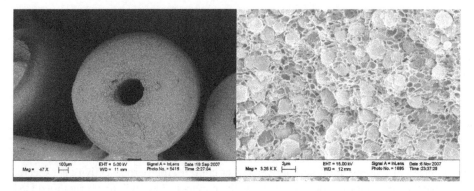

Figure 9.12 SEM images of hollow-fiber composite sorbent synthesized by incorporating zeolites Na-X within CA. Source: reprinted from Ref. [11] with permission from the American Chemical Society.

were adjusted to obtain a highly porous structure with sieve-in-a-cage morphology to ensure a rapid CO_2 transport within the composite hollow fibers. A high filler loading of 75 wt% was achieved with such a porous structure of the composite. Then, the lumen side of the hollow fibers was coated with polyvinylidene chloride (PVDC) acting as a barrier. The composite adsorbent tested under repeated thermal cycles between 100 °C and 45 °C showed good regenerability together with rapid CO_2 adsorption/desorption. In the following analytical study about energy consumption in the CO_2 capture process, the rapid temperature swing of a zeolite/polymer hollow-fiber composite adsorber with heat integration, such as utilizing heat flows in the feed water preheating system and heats of compression of CO_2, was shown to be highly promising [52]. The hollow-fiber composite made with zeolite Na-X and CA was packed into a module and tested under dynamic CO_2 flow condition [53]. Kinetic limitations in the hollow-fiber modules are largely overcome by increasing the superficial gas velocity and fiber packing. In addition, the heat generated by adsorption was successfully dissipated by supplying a coolant to the lumen side of the hollow fiber, such that the CO_2 uptake capacity remained high at any adsorption condition.

Amine/silica/CA composite hollow-fiber adsorbent was produced using a post-spinning infusion of amine, allowing for functionalization of polymer/silica hollow fibers with different types of amines during the solvent-exchange step after fiber spinning (Figure 9.13) [54]. Infusion of two different amines, aminopropyltrimethoxysilane (APS) and polyethyleneimine (PEI), gave similar CO_2 uptake capacity of about 1 mmol/g at CO_2 partial pressure and temperature of 0.1 bar and 35 °C, respectively. Cyclic adsorption/desorption results showed that the hollow-fiber adsorbent derived from APS was stable and did not lose CO_2 uptake capacity over several cyclic runs, while PEI-derived fibers exhibited a 10% decrease in capacity over five adsorption cycles. In contrast, fibers that are spun with pre-synthesized, amine-loaded mesoporous silica powders show negligible CO_2 uptake and low amine loadings because of loss of amines from the silica materials during the fiber-spinning process. In the following study, a thermally stable, mechanically durable, and chemically resistant polyamide-imide(PAI)/silica/PEI hollow-fiber adsorbent was prepared and showed a CO_2 uptake of 1.19 mmol/g at conditions relevant to the flue

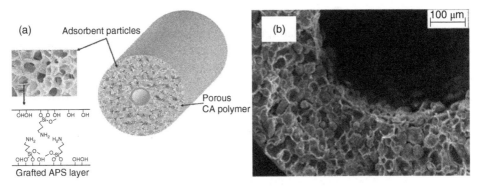

Figure 9.13 (a) Scheme and (b) SEM image of an amine/silica/CA hollow-fiber composite adsorbent. Source: reprinted from Ref. [54] with permission from the American Chemical Society.

gas of coal-fired power plants [55]. The addition of glycerol to the hollow-fiber sorbent increased CO_2 uptake to 2.0 mmol g^{-1} owing to its plasticization effect. Furthermore, it was reported that active cooling by the flow of coolant through the lumen side of hollow fibers can effectively improve the CO_2 uptake property of the PAI/silica/PEI hollow-fiber adsorbent [56].

Instead of solid adsorbent particles, ionic liquids possessing a potential to work in humid flue gas streams typically seen in actual operation have been incorporated into a PAI polymer matrix for application in post-combustion CO_2 capture [57]. In the best result, a PAI sorbent containing 44 wt% of an equimolar of 1-(2-hydroxylethyl)-3-methylimidazolium bis(trifluoromethylsulfonyl)imide ([Im$_{21}$OH][Tf$_2$N]) and 1,8-diazabicyclo-[5.4.0]undec-7-ene (DBU) showed a CO_2 uptake of 0.57 mmol g^{-1} at 0.1 bar of CO_2 partial pressure and 35 °C with decent uptake kinetics.

9.5 Composite Membranes (Mixed-Matrix Membranes)

Mixed-matrix composite membranes are heterogeneous membranes composed of functional fillers embedded in the continuous polymeric matrix; their advantages are introduced in Section 9.1.3. Most widely used fillers for making CO_2-selective membranes are zeolites and MOFs, although other fillers described in Section 9.2 can also be employed. Glassy polymers having a good mechanical stability and film-forming property are more popular than rubbery polymers as matrices. In particular, the polyimide family, including those that are in-house, are extensively employed owing to their good gas-separation properties. In contrast to composite adsorbents in which the separation is solely based on the affinity between CO_2 and adsorbents, both diffusivity and solubility (sorption property) of gases in membrane materials are highly important for composite membranes. In fact, the major role of fillers in mixed-matrix membranes is increasing either or both of them. In most studies, mixed-matrix membranes were fabricated into dense film forms to investigate intrinsic mass-transport properties. Only a few studies demonstrated CO_2 separation properties with asymmetric hollow-fiber membranes, a more desirable configuration for industrial applications.

Zeolites, a robust framework featuring a molecular-sieving property, have been a mainstay for mixed-matrix membrane fabrication. For application in CO_2 separation, zeolites with small pore dimensions, such as LTA, CHA, and MFI have been incorporated in polymer matrices. However, FAU zeolites were often used too, since they can selectively adsorb CO_2 from gas mixtures and thus improve the sorption selectivity of membranes. Controlling zeolite/polymer interfacial morphology has been a key requirement for successful fabrication of composite membranes. The most widely used method is modifying the zeolite surfaces, which are typically covered with hydrophilic hydroxyl groups (silanol or aluminol) with organic molecules [13]. A well-known example is the chemical coupling of zeolite and polyimide using silane chemistry. Creating highly roughened zeolite surfaces by the deposition of an inorganic nanostructure also proved to be highly effective in controlling zeolite/polymer interfaces [20, 29, 58, 59]. Important work on zeolite-based composite membranes for CO_2 capture is summarized in Table 9.3.

Recently, MOFs have been extensively studied as fillers in composite membranes. A wide range of frameworks that can tune the sorption or diffusion properties of membranes are available. Since conventional solvothermal methods give large crystals that are not suitable for membrane fabrications, special techniques including microwave synthesis [93], non-solvent-induced crystallization [94], and facilitated reaction in the presence of an organic base [95] should be employed to obtain nanocrystal forms of MOFs. Indeed, the synthesis of MOF nanocrystals is highly important for the fabrication of asymmetric hollow-fiber membranes in which an ultrathin skin layer involves MOF crystals without forming defects. Furthermore, MOF nanocrystals showed good adhesion with polymers owing to their large external surface areas, which contain organic moieties. MOF-based composite membranes for carbon capture are a hot research area, and thus numerous papers have been published. A recent review paper has summarized all important composite membranes synthesized so far [96]. Among them, the most recent studies that have conducted mixed-gas permeation testing are summarized in Table 9.4. Other fillers can also be employed to design CO_2-selective membranes; studies on such composite membranes are comprehensively reviewed elsewhere [25].

9.6 Conclusion and Outlook

Composite materials that combine the advantages of fillers and polymers have been synthesized for applications in energy-efficient carbon-capture processes such as adsorptive and membrane-based technologies. Hollow-fiber composite adsorbents demonstrated promising performance for rapid TSA that is applicable to post-combustion CO_2 capture. Mixed-matrix membranes also successfully showed improved CO_2 separation performance compared to conventional polymeric membranes. For both cases, however, scaling up the synthesis of materials and testing them in the more realistic conditions remain challenging. For example, the presence of water and other impurities in the flue gas of a coal-fired power plant could not only reduce CO_2 capture performance, but also decrease the effective lifetime of composite materials. For membrane-based carbon capture, mixed-matrix membranes should be fabricated into an asymmetric hollow-fiber configuration to advance from the laboratory scale.

Table 9.3 Summary of zeolite-based composite membranes used for separation of CO_2 from CH_4 and from N_2.

Polymer	Filler	Structural type	Filler loading (wt%)	P_{CO_2} (barrer)	P_{CO_2}/P_{CH_4}	P_{CO_2}/P_{N_2}	Refs.
					CO₂ separation performance		
Matrimid	Na-Y	FAU	10 20	15.02 22.00	39.8 27.6	—	[60]
Poly(amide-b-ethylene oxide)	Na-X	FAU	10 15	103.5 113.0	—	39.0 47.0	[61]
P84	Na-X	FAU	1.5 4.4	4.96 9.07	6.1 5.3	8.9 20.2	[62]
Matrimid	Na-X	FAU	30	17.54	—	31.9	[63]
6FDA-6FpDA-DABA	ZSM-2	FAU-EMT	20	15.96	24.2	13.3[a]	[64]
Polyimide	K-A (3A)	LTA	10 30	5.30 5.78	17.7 15.2	—	[65]
Polyetherimide	K-A (3A)	LTA	10 30	20.30 35.10	9.2 8.01	—	[65]
P84	Na-A (4A)	LTA	8	0.81	—	15.9	[66]
Polyimide	Na-A (4A)	LTA	10 30	6.51 7.38	19.7 27.3	—	[65]
Polyetherimide	Na-A (4A)	LTA	10 30	26.3 39.6	9.5 6.5	—	[65]
Pebax-1657	Na-A (4A)	LTA	10 30	97.1 55.8	26.5 7.9	54.0 12.9	[67]
Polycarbonate	Na-A (4A)	LTA	20 30	7.8 7.0	32.5 37.6	39.0[a] 39.1[a]	[68]
Polycarbonate	Na-A (4A)	LTA	10 30	8.2 7.0	32.8 37.6	38.9 39.1	[69]
Polyethersulfone	Na-A (nano sized)	LTA	20	2.3	31.2	25.6[a]	[70]
Polyethersulfone	Na-A (micro sized)	LTA	20	1.6	31.1	26.9[a]	[70]
Polyimide	Ca-A (5A)	LTA	10 30	7.1 7.9	15.8 16.2	—	[65]
Polyetherimide	Ca-A (5A)	LTA	10 30	24.4 36.3	14.7 11.2	—	[65]
Polyethersulfone	SAPO-34	CHA	20	5.8	37.0	—	[71]
Polyethersulfone	SAPO-34	CHA	20	6.5[b]	33.1	—	[72]
Polyphosphazene	SAPO-34	CHA	22	48.0	17.5	53.0	[73]
Pebax1657	SAPO-34	CHA	9 33 50	100[b] 250[b] 338	17.0[b] 17.0[b] 17.0[b]	52.0[b] 55.0[b] 53.0[b]	[74]
Pebax1074	SAPO-34	CHA	10 30	220[b] 250[b]	31.6[b] 44.5[b]	78.1[b] 83.4[b]	[75]
Crosslinked PDMC	SSZ-13	CHA	25	148	38.9	—	[76]
Matrimid	Silicalite	MFI	20	19.2	—	35.6	[63]
Matrimid	Silicalite	MFI	35	31	—	39	[29]
Matrimid	Silicalite	MFI	10 30	8.3 14.6	67.2 56.5	59.1[a] 47.1[a]	[77]

a) Calculated from single-gas permeability.
b) Extrapolated data.

Table 9.4 Summary of MOF-based composite membranes used for separation of CO_2 from a gas mixture.

Polymer	Filler	Filler loading (wt%)	CO_2/CH_4 ratio in the feed	CO_2/N_2 ratio in the feed	ΔP (bar)	CO_2 separation performance			Refs.
						P_{CO2} (barrer)	P_{CO2}/P_{CH4}	P_{CO2}/P_{N2}	
Polysulfone	NH_2-MIL- 125(Ti)	10 30	50/50	—	3 10 3 10	18.5 15.0 40.0 36.8	28.3 28.5 29.2 5.7	—	[78]
Polysulfone	ZIF-8	16	10/90 50/50 90/10	—	—	—	28.4[a] 31.0[a] 30.0[a]	—	[79]
Polysulfone	NH_2-MIL-53(Al) NH_2-MIL-101(Al)	25	50/50	—	3	5.5[a] 8.6[a]	27.5[a] 28.7[a]	—	[80]
Polysulfone	MIL-68(Al)	4 8	50/50	—	1.7	4.7 4.7	32.2 36.5	—	[81]
Matrimid	Fe-BTC	10 30	50/50	—	5	8.7[a] 12.2[a]	21.8[a] 28.0[a]	—	[82]
Matrimid	NH_2-MIL-53(Al) NH_2-MIL-101(Al)	15	50/50	—	3	8.3[a] 9.5[a]	38.1[a] 36.0[a]	—	[80]
Matrimid	b-CuBDC nc-CuBDC ns-CuBDC	8	50/50	—	2 2 7.5	—	45.0[a] 49.5[a] 88.3[a]	—	[83]
Matrimid	MIL-53(Al) ZIF-8 Cu_3BTC_2	30	50/50	—	2.5	17.5[a] 20.0[a] 13.0[a]	46.3[a] 43.6[a] 51.1[a]	—	[84]
Matrimid	ZIF-7-8(20) ZIF-8-ambz-(15) ZIF-8-ambz-(30)	15	50/50	—	13.9[a] 12.5[a] 14.8[a]	18.8[a] 13.7[a] 10.0[a]	41.8[a] 40.5[a] 43.1[a]	—	[85]
Matrimid	MIL-53 MIL-53-NH_2	15	50/50	—	10.2	—	8.5 2.1	—	[86]

Polymer	Filler								Ref.
Matrimid	TKL-107	5 20	20/80	—	1.9	$7.4^{a)}$ $15.0^{a)}$	$43.2^{a)}$ $50.3^{a)}$	—	[87]
Polyetherimide	MIL-53 MIL-53-NH$_2$	15	50/50	—	10.2	—	42.8 36.1	—	[86]
Polyetherimide	ZIF-8	17,v%	—	20/80	$1.4^{b)}$ $3.4^{b)}$	$26^{c)}$ $26^{c)}$	— —	$30^{a)}$ $32^{a)}$	[88]
Pebax-2533	ZIF-8	5 40	—	10/90	1.5	$273^{a)}$ $1100^{a)}$	— —	$55.4^{a)}$ $35.0^{a)}$	[89]
PMP	NH$_2$-MIL-53(Al)	15 30	10/90	—	2 8 2 8	126.7 244.6 204.4 339.5	13.3 18.2 19.2 22.9	—	[90]
6FDA-durene: DABA, 9:1	ZIF-8	20	50/50	—	$2^{a),b)}$ $35^{a),b)}$	$926^{a)}$ $700^{a)}$	$23.2^{a)}$ $18.4^{a)}$	—	[91]
6FDA-ODA	UiO-66 NH$_2$-UiO-66 MOF-199 NH$_2$-MOF-199 UiO-67	25	50/50	—	10	—	42.3 44.7 50.7 52.4 15.0	—	[92]

a) Extrapolated data.
b) Feed pressure.
c) Gas permeance in GPU.

References

1 Zhang, X. and Caldeira, K. (2015). Time scales and ratios of climate forcing due to thermal versus carbon dioxide emissions from fossil fuels. *Geophysical Research Letters* 42: 4548–4555.

2 Zhang, Y., Sunarso, J., Liu, S., and Wang, R. (2013). Current status and development of membranes for CO2/CH4 separation: a review. *International Journal of Greenhouse Gas Control* 12: 84–107.

3 Chuah, C.Y., Goh, K., Yang, Y. et al. (2018). Harnessing filler materials for enhancing biogas separation membranes. *Chemical Reviews* 118: 8655–8769.

4 Sumida, K., Rogow, D.L., Mason, J.A. et al. (2012). Carbon dioxide capture in metal–organic frameworks. *Chemical Reviews* 112: 724–781.

5 Lee, S.-Y. and Park, S.-J. (2015). A review on solid adsorbents for carbon dioxide capture. *Journal of Industrial and Engineering Chemistry* 23: 1–11.

6 Choi, S., Drese, J.H., and Jones, C.W. (2009). Adsorbent materials for carbon dioxide capture from large anthropogenic point sources. *ChemSusChem* 2: 796–854.

7 Yaumi, A., Bakar, M.A., and Hameed, B. (2017). Recent advances in functionalized composite solid materials for carbon dioxide capture. *Energy* 124: 461–480.

8 Mondal, M.K., Balsora, H.K., and Varshney, P. (2012). Progress and trends in CO_2 capture/separation technologies: a review. *Energy* 46: 431–441.

9 Yang, H., Xu, Z., Fan, M. et al. (2008). Progress in carbon dioxide separation and capture: a review. *Journal of Environmental Sciences* 20: 14–27.

10 Van der Bruggen, B., Isabel, C.E., and Patricia, L. (2011). Analysis of the development of membrane technology for gas separation and CO_2 capture. In: *Modern Applications in Membrane Science and Technology* (eds. I. Escobar and B. Van der Bruggen), 7–26. Washington, D.C.: American Chemical Society.

11 Lively, R.P., Chance, R.R., Kelley, B.T. et al. (2009). Hollow fiber adsorbents for CO2 removal from flue gas. *Industrial & Engineering Chemistry Research* 48: 7314–7324.

12 Keller, L., Ohs, B., Abduly, L., and Wessling, M. (2019). Carbon nanotube silica composite hollow fibers impregnated with polyethylenimine for CO_2 capture. *Chemical Engineering Journal* 359: 476–484.

13 Chung, T.-S., Jiang, L.Y., Li, Y., and Kulprathipanja, S. (2007). Mixed matrix membranes (MMMs) comprising organic polymers with dispersed inorganic fillers for gas separation. *Progress in Polymer Science* 32: 483–507.

14 Han, Y. and Zhang, Z. (2019). Nanostructured membrane materials for CO_2 capture: a critical review. *Journal of Nanoscience and Nanotechnology* 19: 3173–3179.

15 Khalilpour, R., Mumford, K., Zhai, H. et al. (2015). Membrane-based carbon capture from flue gas: a review. *Journal of Cleaner Production* 103: 286–300.

16 Chen, X.Y., Vinh-Thang, H., Ramirez, A.A. et al. (2015). Membrane gas separation technologies for biogas upgrading. *RSC Advances* 5: 24399–24448.

17 Robeson, L.M. (2008). The upper bound revisited. *Journal of Membrane Science* 320: 390–400.

18 Vinh-Thang, H. and Kaliaguine, S. (2013). Predictive models for mixed-matrix membrane performance: a review. *Chemical Reviews* 113: 4980–5028.

19 Keskin, S. and Sholl, D.S. (2010). Selecting metal organic frameworks as enabling materials in mixed matrix membranes for high efficiency natural gas purification. *Energy & Environmental Science* 3: 343–351.

20 Bae, T.-H., Liu, J., Thompson, J.A. et al. (2011). Solvothermal deposition and characterization of magnesium hydroxide nanostructures on zeolite crystals. *Microporous and Mesoporous Materials* 139: 120–129.

21 Giordano, G., Renzo, F.D., Remoueé, F. et al. (1994). Synthesis of high-silica zeolites with unidirectional medium pores systems using nitrogen-free templates. In: *Zeolites and Related Microporous Materials: State of the Art*, Studies in Surface Science and Catalysis, vol. 84 (eds. J. Weitkamp et al.), 141–146. Elsevier Science B.V.

22 Takako, N., Takuji, I., Chie, A. et al. (2012). Solvothermal synthesis of -lit-type zeolite. *Crystal Growth and Design* 12: 1752–1761.

23 Ren, L., Wu, Q., Yang, C. et al. (2012). Solvent-free synthesis of zeolites from solid raw materials. *Journal of the American Chemical Society* 134: 15173–15176.

24 Goh, P.S., Ismail, A.F., Sanip, S.M. et al. (2011). Recent advances of inorganic fillers in mixed matrix membrane for gas separation. *Separation and Purification Technology* 81: 243–264.

25 Rezakazemi, M., Ebadi Amooghin, A., Montazer-Rahmati, M.M. et al. (2014). State-of-the-art membrane based CO_2 separation using mixed matrix membranes (MMMs): an overview on current status and future directions. *Progress in Polymer Science* 39: 817–861.

26 Dyballa, M., Obenaus, U., Lang, S. et al. (2015). Brønsted sites and structural stabilization effect of acidic low-silica zeolite a prepared by partial ammonium exchange. *Microporous and Mesoporous Materials* 212: 110–116.

27 Du, X. and Wu, E. (2007). Porosity of microporous zeolites a, X and ZSM-5 studied by small angle X-ray scattering and nitrogen adsorption. *Journal of Physics and Chemistry of Solids* 68: 1692–1699.

28 Krishna, R. and van Baten, J.M. (2012). A comparison of the CO_2 capture characteristics of zeolites and metal–organic frameworks. *Separation and Purification Technology* 87: 120–126.

29 Bae, T.-H., Liu, J., Lee, J.S. et al. (2009). Facile high-yield Solvothermal deposition of inorganic nanostructures on zeolite crystals for mixed matrix membrane fabrication. *Journal of the American Chemical Society* 131: 14662–14663.

30 Pera-Titus, M. (2014). Porous inorganic membranes for CO_2 capture: present and prospects. *Chemical Reviews* 114: 1413–1492.

31 Rodenas, T., Van Dalen, M., Serra-Crespo, P. et al. (2014). Mixed matrix membranes based on NH2-functionalized MIL-type MOFs: influence of structural and operational parameters on the CO_2/CH4 separation performance. *Microporous and Mesoporous Materials* 192: 35–42.

32 Klimakow, M., Klobes, P., Thünemann, A.F. et al. (2010). Mechanochemical synthesis of metal-organic frameworks: a fast and facile approach toward quantitative yields and high specific surface areas. *Chemistry of Materials* 22: 5216–5221.

33 Basu, S., Khan, A.L., Cano-Odena, A. et al. (2010). Membrane-based technologies for biogas separations. *Chemical Society Reviews* 39: 750–768.

34 Wahby, A., Silvestre-Albero, J., Sepúlveda-Escribano, A., and Rodríguez-Reinoso, F. (2012). CO2 adsorption on carbon molecular sieves. *Microporous and Mesoporous Materials* 164: 280–287.

35 Carta, D., Casula, M.F., Bullita, S. et al. (2014). Direct sol–gel synthesis of doped cubic mesoporous SBA-16 monoliths. *Microporous and Mesoporous Materials* 194: 157–166.

36 Bakhtiari, N. and Azizian, S. (2015). Adsorption of copper ion from aqueous solution by nanoporous MOF-5: a kinetic and equilibrium study. *Journal of Molecular Liquids* 206: 114–118.

37 Lin, K.S., Adhikari, A.K., Ku, C.N. et al. (2012). Synthesis and characterization of porous HKUST-1 metal organic frameworks for hydrogen storage. *International Journal of Hydrogen Energy* 37: 13865–13871.

38 Valero, M., Zornoza, B., Téllez, C., and Coronas, J. (2014). Mixed matrix membranes for gas separation by combination of silica MCM-41 and MOF NH2-MIL-53(Al) in glassy polymers. *Microporous and Mesoporous Materials* 192: 23–28.

39 Ren, Z.F. and Huang, Z.P. (1998). Synthesis of large arrays of well-aligned carbon nanotubes on glass. *Science* 282: 1105–1107.

40 He, M., Zhang, S., Wu, Q. et al. (2018). Designing catalysts for chirality-selective synthesis of single-walled carbon nanotubes: past success and future opportunity. *Advanced Materials* 31: 1800805.

41 Chen, Y. and Zhang, J. (2014). Chemical vapor deposition growth of single-walled carbon nanotubes with controlled structures for nanodevice applications. *Accounts of Chemical Research* 47: 2273–2281.

42 Kumar, S., Ahlawat, W., Kumar, R., and Dilbaghi, N. (2015). Graphene, carbon nanotubes, zinc oxide and gold as elite nanomaterials for fabrication of biosensors for healthcare. *Biosensors and Bioelectronics* 70: 498–503.

43 Cazorla, C., Shevlin, S.A., and Guo, Z.X. (2011). Calcium-based functionalization of carbon materials for CO2 capture: a first-principles computational study. *Journal of Physical Chemistry C* 115: 10990–10995.

44 Gadipelli, S. and Guo, Z.X. (2015). Graphene-based materials: synthesis and gas sorption, storage and separation. *Progress in Materials Science* 69: 1–60.

45 Qiao, Z., Zhao, S., Wang, J. et al. (2016). A highly permeable aligned montmorillonite mixed-matrix membrane for CO$_2$ separation. *Angewandte Chemie* 128: 9467–9471.

46 Liu, G., Jin, W., and Xu, N. (2015). Graphene-based membranes. *Chemical Society Reviews* 44: 5016–5030.

47 Moore, T.T. and Koros, W.J. (2005). Non-ideal effects in organic-inorganic materials for gas separation membranes. *Journal of Molecular Structure* 739: 87–98.

48 Dai, H. (2002). Carbon nanotubes: opportunities and challenges. *Surface Science* 500: 218–241.

49 Li, Y., Chung, T.-S., Cao, C., and Kulprathipanja, S. (2005). The effects of polymer chain rigidification, zeolite pore size and pore blockage on polyethersulfone (PES)-zeolite a mixed matrix membranes. *Journal of Membrane Science* 260: 45–55.

50 Moaddeb, M. and Koros, W.J. (1997). Gas transport properties of thin polymeric membranes in the presence of silicon dioxide particles. *Journal of Membrane Science* 125: 143–163.

51 Mahajan, R. and Koros, W.J. (2002). Mixed matrix membrane materials with glassy polymers. Part 2. *Polymer Engineering & Science* 42: 1432–1441.

52 Lively, R.P., Chance, R.R., and Koros, W.J. (2010). Enabling low-cost CO2 capture via heat integration. *Industrial & Engineering Chemistry Research* 49: 7550–7562.

53 Lively, R.P., Leta, D.P., DeRites, B.A. et al. (2011). Hollow fiber adsorbents for CO2 capture: kinetic sorption performance. *Chemical Engineering Journal* 171: 801–810.

54 Rezaei, F., Lively, R.P., Labreche, Y. et al. (2013). Aminosilane-grafted polymer/silica hollow fiber adsorbents for CO2 capture from flue gas. *ACS Applied Materials & Interfaces* 5: 3921–3931.

55 Labreche, Y., Fan, Y., Rezaei, F. et al. (2014). Poly(amide-imide)/silica supported PEI hollow fiber sorbents for Postcombustion CO2 capture by RTSA. *ACS Applied Materials & Interfaces* 6: 19336–19346.

56 Fan, Y., Labreche, Y., Lively, R.P. et al. (2014). Dynamic CO2 adsorption performance of internally cooled silica-supported poly(ethylenimine) hollow fiber sorbents. *AIChE Journal* 60: 3878–3887.

57 Lee, J.S., Lively, R.P., Huang, D. et al. (2012). A new approach of ionic liquid containing polymer sorbents for post-combustion CO2 scrubbing. *Polymer* 53: 891–894.

58 Shu, S. and Husain, W.J. (2007). Koros, a general strategy for adhesion enhancement in polymeric composites by formation of nanostructured particle surfaces. *The Journal of Physical Chemistry C* 111: 652–657.

59 Lydon, M.E., Unocic, K.A., Bae, T.-H. et al. (2012). Structure–property relationships of inorganically surface-modified zeolite molecular sieves for nanocomposite membrane fabrication. *The Journal of Physical Chemistry C* 116: 9636–9645.

60 Ebadi Amooghin, A., Omidkhah, M., and Kargari, A. (2015). Enhanced CO_2 transport properties of membranes by embedding nano-porous zeolite particles into Matrimid® 5218 matrix. *RSC Advances* 5: 8552–8565.

61 Bryan, N., Lasseuguette, E., van Dalen, M. et al. (2014). Development of mixed matrix membranes containing zeolites for post-combustion carbon capture. *Energy Procedia* 63: 160–166.

62 Karkhanechi, H., Kazemian, H., Nazockdast, H. et al. (2012). Fabrication of homogenous polymer-zeolite nanocomposites as mixed-matrix membranes for gas separation. *Chemical Engineering & Technology* 35: 885–892.

63 Chaidou, C.I., Pantoleontos, G., Koutsonikolas, D.E. et al. (2012). Gas separation properties of polyimide-zeolite mixed matrix membranes. *Separation Science and Technology* 47: 950–962.

64 Pechar, T.W., Tsapatsis, M., Marand, E., and Davis, R. (2002). Preparation and characterization of a glassy fluorinated polyimide zeolite-mixed matrix membrane. *Desalination* 146: 3–9.

65 Ozturk, B. and Demirciyeva, F. (2013). Comparison of biogas upgrading performances of different mixed matrix membranes. *Chemical Engineering Journal* 222: 209–217.

66 Chong Lua, A. and Shen, Y. (2013). Influence of inorganic fillers on the structural and transport properties of mixed matrix membranes. *Journal of Applied Polymer Science* 128: 4058–4066.

67 Surya Murali, R., Ismail, A.F., Rahman, M.A., and Sridhar, S. (2014). Mixed matrix membranes of Pebax-1657 loaded with 4A zeolite for gaseous separations. *Separation and Purification Technology* 129: 1–8.

68 Sen, D., Kalipcilar, H., and Yilmaz, L. (2006). Development of zeolite filled polycarbonate mixed matrix gas separation membranes. *Desalination* 200: 222–224.

69 Şen, D., Kalıpçılar, H., and Yilmaz, L. (2007). Development of polycarbonate based zeolite 4A filled mixed matrix gas separation membranes. *Journal of Membrane Science* 303: 194–203.

70 Huang, Z., Li, Y., Wen, R. et al. (2006). Enhanced gas separation properties by using nanostructured PES-zeolite 4A mixed matrix membranes. *Journal of Applied Polymer Science* 101: 3800–3805.

71 Cakal, U., Yilmaz, L., and Kalipcilar, H. (2012). Effect of feed gas composition on the separation of CO_2/CH_4 mixtures by PES-SAPO 34-HMA mixed matrix membranes. *Journal of Membrane Science* 417–418: 45–51.

72 Oral, E.E., Yilmaz, L., and Kalipcilar, H. (2014). Effect of gas permeation temperature and annealing procedure on the performance of binary and ternary mixed matrix membranes of polyethersulfone, SAPO-34, and 2-hydroxy 5-methyl aniline. *Journal of Applied Polymer Science* 131: 40679.

73 Jha, P. and Way, J.D. (2008). Carbon dioxide selective mixed-matrix membranes formulation and characterization using rubbery substituted polyphosphazene. *Journal of Membrane Science* 324: 151–161.

74 Zhao, D., Ren, J., Li, H. et al. (2014). Poly(amide-6-b-ethylene oxide)/SAPO-34 mixed matrix membrane for CO_2 separation. *Journal of Energy Chemistry* 23: 227–234.

75 Rabiee, H., Meshkat Alsadat, S., Soltanieh, M. et al. (2015). Gas permeation and sorption properties of poly(amide-12-b-ethyleneoxide)(Pebax1074)/SAPO-34 mixed matrix membrane for CO_2/CH_4 and CO_2/N_2 separation. *Journal of Industrial and Engineering Chemistry* 27: 223–239.

76 Ward, J.K. and Koros, W.J. (2011). Crosslinkable mixed matrix membranes with surface modified molecular sieves for natural gas purification: I. preparation and experimental results. *Journal of Membrane Science* 377: 75–81.

77 Zhang, Y., Balkus, K.J. Jr.,, Musselman, I.H., and Ferraris, J.P. (2008). Mixed-matrix membranes composed of Matrimid® and mesoporous ZSM-5 nanoparticles. *Journal of Membrane Science* 325: 28–39.

78 Guo, X., Huang, H., Ban, Y. et al. (2015). Mixed matrix membranes incorporated with amine-functionalized titanium-based metal-organic framework for CO_2/CH_4 separation. *Journal of Membrane Science* 478: 130–139.

79 Sorribas, S., Zornoza, B., Téllez, C., and Coronas, J. (2014). Mixed matrix membranes comprising silica-(ZIF-8) core–shell spheres with ordered meso–microporosity for natural- and bio-gas upgrading. *Journal of Membrane Science* 452: 184–192.

80 Rodenas, T., van Dalen, M., Serra-Crespo, P. et al. (2014). Mixed matrix membranes based on NH_2-functionalized MIL-type MOFs: influence of structural and operational parameters on the CO_2/CH_4 separation performance. *Microporous and Mesoporous Materials* 192: 35–42.

81 Seoane, B., Sebastian, V., Tellez, C., and Coronas, J. (2013). Crystallization in THF: the possibility of one-pot synthesis of mixed matrix membranes containing MOF MIL-68(Al). *CrystEngComm* 15: 9483–9490.

82 Shahid, S. and Nijmeijer, K. (2014). High pressure gas separation performance of mixed-matrix polymer membranes containing mesoporous Fe(BTC). *Journal of Membrane Science* 459: 33–44.

83 Rodenas, T., Luz, I., Prieto, G. et al. (2015). Metal–organic framework nanosheets in polymer composite materials for gas separation. *Nature Materials* 14: 48–55.

84 Shahid, S. and Nijmeijer, K. (2014). Performance and plasticization behavior of polymer–MOF membranes for gas separation at elevated pressures. *Journal of Membrane Science* 470: 166–177.

85 Thompson, J.A., Vaughn, J.T., Brunelli, N.A. et al. (2014). Mixed-linker zeolitic imidazolate framework mixed-matrix membranes for aggressive CO_2 separation from natural gas. *Microporous and Mesoporous Materials* 192: 43–51.

86 Chen, X.Y., Hoang, V.-T., Rodrigue, D., and Kaliaguine, S. (2013). Optimization of continuous phase in amino-functionalized metal-organic framework (MIL-53) based co-polyimide mixed matrix membranes for CO_2/CH_4 separation. *RSC Advances* 3: 24266–24279.

87 Zhang, D.-S., Chang, Z., Li, Y.-F. et al. (2013). Fluorous metal-organic frameworks with enhanced stability and high H_2/CO_2 storage capacities. *Scientific Reports* 3: 3312.

88 Dai, Y., Johnson, J.R., Karvan, O. et al. (2012). Ultem®/ZIF-8 mixed matrix hollow fiber membranes for CO_2/N_2 separations. *Journal of Membrane Science* 401-402: 76–82.

89 Nafisi, V. and Hägg, M.-B. (2014). Development of dual layer of ZIF-8/PEBAX-2533 mixed matrix membrane for CO_2 capture. *Journal of Membrane Science* 459: 244–255.

90 Abedini, R., Omidkhah, M., and Dorosti, F. (2014). Highly permeable poly(4-methyl-1-pentyne)/NH_2-MIL 53 (Al) mixed matrix membrane for CO_2/CH_4 separation. *RSC Advances* 4: 36522–36537.

91 Askari, M. and Chung, T.S. (2013). Natural gas purification and olefin/paraffin separation using thermal cross-linkable co-polyimide/ZIF-8 mixed matrix membranes. *Journal of Membrane Science* 444: 173–183.

92 Nik, O.G., Chen, X.Y., and Kaliaguine, S. (2012). Functionalized metal organic framework-polyimide mixed matrix membranes for CO_2/CH_4 separation. *Journal of Membrane Science* 413-414: 48–61.

93 Ni, Z. and Masel, R.I. (2006). Rapid production of metal–organic frameworks via microwave-assisted Solvothermal synthesis. *Journal of the American Chemical Society* 128: 12394–12395.

94 Bae, T.-H., Lee, J.S., Qiu, W. et al. (2010). A high-performance gas-separation membrane containing submicrometer-sized metal–organic framework crystals. *Angewandte Chemie International Edition* 49: 9863–9866.

95 Bae, T.-H. and Long, J.R. (2013). CO2/N2 separations with mixed-matrix membranes containing Mg2(dobdc) nanocrystals. *Energy & Environmental Science* 6: 3565–3569.

96 Seoane, B., Coronas, J., Gascon, I. et al. (2015). Metal-organic framework based mixed matrix membranes: a solution for highly efficient CO2 capture? *Chemical Society Reviews* 44: 2421–2454.

10

Poly(Amidoamine) Dendrimers for Carbon Capture

Ikuo Taniguchi

International Institute for Carbon-Neutral Energy Research (WPI-I2CNER), Kyushu University, Fukuoka, Japan

10.1 Introduction

This chapter describes the pervasive potential of dendrimers, especially poly(amidoamine) (PAMAM) dendrimers, as promising materials for effective carbon capture. The word *dendrimer* originates from *dendron* in Greek for tree or branch and postulates highly branched, monodisperse macromolecular compounds with well-defined chemical structures and physicochemical properties of both the interior and exterior surface of the nanostructures [1, 2].

In 1978, Vögtle and co-workers reported the first synthesis of polyamine dendritic structure as cascade molecules by the Michael addition of acrylonitrile to benzylamine to give a bisnitrile and the following reduction of the nitrile groups to primary amines, which corresponds to generation 0 (*G* 0) dendrimer. Reiteration of the stepwise reactions gives a hyperbranched or starburst architecture of the nanoscopic compounds or higher-generation dendrimer, as shown in Figure 10.1a [3]. The building block chemistry

Materials for Carbon Capture, First Edition. Edited by De-en Jiang, Shannon M. Mahurin and Sheng Dai.
© 2020 John Wiley & Sons Ltd. Published 2020 by John Wiley & Sons Ltd.

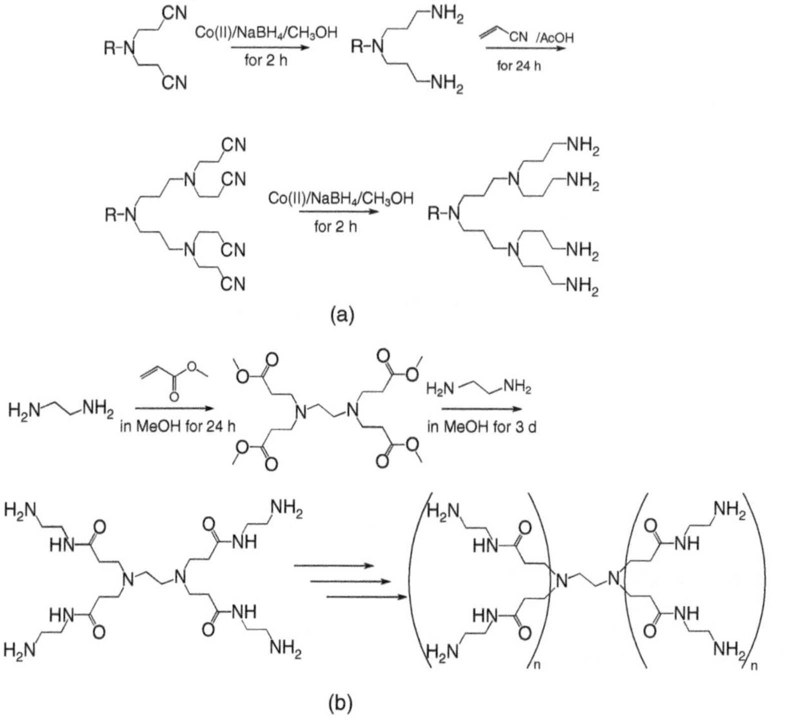

Figure 10.1 Synthetic schemes of (a) Vögtle's cascade molecules and (b) poly(amidoamine) (PAMAM) dendrimers.

is referred to as the *divergent method*, and key contributions to the method have been made by Denkewalter [4], Newkome [5], and Tomalia [6] to introduce a diverse family of dendrimers including poly(amines), PAMAMs, poly(ethers), poly(siloxanes), poly(thioethers), poly(amidoalcohols), poly(phosphoniums), poly(alkanes), and poly(nucleic acids) [1]. The chemistry of dendrimers as quantized building blocks and the many potential applications have been extensively investigated in the past couple of decades, and a number of articles and reviews have been seen in the scientific literature [1–15]. Among the diverse dendrimer families, PAMAM dendrimers are the most intensively studied and are developed by repetition of the two-step sequential reactions of methylacrylate to primary amine and ethylenediamine to the resulting methoxy group, as shown in Figure 10.1b [6].

With increase of the dendrimer generation, the number of surface terminal functional groups is exponentially increased, and an increasing number of branching reactions is required on the molecule to extend the generation [16, 17], which sometimes causes incomplete growth of the branching reactions or defects in the higher-generation dendrimer by the so-called *de Gennes dense-packed stage* [18]. On the other hand, a solution associated with the divergent method was first reported by Fréchet and Hawker [19] and by Neenan and Miller [20], independently. The methodology is called a *convergent approach*, and as in the divergent approach, basically each generation is formed by two sequential reactions. Dendrimer formation proceeds from surface to core in the convergent approach, while the dendritic architectures grow from core to surface in the divergent approach.

Significant defects of the surface functional groups are suppressed by the convergent method; however, the reactive group at the focal point (core) has less reactivity in the higher-generation synthesis due to steric hindrance, which results in low reactivity and efficiency [7].

In both the divergent and convergent approaches, difficulties are often found in the syntheses of higher-generation dendrimers with well-defined or precisely controlled nanoarchitectures as described previously. However, covalent assembly of different dendrons developed by the divergent method explores the potential of dendrimer chemistry with tuning of the versatile functionalities. Nanostructure beyond the dendrimer has also been a topic, and more complex nanoscale architectures have been developed by assembling dendrimers as reactive modules [11].

The spherical structure of dendrimers displays unusual physical properties. The solution viscosity is much lower than that of linear polymers with the same molecular weights [21]. Thus, the dendritic structures can be used for rheology modifiers, surface-modifying agents, lubricants, and so on [2]. However, the fascinating nanoscopic architectures of dendrimers have mostly stimulated imaginations for biomedical application for many years, such as bioimaging, drug and gene carriers, drugs/vaccines, and scaffolds for tissue repair [22]. A typical example is encapsulation of the bioactive agents in the interior nanospace of the dendrimers, and the controlled release of the agents has been also studied.

10.2 Poly(Amidoamine) in CO$_2$ Capture

Apart from such biomedical and pharmaceutical applications, an interesting feature of PAMAM dendrimers was first reported by Sirkar and co-workers at the New Jersey Institute of Technology at the very beginning of this century [23–25], in which the promising potential of the dendrimers was introduced as materials for CO$_2$ capture. PAMAM dendrimers are one of the most important and well-characterized/used dendrimer families in the history of dendritic molecules.

10.2.1 A Brief History

In addition to the synthetic chemistries, the physical properties of PAMAM dendrimers have been also studied. The nanoscale dimensions have been reported by Tomalia and co-workers: for example, the sizes of ammonia core-PAMAM dendrimers increases from 3.1, 4.0, 5.3, 6.7, and 8.0 nm with increase of the dendrimer generation G from 3.0 to 7.0 [26]. The size-scaling properties of PAMAM dendrimers have captivated the potential application of dendrimers in the field of size-exclusion chromatography (SEC). Sirkar and co-workers tried to use the dendrimers with various generations for calibration agents in SEC. It was the first attempt for them to use PAMAM dendrimers, and due to the high density of primary amines of the dendrimers, they found unusually high CO$_2$ separation performance of the dendrimers, when the dendrimers were immobilized in a porous poly(vinylidene fluoride) (PVDF) substrate to form an immobilized liquid membrane (ILM) [23]. This serendipity gave rise to increased attention on dendritic molecules as promising for CO$_2$ capture.

In comparison to glassy polymeric membranes, the PAMAM ILMs gave a higher separation factor over N_2, especially at low CO_2 partial pressures under isobaric conditions of atmospheric pressure. With such dense membrane systems, the permeability P of gas is expressed by multiplication of the solubility S in the membrane and the diffusivity D in the membrane as $P = S \times D$. Due to specific interaction between CO_2 and the amino groups of PAMAM, the solubility of CO_2 is considerably elevated to give much higher permeability.

The PAMAM ILMs are preferable for CO_2 separation at coal-fired power plants (post-combustion carbon capture) or at steel works. The target gas containing saturated water vapor is exhausted at ambient temperature, and the CO_2 partial pressure is low and mostly between 10 and 25 vol.% depending on the sources, although the gas permeability should be improved for implementation. Mechanism of the attractive preferential CO_2 permeation through PAMAM ILMs is explained by a carrier-mediated facilitated transportation mechanism, often seen on the biomembranes by transmembrane proteins [27].

For preparation of the PAMAM ILMs, PAMAM/methanol solution was added on a porous PVDF support, and then the solvent was removed under vacuum. They confirmed no significant leakage of PAMAM dendrimers from the ILMs over a month at total pressure of 121.6 kPa, which results in decrease of the CO_2 separation properties. The CO_2/N_2 separation factor of the PAMAM ($G\,0$) ILMs at CO_2 partial pressure $p(CO_2)$ of 0.36 cmHg or 480 Pa was around 18 000 with the permeability $P(CO_2)$ or permeance $Q(CO_2)$ of 3600 barrer and 9.8 GPU, respectively, under highly humidified conditions [24]. They have concluded that the excellent CO_2 selectivity over N_2 was a result of a CO_2-selective molecular gate function, although detailed preferential CO_2 permeation mechanisms have not been postulated. With increase of $p(CO_2)$, the separation performance goes down, which is often found in the carrier-mediated facilitated transportation manner. The $\alpha(CO_2/N_2)$ and $Q(CO_2)$ drop to 720 and 0.4 GPU, respectively, at $p(CO_2)$ of 30 cmHg or 40 kPa. However, the CO_2 selectivity over N_2 is still much higher than that of other polymeric materials [28], which stimulates further manipulation of PAMAM dendrimers as effective CO_2 capture materials.

10.2.2 Immobilization of PAMAM Dendrimers

PAMAM dendrimers with lower generations flow at ambient conditions, and the CO_2 separation performance of the dendrimer ($G\,0$) ILMs has been studied in the form of flat membranes for the gas permeation test at ambient pressure. Thinking of fabrication of membrane modules or use under pressure, stable immobilization of the fascinating dendritic molecules should be an inevitable issue for CO_2 capture. Effective immobilization of PAMAM dendrimers to suitable polymer matrices has been studied by Kazama and co-workers at the Research Institute of Innovative Technology for the Earth (RITE, Kyoto, Japan).

10.2.2.1 Immobilization in Crosslinked Chitosan

They succeeded in the immobilization of PAMAM dendrimers ($G\,0$) in a crosslinked chitosan on a lumen side of a commercial polysulfone (PSF) hollow-fiber membrane for ultra-filtration to form a composite membrane configuration by the in situ modification (IM) method, as shown in Figure 10.2 [29, 30]. In brief, chitosan/acetic acid aqueous solution was circulated in the lumen side of the hollow-fiber while the shell side was

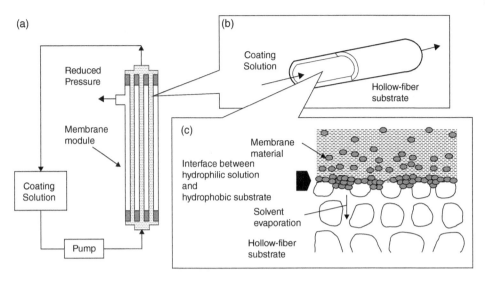

Figure 10.2 Schematic illustration of the in situ modification (IM) method. (a) Flow diagram; (b) solution flow in a hollow-fiber; (c) accumulation of solutes on the inner surface of a hollow-fiber. Source: reprinted from Ref. [30] with permission from Elsevier Science.

evacuated as schematized in Figure 10.2. Then, ethylene glycol diglycidyl ether (EGDGE) was added to crosslink the polysaccharide by the reaction between primary amines of chitosan and epoxide rings of EGDGE.

The circulation of chitosan solution followed by EGDGE crosslinking resulted in the formation of a chitosan gutter layer on the inner surface of the hollow-fiber with thicknesses of a couple of hundred nanometers. The thickness was controlled by the circulation period of the chitosan solution. Then aqueous PAMAM dendrimer solution was also circulated in the lumen side of the resulting hollow-fiber membrane to immobilize the dendrimer into the crosslinked chitosan. The SEM images in Figure 10.3 revealed the formation of a composite membrane configuration prepared by the IM method. After circulation of the dendrimer solution, increase of the thickness of the chitosan layer was observed, which indicated dendrimer loading in the polysaccharide matrix. The CO_2 separation properties over N_2 of the composite hollow-fiber membranes were examined at $p(CO_2)$ of 5 kPa and 313 K under atmospheric pressure, and the results are listed in Table 10.1. The pristine PSF exhibits high gas permeability, 1.5×10^{-7} m³(STP)/(m² s Pa) or 2.0×10^4 GPU in CO_2 permeance, with no selectivity due to its porous nature. By coating the inner surface with crosslinked chitosan, CO_2 permeance indeed dropped down by one order. However, CO_2 and N_2 penetrated through a pore or defect to give equivalent permeance, even though the surface pore was filled by the chitosan coating in the SEM images (Figure 10.3). On the other hand, after PAMAM dendrimer immobilization, the CO_2 selectivity rose to 400, while CO_2 permeance decreased to 1.6×10^{-10} m³(STP)/(m² s Pa) or 21.3 GPU under the same operation conditions. The significant enhancement in the selectivity was due to specific interaction between CO_2 and dendrimers immobilized physically in the crosslinked chitosan. The CO_2 separation performance of the hollow-fiber membranes did not change over 1000 hours under the operation conditions, although the results were unpublished. The long-term durability

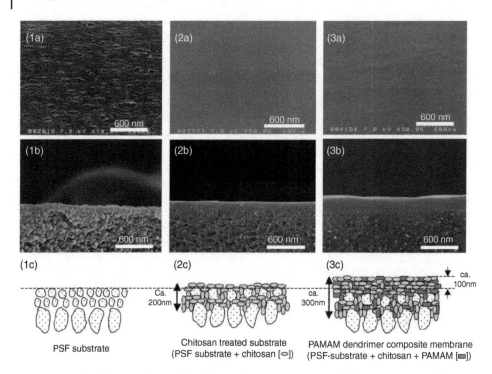

Figure 10.3 SEM images of poly(amidoamine) (PAMAM) dendrimer composite membrane and substrates. (1) polysulfone (PSF) support, (2) PSF support + chitosan and (3) PSF support + chitosan + PAMAM dendrimer, where a, b, and c are surface, cross-sectional, and schematic images, respectively. Source: reprinted from Ref. [30] with permission from Elsevier Science.

Table 10.1 CO_2 separation properties over N_2 of various composite hollow-fiber membranes.

	CO_2 permeance		
	$m^3(STP)/(m^2 \, s \, Pa)$	GPU	$\alpha(CO_2/H_2)$
Polysulfone (PSF) substrate	1.5×10^{-7}	20 000	1
PSF substrate + Chitosan	4.3×10^{-8}	5700	1
PSF substrate + Chitosan + poly(amidoamine) (PAMAM)	1.6×10^{-10}	21	400

CO_2 partial pressure: 5 kPa at atmospheric pressure. Feed (CO_2/N_2 5/95 by vol%) and sweep (He) gas-flow rates: 100 and 10 ml min^{-1}, respectively at 313 K under saturated humid conditions.

indicated that PAMAM dendrimers were stably incorporated into the crosslinked chitosan gutter layer. The immobilization of PAMAM dendrimers by the IM method is straightforward and readily scalable. These simple procedures are of great importance in processing a membrane module for demonstration.

X-ray photoelectron spectroscopy (XPS) confirmed a peak at 410 eV (N_{1s}) on the inner surface of the hollow-fiber after the dendrimer circulation, which indicated dendrimer immobilization in the crosslinked chitosan. The PAMAM dendrimer loading efficiency was limited by the IM method, and the maximum loading was ca. 30 wt% in the CO_2 selective layer. Thus, an alternative immobilization strategy of PAMAM dendrimers should be developed to improve the loading efficiency for higher CO_2 separation performance.

10.2.2.2 Immobilization in Crosslinked Poly(Vinyl Alcohol)

Poly(vinyl alcohol) (PVA) is one of the potential candidates as a polymer matrix for stable incorporation of PAMAM dendrimers. In addition to good membrane formability, the high mechanical properties and thermal stability of PVA are suitable for use under pressure. Ho and co-workers have been intensively investigating PVA membranes for use under pressure, such as CO_2 capture at an integrated coal gasification combined cycle (IGCC) plant (pre-combustion carbon capture), where CO_2 has to be separated over H_2 [31–39]. They have developed sterically hindered amine-blended crosslinked PVA membranes, in which the amines work as fixed carriers for CO_2 by a facilitated transportation manner. The PVA-based membranes having N-isopropylamine moiety exhibit high CO_2 separation properties under pressure, $p(CO_2)$ of 40.5 kPa. The CO_2 permeability and separation factor over H_2 are 300 barrer (14 GPU in the permeance) and 40, respectively, at 383 K under humidified conditions [33]. The CO_2 separation performance can be further extended by the addition of fumed silica to the PVA-based membranes to form mixed matrix membranes. CO_2 permeability and selectivity are improved to 1300 barrer (41 GPU) and 87, respectively, at total pressure of 1.5 MPa ($p(CO_2)$ of 303 kPa) and 380 K under 73% relative humidity [35]. Thus, PVA-derivative membranes hold promising potential for pre-combustion carbon capture.

PAMAM dendrimer (G 0) immobilization in a crosslinked PVA has also been investigated at RITE [40, 41]. Commercial PVA is dissolved in water and crosslinked by a titanium crosslinker, diisopropoxybis(triethanolaminato)titanium. The crosslinking reaction is carried out in water, and the Ti crosslinker selectively reacts to the hydroxyl group of PVA in the presence of PAMAM dendrimers. A commercial Ti crosslinker/isopropanol solution from Matsumoto Fine Chemical (Chiba, Japan) is added to a PVA aqueous solution with a predetermined reaction stoichiometry. After agitation for a half hour, the viscosity of the reaction mixture is increased due to the partial crosslinking. The reaction mixture is cast on a Teflon dish, and the PAMAM/PVA membranes are formed during slow evaporation of water for a couple of days at ambient conditions to avoid defect formation. The residual water is then removed under vacuum, and the following heating process completes the crosslinking reaction. The amount of the casting solution to a Teflon dish determines the thickness of the resulting polymeric membranes. PAMAM dendrimer incorporation can be readily tuned by controlling the mixture ratio of the amine to PVA, and a self-standing membrane containing up to 82.8 wt% of PAMAM dendrimer is available. The PAMAM/PVA membranes show CO_2-selective permeation over H_2 as well as the PVA derivative membranes developed by Ho and co-workers. At $p(CO_2)$ of 5 kPa and 313 K under 80% relative humidity, CO_2 permeation is much higher than H_2 permeation, as shown in Figure 10.4a. While H_2 permeance decreases from 4.71×10^{-14} to 3.86×10^{-14} m³(STP)/(m² s Pa) or 6.28×10^{-3} to 5.15×10^{-3} GPU, CO_2 permeance increases from 1.00×10^{-10} to 4.71×10^{-10} m³(STP)/(m² s Pa) or 13.3 to 62.8 GPU with increase of the PAMAM dendrimer fraction from 22.7 to 82.8 wt% in the polymeric membranes. As a result, the separation factor reached 1200 at the

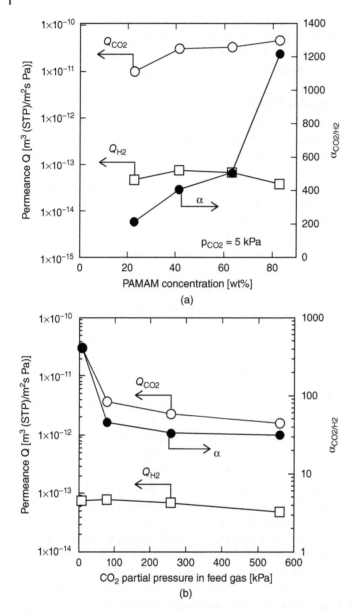

Figure 10.4 CO_2 separation properties of poly(amidoamine)/poly(vinyl alcohol (PAMAM/PVA) membranes under 80% relative humidity at 40 °C. (a) Effect of PAMAM dendrimer concentration at $p(CO_2)$ of 5 kPa; (b) effect of $p(CO_2)$ in the feed gas of PAMAM/PVA membrane (41.6/52.0 by wt). Source: reprinted from Ref. [40] with permission from Elsevier Science.

operation conditions. The obtained results also indicate selective CO_2 permeation based on the amine-mediated facilitated transportation mechanism. The CO_2 separation test is also conducted under pressurized conditions by varying $p(CO_2)$ from 5 to 560 kPa, as shown in Figure 10.4b. CO_2 permeance decreases as expected due to the facilitated transportation manner, in which permeance is inversely proportional to transmembrane CO_2 partial pressure. However, $\alpha(CO_2/H_2)$ does not change much over $p(CO_2)$ of 240 kPa. In a typical carrier-mediated facilitated transportation, CO_2 selectivity over H_2 also should be decreased with increase of $p(CO_2)$. The obtained results have often been found in the gas-separation test of PAMAM membranes under pressure, although the reason has not been made clear. The CO_2 permeance and selectivity are 1.2×10^{-11} m³(STP)/(m² s Pa) or 1.6 GPU and 42, respectively, at $p(CO_2)$ of 560 kPa and 333 K under 80% relative humidity. In comparison to the PVA-based membranes developed (25 μm in thickness) by Ho and co-workers, the thickness of PAMAM/PVA membranes is about 400 μm [40]. The lower CO_2 permeance would be improved by reducing the membrane thickness.

Another advantage of PVA is good compatibility with amines, which were homogeneously distributed over the PVA matrix. Transparency of the resulting membrane suggests no phase separation between amines and PVA on microns or higher scale. In addition, small-angle X-ray scattering (SAXS) measurement gives no peaks in the region from 0.2 to $1.0\,nm^{-1}$ in the scattering vector [41]. PAMAM dendrimers and the polymer matrix are miscible, while poly(amidoamine)/poly(ethylene glycol) (PAMAM/PEG) membranes show a phase-separated structure on a couple of microns scale as described shortly. Thus, fabrication of a thinner membrane would be one of the effective approaches to elevate CO_2 permeance. The CO_2 separation performance of PAMAM/PVA membranes did not change for over a week in the operating conditions. The leakage of dendrimers from the membranes is negligible under pressurized conditions, indicating stable immobilization of the dendrimers in the crosslinked PVA. The excellent mechanical and thermal properties of PVA represent a promising membrane matrix for PAMAM dendrimers. However, the relatively complex and time-consuming fabrication procedures of the PAMAM/PVA membranes should be improved to scale up for pilot and demonstration studies.

10.2.2.3 Immobilization in Crosslinked PEG

PEG or poly(ethylene oxide) (PEO) shows excellent CO_2 solubility due to quadruple-pole interaction between the EO unit and CO_2 and has been well known as a CO_2-philic polymer. Various PEG-containing polymers have been developed, and some of them are commercialized, such as copolymer of PEG and poly(butylene terephthalate) (Polyactive®) and PEG and polyamide (Pebax®). While high-molecular-weight crystalline PEGs often suppress gas transport, amorphous and flexible short PEG chains are suitable for preferential CO_2 permeation. Freeman and co-workers reported CO_2 separation with a crosslinked PEG membrane obtained by photopolymerization of PEG diacrylates with good CO_2 permeability under pressurized conditions [42–44]. The CO_2 permeance and selectivity over H_2 are 6.3 GPU and 9.4, respectively, at 308 K and $p(CO_2)$ of 1.7 MPa under dry conditions [45].

The pressure tolerance of the crosslinked PEG membranes is quite attractive for gas separation under pressurized conditions, such as in pre-combustion carbon capture. Incorporation of PAMAM dendrimers in a crosslinked PEG has been investigated by Taniguchi and co-workers [46]. PAMAM dendrimers can be entrapped by photopolymerization of PEG dimethacrylate (PEGDMA) in the presence of the dendrimers in ethanol. UV curing is initiated by 1-hydroxycyclohexyl phenyl ketone (Irgacure 184), where the molar ratio of the photoinitiator to methacrylate is 60–100. Macroscopic homogeneous membranes can be readily obtained by UV irradiation for a few minutes. A self-standing membrane is also obtained in the same manner when the dendrimer fraction is below 72 wt% in the resulting polymeric membranes. The PAMAM/PEG membranes exhibit preferential CO_2 permeation over H_2 as well as PAMAM/PVA membranes, and the separation factor is much higher than the crosslinked PEG membranes without PAMAM dendrimers. As shown in Figure 10.5a, the $\alpha(CO_2/H_2)$ of PAMAM/PEG membranes goes up with increase of the dendrimer (G 0) weight fraction at $p(CO_2)$ of 5 kPa and 298 K under 80% relative humidity. CO_2 permeance actually rises from 2.17×10^{-3} to 3.65×10^{-3} GPU with increase of the dendrimer fraction. However, a significant decrease in H_2 permeance from 3.19×10^{-4} to 2.96×10^{-6} GPU results in providing excellent CO_2 selectivity over H_2 under the operation conditions. Thus, the dendrimers suppress H_2 permeation rather than accelerating CO_2 permeation. The gas permeances are mostly lower than the other PAMAM ILMs mentioned earlier because the PAMAM/PEG membrane thickness is much greater and about 400–650 µm.

PAMAM dendrimers with various generations are commercially available, and the effect of CO_2 separation performance on the generation is studied under the same experimental conditions. The obtained results are displayed in Figure 10.5b, where the dendrimer weight fraction is kept to 50 wt%. Both the CO_2 and H_2 permeances increase with increase of the dendrimer generation. The number of primary amines indeed increases in one dendritic molecule as the generation grows, but the amine density decreases because of the fixed PAMAM dendrimer weight fraction in the membrane in Table 10.2. The amine density of PAMAM dendrimer (G 0) is 3.87 mmol g^{-1}-membrane and much higher than that of the other dendrimers with higher generations. In addition, increase in gas permeation results from non-specific gas permeation through the interior nanospace of the dendrimer. As a consequence, the smallest PAMAM dendrimer (G 0) gives the highest CO_2 selectivity among the dendrimers examined in crosslinked PEGs. Thinking of the de Gennes dense-packed stage [18], the branches of PAMAM dendrimers used in this study would be flexible enough not to suppress gas permeation through the inside of the dendrimers. Hereafter, the PAMAM dendrimer (G 0) is used for the PAMAM/PEG membrane preparation.

Facile immobilization of PAMAM dendrimers into a crosslinked PEG is introduced. PAMAM/PEG membranes can be readily prepared by the UV curing method, and the resulting polymeric membranes exhibit very high CO_2 selectivity over H_2. As well as PAMAM/PVA membranes, the leakage of PAMAM dendrimers is not confirmed during gas-separation experiments. The CO_2 separation performance of PAMAM/PEG membranes is strongly dependent on the dendrimer weight fraction and generation. Interplay between the morphology and CO_2 separation properties is studied to understand the key factors for preferential CO_2 permeation in the following section.

10.3 Factors to Determine CO_2 Separation Properties

10.3.1 Visualization of Phase-Separated Structure

The photopolymerization procedure to incorporate PAMAM dendrimers into a crosslinked PEG is quite effective for preparation of CO_2 separation membranes, and the dendrimer weight fraction also can be controlled by the UV curing method. The resulting PAMAM/PEG membranes are translucent, whereas a crosslinked PEG prepared in the absence of the dendrimer is transparent, as shown in Figure 10.6a. PEGDMA and PAMAM

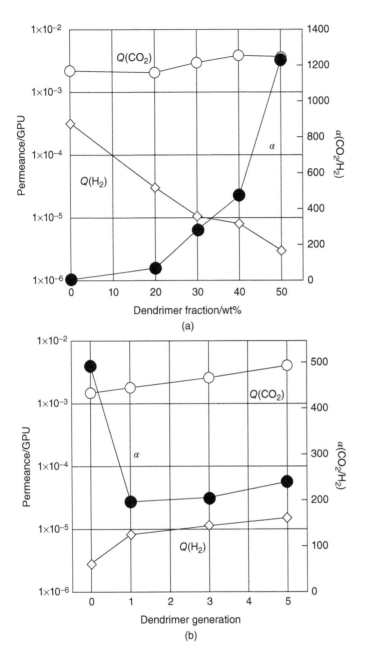

Figure 10.5 CO$_2$ separation properties of poly(amidoamine)/poly(ethylene glycol) (PAMAM/PEG) membranes over H$_2$ at 25 °C under 80% relative humidity and at p(CO$_2$) of 5 kPa. Average PEG unit is 14. (a) Effect of dendrimer (G 0) fraction; (b) effect of dendrimer generation with constant dendrimer fraction (50 wt%).

Table 10.2 Amine density in poly(amidoamine)/poly(ethylene glycol) (PAMAM/PEG) membranes with various dendrimer generations.

Dendrimer generation	Amines	MW	Amines/mmol/g-membrane[a]
0.0	4	517	3.87
1.0	8	1430	2.80
3.0	32	6909	2.32
5.0	128	28 826	2.22

a) PAMAM/PEG membranes are prepared by photopolymerization of photopolymerization of PEG dimethacrylate (PEGDMA) in the presence of PAMAM dendrimers, where the dendrimer weight fraction in the resulting membranes is kept to 50 wt%.

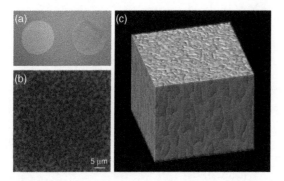

Figure 10.6 Structural analysis of poly(amidoamine)/poly(ethylene glycol) (PAMAM/PEG) membranes prepared by photopolymerization. Average PEG unit is 14. (a) Left: with PAMAM dendrimer; right: without the dendrimer; $\phi = 1.0$ cm. (b) A fluorescent image by LSCM of PAMAM/PEG membrane (50/50 by wt.) at 50 μm depth from the surface. (c) A reconstructed 3D image of PAMAM/PEG membrane (50/50 by wt.) in $35 \times 35 \times 30$ μm. PEG-rich, and the dendrimer-rich phases are colored in green and yellow, respectively. (*See color plate section for color representation of this figure*).

dendrimers are dissolved in ethanol in the photopolymerization due to immiscibility of the solutes. Thus, the turbidity of PAMAM/PEG membranes results from phase separation between the PAMAM dendrimer and the polymer matrix on a micron scale. The detailed morphology is studied on a laser scanning confocal microscope (LSCM) [47–49].

To visualize a phase separation under LSCM, the fluorescein moiety is tagged covalently to the hydroxyl group of hydroxyl-terminated PEG methacrylate to form fluorescein-tagged PEGMA (FITC-PEGMA), which is copolymerized with PEGDMA in the presence of PAMAM dendrimers to stain only the polymer matrix [50]. Figure 10.6b exhibits a typical LSCM image of PAMAM/PEG membrane with PAMAM weight fraction of 50% at 50 μm from the membrane surface. A phase-separated structure on a couple of microns scale is seen as expected: so-called *macrophase separation*, where the bright and dark phases indicate the PEG-rich and dendrimer-rich domains, respectively, under fluorescent LSCM. The fluorescent images are processed computationally to extract the interface properly between the PEG- and PAMAM-rich domains [48, 50–52]. The 3D structure is reconstructed in $35 \times 35 \times 30$ μm by the processed images at each depth as shown in

Figure 10.6c, where the PEG-rich and PAMAM dendrimer-rich domains are colored green and yellow, respectively. Formation of a bicontinuous structure is confirmed upon the photopolymerization-induced phase separation on a couple of microns scale. The phase-separated structure is dependent on the dendrimer weight fraction, and an average spacing of PEG-rich domain (Λ_{ave}) is determined from the fluorescent images, which corresponds to an average PAMAM domain size [48]. A plot of the magnitude of the Fourier transformation as a function of the wavenumber (q) shows a peak derived from a periodical structure of the PEG-rich phase, and the average PAMAM domain size (Λ_{ave}) is calculated from the q at the peak top (q_{max}) by the following equation:

$$\Lambda_{ave} = \frac{2\pi}{q_{max}}$$

In general, an immiscible polymer blend would give such macrophase-separated structure, and the average domain size would increase with incubation period. However, PEG is crosslinked in the PAMAM/PEG membranes, thus the phase separation is stationary or fixed. In the polymeric membrane formulations, PEGDMA with 750 average molecular weight has been used, and the average ethylene glycol (EG) unit is 14. Since PEG shows poor compatibility with PAMAM dendrimers, the longer PEG receives higher repulsive force from the PAMAM dendrimers, which results in providing different Λ_{ave} values. PEGDMAs with various EG units are used to prepare the PAMAM/PEG membranes, and the polymerization-induced phase separation is examined under LSCM by the same procedures. The results of the calculations are summarized in Table 10.3.

The characteristic values of the phase separation are related to the dendrimer weight fraction and number of EG units. When the dendrimer fraction is 50 wt%, the volume fractions are close to 50 vol%. However, the values of the volume fractions are not equivalent to those of the weight fraction when the values are far from symmetric, which indicates partial mixing of the minor domain with the major one. With increase of the PAMAM weight fraction, Λ_{ave} increases from 2.0 ± 0.1 to 4.1 ± 0.1 μm, and the interface area per unit volume decreases from 3.1 ± 0.2 to 1.6 ± 0.1 μm^2 μm^{-3}. As the dendrimer-rich domain grows, the curvature of the interface becomes smoother, with a decrease in the interface/volume.

Table 10.3 Effect of poly(amidoamine) (PAMAM) dendrimer weight fraction and ethylene glycol (EG) unit length on photopolymerization-induced phase separation between PAMAM dendrimer and poly(ethylene glycol) (PEG) matrix.

PAMAM weight fraction/wt%	Average EG unit	Λ_{ave} μm^{-1}	PAMAM volume fraction/vol%	Interface area/volume/ μm^2 μm^{-3}
30	14	2.0 ± 0.1	44.3 ± 0.2	3.1 ± 0.2
50	3	2.8 ± 0.1	47.5 ± 2.4	2.0 ± 0.1
	9	2.5 ± 0.2	49.9 ± 2.0	2.3 ± 0.1
	14	2.2 ± 0.2	49.2 ± 1.0	2.5 ± 0.1
	23	2.7 ± 0.1	50.9 ± 0.5	2.2 ± 0.1
70	128	4.1 ± 0.0	56.0 ± 0.3	1.6 ± 0.0

a) $\pm$ denotes standard deviation (n = at least 3).

On the other hand, when the number of EG units of PEGDMA changes from 3 to 23, Λ_{ave} first goes up and then goes down [53]. In comparison to the other polymerization manners of methacrylates, this polymerization requires a much shorter reaction period, and thus both kinetics and thermodynamics should be taken into consideration. The rate of phase separation R is expressed by multiplying the diffusion coefficient of monomer D and thermodynamic driving force ε. The longer PEG chain or higher number of EG units gives rise to the ε due to incompatibility of the dendrimer and the matrix, although the D of PEGDMA decreases. PAMAM/PEGDMA membranes with EG units of 3 and 23 represent a lower interface area/volume and a larger Λ_{ave} than those with EG units of 9 and 14. PEGDMA (EG 23) gives the smallest D among used in the reaction conditions; however, the ε is the greatest, and the onset of phase separation should be the earliest. As a consequence, the photopolymerization reaction reaches the late stage of phase separation even during the short reaction period. On the other hand, with PEGDMA (EG 3), although the onset of the phase separation is the slowest due to the smallest ε, the fastest diffusion is also enough to reach a late stage of phase separation in the short polymerization period, providing a lower interface area/volume and smoother interface than others.

With PEGDMA (EG 9 and 14), in contrast, the relatively higher ε than that of PEGDMA (EG 3) induces rapid onset of phase separation; however, the lower D does not allow further growth of the phase-separated structure. The polymerization-induced phase separation ceases at an early or intermediate stage of the phase separation, which results in giving the higher interface area/volume and the lower Λ_{ave} found in Table 10.3.

10.3.2 Effect of Humidity

Since PAMAM dendrimer immobilized polymeric membranes exhibit excellent CO_2 selectivity, the PAMAM membranes will be applicable for CO_2 capture at mass-emission sources, such as thermal power stations. The effect of humidity on the CO_2 separation performance of PAMAM/PEG membranes should be examined because CO_2-containing exhaust gas is highly humidified. Change in gas-transport properties of PAMAM/PEG membranes is studied as a function of time at 298 K and $p(CO_2)$ of 5 kPa under 80% relative humidity in Figure 10.7a [53]. The dendrimer weight fraction and average number of EG units of PEGDMA are 50% and 14, respectively. The polymeric membranes are well dried under vacuum to eliminate reaction solvent and to avoid humid absorption in atmosphere before use. The CO_2 and H_2 permeances are ca. 0.6 and 1.4×10^{-2} GPU, respectively, to provide 40 in $\alpha(CO_2/H_2)$ for the first several hours. However, CO_2 permeance increases drastically from 0.6 to 6.5 GPU during the next 10–15 hours of incubation, while H_2 permeance does not change much or rather decreases, and then the selectivity reaches 500. PAMAM/PEG membranes as prepared are translucent due to macrophase separation between PAMAM dendrimers and the polymer matrix. However, the membranes become transparent under humidified conditions by absorbing water. The swelling behavior is equilibrated for 10–16 hours of incubation under 80% relative humidity. Significant increase in CO_2 permeance is found during that incubation period in Figure 10.7b [53].

The CO_2 separation performance of PAMAM/PEG and PEG membranes is examined under different relative humidity at the same experimental conditions, as displayed in Figure 10.8 [53]. Without the PAMAM dendrimer, both gas permeances are independent of

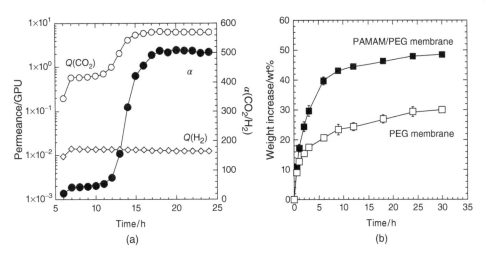

Figure 10.7 (a) Change in CO$_2$ separation properties over H$_2$ of poly(amidoamine)/poly(ethylene glycol) (PAMAM/PEG) membranes (50/50 by wt) as a function of time at 25 °C under 80% relative humidity and at p(CO$_2$) of 5 kPa. Average PEG unit is 14. (b) Time course of swelling of PAMAM/PEG (50/50 by wt) and PEG membranes at 25 °C under 80% relative humidity. Average PEG unit is 14.

relative humidity, and thus CO$_2$ selectivity is constant and around 9 in Figure 10.8a. However, the CO$_2$ permeance of the PAMAM/PEG membrane rises with increase of relative humidity from 0.05 under dry conditions to 1.84 GPU at 80% relative humidity, while the H$_2$ permeance is not dependent on the humidity from 0–50% and slightly increases at 80%. As a consequence, the increase in CO$_2$ permeance is most likely related to the increase in CO$_2$/H$_2$ selectivity at higher relative humidity in Figure 10.8b. While crosslinked PEG membranes display good CO$_2$ separation performance [42–44], it is apparent that the PAMAM dendrimer gives the high CO$_2$ separation properties in the polymeric membrane. In Figure 10.8, the membrane thickness of the crosslinked PEG membrane is ca. 150 μm, while that of the PAMAM/PEG membrane is 400 μm, which results in the difference in gas permeances.

10.3.3 Effect of Phase-Separated Structure

Since PAMAM dendrimers elevate CO$_2$ solubility in the PAMAM/PEG membranes rather than the PEG matrix, the phase-separated structure or distribution of PAMAM dendrimers discussed previously plays an important role in the gas transportation of the polymeric membranes. When the PAMAM dendrimer weight fraction is kept constant at 50 wt%, the polymeric membranes give different phase separations depending on the length of the PEG chains in Table 10.3. The CO$_2$ separation properties of the PAMAM/PEG membranes prepared by various PEGDMAs are examined as a function of EG units, and the results are represented in Figure 10.9. Although the deviation is not small, the CO$_2$ permeation is more sensitive to the PEG length and much higher than the H$_2$ permeation. PAMAM/PEG membranes with PEGDMA (EG 9 and 14) show CO$_2$ permeances above 1.0 GPU, while those with PEGDMA (EG 3 and 23) show 0.34 and 7.6 × 10^{-2} GPU, respectively. In contrast, the H$_2$ permeances do not change much among the polymeric membranes with various PEGDMAs

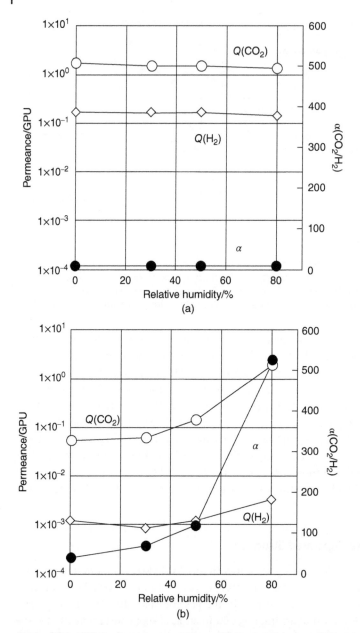

Figure 10.8 Effect of relative humidity on CO_2 separation properties over H_2 of (a) PEG and (b) poly(amidoamine)/poly(ethylene glycol) (PAMAM)/PEG (50/50 by wt) membranes at 25 °C under 80% relative humidity and at $p(CO_2)$ of 5 kPa. Average PEG unit is 14.

tested and are between 1.0×10^{-2} and 6.6×10^{-2} GPU. CO_2/H_2 selectivity is determined by the ratio of CO_2 and H_2 permeances in Figure 10.9, and PEGDMAs (EG 9 and 14) provide higher selectivity than PEGDMAs (EG 3 and 23). When the polymeric membrane is prepared with PEGDMA (EG 14), the separation factor is 550 ± 150 (thickness ca. 400 μm).

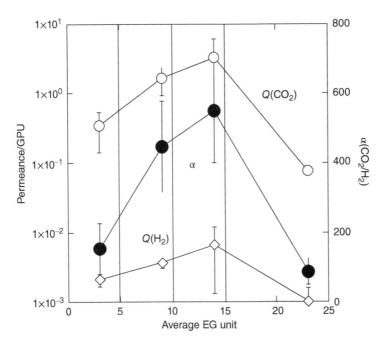

Figure 10.9 Effect of the average ethylene glycol unit number of photopolymerization of PEG dimethacrylate (PEGDMA) on CO_2 separation properties over H_2 of poly(amidoamine)/poly(ethylene glycol) (PAMAM/PEG) (50/50 by wt) membrane at 25 °C under 80% relative humidity and at $p(CO_2)$ of 5 kPa.

With the lowest case, the factor is 87 ± 38 with PEGDMA (EG 23); however, the value is still much higher than that of crosslinked PEG membranes under the same experimental conditions [53].

With the phase-separated structure in Table 10.3, it is clearly found that Λ_{ave} is responsible for the $\alpha(CO_2/H_2)$ or CO_2 permeance. A smaller PAMAM dendrimer-rich domain size gives higher CO_2 separation performance. The gas-separation experiments are conducted under humidified conditions, in which the polymeric membranes are swollen by absorbing humidity and become transparent. On the other hand, phase separation is observed under dry conditions on a LSCM. PAMAM/PEG membranes equilibrated in a humidifier are readily dried during LSCM observations due to rapid water evaporation from the polymeric membranes, and the detailed morphology of the polymeric membranes under humidified conditions cannot be captured by the LSCM.

In dry conditions, the gas-permeation mechanism through a dense polymeric membrane is explained by diffusion in a free volume of the membrane driven by the pressure difference between the feed and permeate sides [54]. Here, CO_2 permeates preferentially through the PAMAM dendrimer-rich domain due to the specific interaction with amino groups of the dendrimer, and thus the free volume in the domain is occupied by CO_2. As a result, H_2 has to penetrate mostly in the PEG-rich domain, where CO_2 still shows higher permeability than H_2 in Figure 10.8a.

On the other hand, in humidified conditions, PAMAM/PEG membranes become transparent by absorbing water, where water works as a compatibilizer and facilitates diffusion of the dendritic molecules into the PEG-rich domain, although the diffusion is not accurately traced by LSCM observations. The polymeric membranes are in a hydrogel state under humidified conditions, and the resulting hydrogel becomes basic due to diffusion of the amine. Thus, the CO_2 solubility of the membrane increases, which results in a significant increase in the CO_2 permeance and separation factor, while H_2 permeance is mostly independent of humidity in Figure 10.8a. Bicarbonate ion formation is confirmed when the dendrimer/water mixture is kept under CO_2 atmosphere, and thus bicarbonate ion is a major migrating species through the polymeric membrane in CO_2 separation over H_2 [55]. The detail is discussed in the following section.

With the results in Table 10.3 and Figures 10.8 and 10.9, effect of the phase separation on CO_2 separation performance can be speculated as follows. When the polymeric membranes are prepared with PEGDMAs (EG 9 and 14), the polymerization-induced phase separation is suspended in an early or intermediate stage of the phase separation. The resulting average PAMAM dendrimer-rich domain is smaller than the membranes with PEGDMAs (EG 3 and 23). In this case, PAMAM dendrimers could diffuse more readily into the PEG-rich domain under humidified conditions, and the dendrimer diffusion is more homogeneous in comparison to polymeric membranes with PEGDMAs (EG 3 and 23), while the diffusion of the PEG-rich domain into the dendrimer-rich domain is limited due to crosslinking of PEG chains. Polymeric membranes with PEGDMAs (EG 9 and 14) thus provide CO_2 with higher chance to permeate through the membrane and, as a result, exhibit higher CO_2 separation performance than the other membranes. This assumption is supported by measuring the optical density of the PAMAM/PEG membranes. Polymeric membranes with PEGDMAs (EG 9 and 14) show lower optical density or less scattering intensity than those with PEGDMAs (EG 3 and 23), which indicates more homogeneous diffusion of the dendrimers in the polymeric membranes [53].

With these investigations, the important factors to control the CO_2 separation properties of the PAMAM/PEG membranes are the dendrimer fraction, generation of the dendrimer, the phase-separated structure between the dendrimer and PEG matrix, and humidity in the feed gas.

10.4 CO_2-Selective Molecular Gate

PAMAM/PEG membranes exhibit preferential CO_2 permeation over even smaller H_2. As shown in Figure 10.5a, increase of the PAMAM dendrimer weight fraction in the polymeric membranes gives rise to $\alpha(CO_2/H_2)$, which results from a significant decrease in H_2 permeation as the dendrimer fraction increases. Thus, PAMAM dendrimers suppress H_2 permeation rather than elevating CO_2 permeation as described previously. Sirkar and co-workers introduced a CO_2-selective molecular gate mechanism to explain the unusually high CO_2 separation performance over N_2 [23], although further study should be made to explain the preferential CO_2 permeation of PAMAM dendrimers.

The ^{13}C NMR technique can identify CO_2 in PAMAM/PEG membranes. When PAMAM/PEG membranes with 50% of the dendrimer weight fraction are equilibrated under 80% relative humidity, 2.0 g of the polymeric membrane absorbs 1.0 g of humidity [53]. Figure 10.8a demonstrates that the gas-permeation properties of crosslinked PEG membrane is insensitive to relative humidity. Thus, the PAMAM dendrimer/water mixture determines the CO_2 separation performance of polymeric membranes under humidified conditions, and the ^{13}C NMR study is carried out with the PAMAM/D_2O mixture. The PAMAM dendrimer is mixed with D_2O with the same molar ratio of PAMAM/H_2O equilibrated under humidity, and humidified CO_2 is introduced for more than 24 hours to allow reaching an interaction equilibrium between the amine and CO_2.

In the resulting PAMAM/D_2O mixture, the carbonyl carbons show the longest relaxation period with 3 seconds > in a T_1 measurement, while the T_1 of other carbons are less than one second. The radio pulse is irradiated at intervals of 30 seconds to eliminate the nuclear overhauser effect for an inverse-gated decoupling ^{13}C NMR. Integration of the carbon resonances gives quantitative information for the peaks found in the NMR measurement [55]. As a control, the NMR measurement of $NaHCO_3$ aqueous solution (pH 8.0) was conducted in the same NMR measurement conditions. Figure 10.10 represents the results of the ^{13}C NMR measurement of PAMAM/D_2O mixture before and after CO_2 treatment. The absence of a peak at 125 ppm after CO_2 treatment suggests that CO_2 does not exist as a gaseous molecule in the dendrimer/D_2O mixture. The peaks at 161 and 164 ppm found only after CO_2 treatment are assigned to carbonyl carbons of bicarbonate and carbamate, respectively, as expressed by the following reactions:

$$CO_2 + 2R - NH_2 \rightarrow R - NH - COO^- \cdots N^+H_3 - R$$

$$CO_2 + OD^- \rightarrow DCO_3^-$$

The $NaHCO_3$ aqueous solution gives a peak at 161 ppm, and this result suggests that the bicarbonate ion in the PAMAM/D_2O mixture after CO_2 treatment is well hydrated with

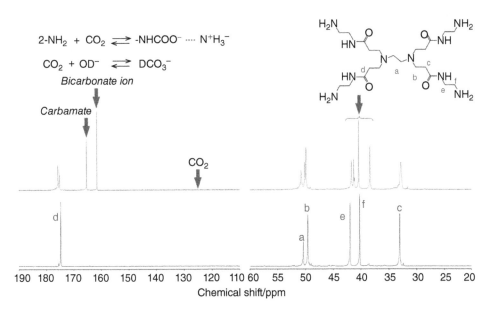

Figure 10.10 Inverse gate decoupling ^{13}C NMR spectra of poly(amidoamine) (PAMAM) dendrimer/D_2O mixture (50/50 by wt) before (bottom) and after (top) CO_2 equilibration at 25 °C.

deuterated water away from the amines of the dendrimer. By equilibrating the mixture under CO_2 atmosphere, a peak at 175 ppm assigned to amide carbon of the dendrimer is split in two peaks because of exchanging a proton for deuterium.

Here, the bicarbonate ion can be formed only in the presence of water, and as described in Figure 10.8b, the CO_2 permeance of PAMAM/PEG membranes is drastically enhanced under highly humidified conditions. Thus, the bicarbonate ion is a major migrating species of CO_2 in polymeric membranes. In addition, the peak shift of methylene carbons of the branching end of the dendrimer is seen after CO_2 treatment in Figure 10.10, which indicates that the secondary and tertiary amino groups of the PAMAM dendrimer do not participate in the interaction with CO_2.

The inverse-gate decoupling ^{13}C NMR allows quantitative analyses of the interaction between the PAMAM dendrimer and absorbed CO_2. From the ^{13}C NMR results in Figure 10.10, 5.6 mmol of CO_2 is interacted with 1.0 g of the dendrimer, which indicates 2.9 mol of CO_2 interacts with one mol of the PAMAM dendrimer in the experimental conditions, and the ratio of bicarbonate to carbamate was 39 : 61 by mol.

Interaction between CO_2 and a primary amino group of PAMAM dendrimers gives a carbamate, which interacts with a free primary amino group, resulting in the formation of quasi-crosslinking of the dendrimer. The quasi-crosslinking formation is confirmed by SAXS. The polymeric membranes are first equilibrated under 80% relative humidity in a humidified chamber cell, and then the atmosphere in the cell is changed from N_2 to CO_2. The scattering intensity increases, while no peaks derived from periodical structures are found. The intensity decreases when the atmosphere is switched back from CO_2 to N_2. The quasi-crosslinking of the dendrimer is confirmed by different techniques such as differential scanning calorimetry (DSC) and attenuated total reflection (ATR)-IR [55]. Tensile testing of the PAMAM/PEG membrane also supports formation of quasi-crosslinking. When the membranes are equilibrated in CO_2 atmosphere, the Young's modulus of the polymeric membranes increased more than 30 times in comparison to prepared membranes, from 0.22 ± 0.05 to 7.33 ± 1.00 MPa. In general, increase in Young's modulus accompanies a decrease in the elongation-to-break. However, with PAMAM/PEG membranes, the elongation becomes more than double under CO_2 atmosphere from 24.9 ± 8.7 to $55.5 \pm 1.0\%$ [55]. This is due to reversible and rearrangeable crosslinking between the PAMAM dendrimer and CO_2. This is similar to slime formed by the crosslinking between borax and hydroxyl groups of PVA [56].

The obtained result suggests that an increased crosslinking density from the quasi-crosslinking between the dendrimer and CO_2 suppresses H_2 permeation. In addition, CO_2 absorbed in water forms carbamate and bicarbonate ion, and the resulting highly charged species reduce the solubility of nonpolar H_2 in the membranes by the salting-out effect to give high CO_2 separation factors. This can be a "CO_2-selective molecular gate" effect [23].

10.5 Enhancement of CO_2 Separation Performance

A number of CO_2 separation membranes have been investigated for CO_2 separation over N_2, which can be utilized for carbon capture at CO_2 mass-emission source, such as thermal power stations and steel works [57–69]. The CO_2-containing flue gas and blast furnace are exhausted at ambient pressure, and the CO_2 partial pressure is 10–25% depending on the source. Deference of transmembrane CO_2 partial pressure is usually small, which has been making CO_2 separation with membranes challenging. When CO_2 permeance is 2000 GPU, the separation factor should be greater than 50 to compete with the current solution absorption technology, whose CO_2 capture costs are close to $40 USD/t-$CO_2$ [70]. For further cost reduction, the permeance should be improved rather than elevating CO_2 selectivity. On the other hand, in pre-combustion carbon capture, CO_2 separation over H_2 is rather difficult. However, the syngas after the water-gas shift reaction is pressurized, and the CO_2 concentration is ca. 36% in an oxygen-blown plant. The much larger difference in CO_2 partial pressure between feed and permeate sides can be advantageous in comparison to post-combustion carbon capture. RITE examined the required CO_2 separation performance for pre-combustion CO_2 capture, where the separation cost was less than $\overline{Y}$1500 JPY/t-CO_2. The CO_2 permeance and selectivity is greater than 100 GPU and 30 or 40 GPU and 125, respectively, at CO_2 partial pressure of 1.0 MPa. Poly(vinylamine)-based membranes developed by Ho and coworkers are currently close to the target values [71]. For most CO_2 separation membranes, elevating CO_2 permeance is the prime target beyond the upper bound.

PAMAM-containing membranes have been developed for pre-combustion CO_2 capture because they showed quite high CO_2 separation factors, especially at relatively lower CO_2 partial pressure. However, CO_2 permeance should be improved, and one plausible approach can be reducing the membrane thickness. However, through intensive investigations of PAMAM/PEG membranes, the formation of a bicontinuous structure of PAMAM dendrimer-rich and PEG-rich phases is found upon photopolymerization-induced phase separation on a couple of microns scale [50, 55]. When the membrane thickness is below 100 μm, the CO_2 separation properties over H_2 drop down severely.

To suppress the macrophase separation between PAMAM-rich and PEG-rich phases, a compatible crosslinker, 4GMAP (4-arm glycidyl methacrylate-modified PAMAM) [72], is developed from the dendrimer (*G* 0). Figure 10.11 shows the reaction scheme. Glycidyl methacrylate is reacted to PAMAM with a precise molar ratio of 4 to 1. The reaction is conducted in ethanol at ambient conditions to suppress the Michael addition reaction between the methacryl group and the primary amino group. Copolymerization of the crosslinker with PEGDMA prepares transparent membranes in the presence of PAMAM, when 4GMAP content is greater than 7.5 wt%. Suppression of the phase separation is confirmed by increasing compatibility between PAMAM and the polymer matrix, as shown in Figure 10.11, and as a consequence, leakage of PAMAM is suppressed in reducing the membrane thickness.

The polymeric membranes are prepared by the UV curing method with or without the crosslinker, and the thicknesses are ca. 70 μm. The CO_2 separation properties are examined under different pressures, as depicted in Figure 10.12. With 4GMAP (7.5 wt% relative to the membrane), the CO_2 permeance and separation factor are 2 GPU and 50, respectively, at $p(CO_2)$ of 0.082 MPa. Those values decrease with elevating $p(CO_2)$ and reach 0.7 GPU and 17.4, respectively, at $p(CO_2)$ of 0.56 MPa, while the H_2 permeance is almost constant over CO_2 pressure differences. The preferential CO_2 permeation is explained by a facilitated transportation via carrier-mediated diffusion mechanism [27]. On the other hand, without 4GMAP, the polymeric membranes lose selectivity at and above $p(CO_2)$ of 0.23 MPa or total

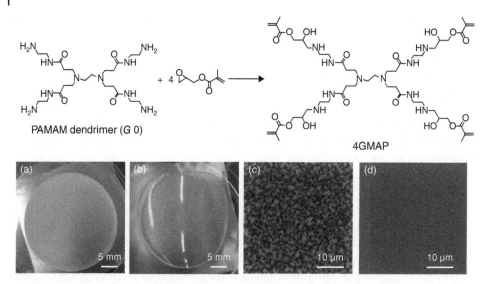

Figure 10.11 Top: Synthetic scheme of a compatible crosslinker 4-arm glycidyl methacrylate-modified PAMAM (4GMAP). Bottom: Poly(amidoamine)/poly(ethylene glycol) (PAMAM/PEG) membranes (PAMAM dendrimer: 50 wt%) (a) without and (b) with 4GMAP (7.5 wt%); and laser scanning confocal microscope (LSCM) images of the membranes (c) without and (d) with 4GMAP (7.5 wt%). Average PEG unit is 14.

pressure of 0.29 MPa. Introduction of the compatible crosslinker also results in enhanced mechanical properties and pressure tolerance [72].

The CO_2 separation performance of 4GMAP-containing membranes is examined with various membrane thicknesses at $p(CO_2)$ of 0.56 MPa, as shown in Figure 10.13. Both the CO_2 and H_2 permeances are elevated by thinning the membrane thickness. CO_2 selectivity is mostly above 10 in the experimental conditions with the thickness between 10 and 100 μm, which is higher than the other CO_2-philic polymers, such as polydimethylsiloxane [73]. However, CO_2 permeance is low, 2.5 GPU with 11 in CO_2 selectivity when the thickness is 18 μm. Those values are far below the target for pre-combustion carbon capture. Thus, further improvement is required in both CO_2 permeance and selectivity, although the quite-high CO_2 separation properties under low CO_2 partial pressure are still attractive. A number of gas-separation membranes have been developed, and some of them have been used, such as N_2-enriching membranes. With CO_2 capture, while CO_2-separation membranes have been investigated [74–76], only a few successful applications can be seen, such as cellulose triacetate membranes for natural gas sweetening by UOP and Cameron (Schlumberger).

10.6 Conclusion and Perspectives

In this chapter, promising CO_2 capture properties of PAMAM dendrimers have been introduced. The dendrimers can be readily incorporated into polymer matrices, such as crosslinked chitosan, PVA, and PEG. The amine-containing polymeric membranes show

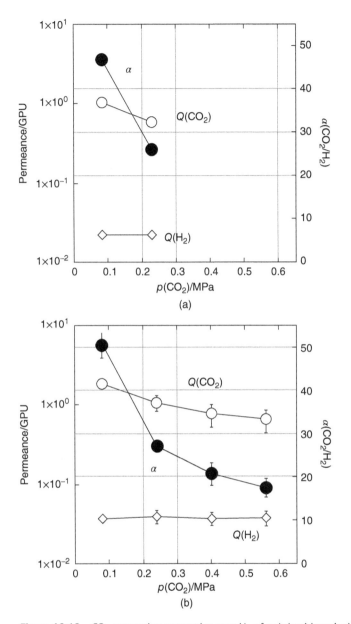

Figure 10.12 CO_2 separation properties over H_2 of poly(amidoamine)/poly(ethylene glycol) (PAMAM/PEG) membranes (PAMAM dendrimer: 50 wt%) (a) without (75 μm in thickness) and (b) with 4-arm glycidyl methacrylate-modified PAMAM (4GMAP) (7.5 wt%) (71 μm in thickness) as a function of $p(CO_2)$ at 40 °C under 80% relative humidity. Average PEG unit is 14.

high CO_2 separation properties over even smaller H_2 by a carrier-mediated facilitated transportation. With the IM method, a hollow-fiber membrane module is readily prepared, although the amine immobilization is limited. Immobilization of PAMAM dendrimers into crosslinked PVA is also straightforward and scalable. The membrane-forming

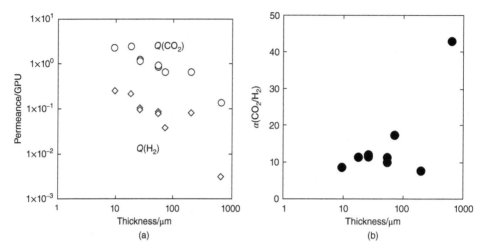

Figure 10.13 Effect of membrane thickness on CO_2 separation properties over H_2 of poly(amidoamine)/poly(ethylene glycol) (PAMAM/PEG) membranes with 4-arm glycidyl methacrylate-modified PAMAM (4GMAP) (PAMAM/PEG/4GMAP: 50/42.5/7.5 wt%) at 40 °C under 90% relative humidity under $p(CO_2)$ of 0.56 MPa and 0.70 MPa of total pressure. Average PEG unit is 14.

process is indeed time-consuming, to evaporate water slowly, but allows fabrication of both hollow-fiber and spiral-wound membranes for demonstration. Miscibility between the amine and polymer matrix also makes it possible to reduce the membrane thickness to enhance gas permeability. On the other hand, crosslinked PEG is promising due to its CO_2-philic nature; the amines are readily immobilized and stable in a couple of minutes UV curing, while incompatibility between amines and PEG results in macrophase separation. However, with a compatible crosslinker, a thinner membrane formulation can be available by suppressing phase separation. PAMAM dendrimers are also mixed with PEG-grafted comb copolymer to improve CO_2 separation performance [77]. A recent review summarizes developments of CO_2-selective membranes over H_2 [78].

The CO_2 permeability of PAMAM membranes should be further improved for use, although they display high CO_2 selectivity. A plausible approach can be to accelerate bicarbonate formation in the polymeric membranes under humidity, which is the major migrating species of CO_2 in the dendrimer-containing membranes. CO_2 interacts with the primary amino groups on the branching end of the dendrimers in Figure 10.10. Basically, primary amino groups show stronger interaction with CO_2 than secondary and tertiary amines, which indicates that dissociation between CO_2 and primary amino groups requires higher energy in comparison to secondary and tertiary amino groups for diffusion. Thus, CO_2 permeability is enhanced by replacing primary amino group with secondary and tertiary amino groups of the dendritic molecules. In addition, use of biocatalysts, carbonic anhydrase and the derivatives, is also effective, which has been investigated in recent decades. The enzymes catalyze the bicarbonate formation of dissolved [79], and Codexis and Akermin have examined pilot testing of enzyme-assisted CO_2 capture at the National Carbon Capture Center (AL, USA).

Energy production has been based on fossil resources in the United States and most other developed countries. To mitigate global warming and climate change, we have to switch the energy system from fossil-based to carbon-neutral or renewable, where H_2 can be used as an energy source. However, H_2 is mostly produced by steam reforming of light hydrocarbons, and the produced H_2 is purified by pressure-swing adsorption or Pd membrane separation. The off-gas after the purification consists mainly of H_2 and CO_2 and is led to the steam reformer as fuel to keep it at elevated temperature. Thus, CO_2 is eventually exhausted. Since the off-gas comes out at a moderate temperature and ambient pressure with $p(CO_2)$ of 40–50 kPa, PAMAM dendrimer-containing membranes are suitable for CO_2 separation over H_2 to make the H_2 production system carbon-free. Kimura and co-workers reported a feasibility study of CO_2-free H_2 production at an on-site H_2 refilling station (HRS) by membrane separation, where CO_2 was captured by the dendrimer-containing membranes and then injected into an underground aquifer shallower than 1 km depth, so called "small-scale decentralized CCS" [80]. The cost simulation indicates that steam reforming with small-scale CCS provides H_2 less expensively by $0.38 USD Nm^{-3}-H_2 in comparison to water hydrolysis with renewable energy in Japan. Thus, PAMAM dendrimer-containing membranes hold potential for CO_2 capture in H_2 purification, such as CO_2-free H_2 production by steam-reforming processes at HRSs.

Acknowledgments

This work was supported in part by the Japanese Ministry of Economy, Trade and Industry (METI); Nippon Steel Engineering Co., Ltd.; JST A-STEP (AS251Z01541M); KRI Inc.; and Fukuoka Strategy Conference for Hydrogen Energy. The author is thankful to Dr. Shingo Kazama and his colleagues at the Chemical Research Group of RITE, especially Ms. Hiromi Urai and Rie Sugimoto, for assistance with the membrane fabrication processes and gas chromatography measurements. The author also wishes to acknowledge Prof. Hiroshi Jinnai at Tohoku University for helping with the LSCM measurements and 3D analyses of membrane morphology.

References

1 Tomalia, D.A. (1994). Starburst/cascade dendrimers: fundamental building blocks for a new nanoscopic chemistry set. *Adv. Mater.* 7: 529–539.

2 Matthews, O.A., Shipway, A.N., and Stoddart, J.F. (1998). Dendrimers-branching out from curiousities into new technologies. *Prog. Polym. Sci.* 23: 1–56.

3 Vögtle, F., Gestermann, S., Hesse, R. et al. (2000). Functional dendrimers. *Prog. Polym. Sci.* 25: 987–1041.

4 Grate, J.W. (2000). Acoustic wave microsensor arrays for vapor sensing. *Chem. Rev.* 100: 2627–2648.

5 Voit, B. (2000). New developments in hyperbranched polymers. *J. Polym. Sci. Part A: Polym. Chem.* 38: 2505–2525.

6 Esfand, R. and Tomalia, D.A. (2001). Poly(amidoamine) (PAMAM) dendrimers: from biomimicry to drug delivery and biomedical applications. *Drug Deliv. Today* 6: 427–436.

7 Dykes, G.M. (2001). Dendrimers: a review of their appeal and applications. *J. Chem. Technol. Biotechnol.* 76: 903–918.

8 Jikei, M. and Kakimoto, M. (2001). Hyperbranched polymers: a promising new class of materials. *Prog. Polym. Sci.* 26: 1233–1285.

9 Hecht, S. (2003). Functionalizing the interior of dendrimers: synthetic challenges and applications. *J. Polym. Sci. Part A: Polym. Chem.* 41: 1047–1058.

10 Yates, C.R. and Hayes, W. (2004). Synthesis and applications of hyperbranched polymers. *Eur. Polym. J.* 40: 1257–1281.

11 Tomalia, D.A. (2005). Birth of a new macromolecular architecture: dendrimers as quantized building blocks for nanoscale synthetic polymer chemistry. *Prog. Polym. Sci.* 30: 294–324.

12 Svenson, S. and Tomalia, D.A. (2005). Dendrimers in biomedical applications-reflections on the field. *Adv. Drug Deliv. Rev.* 57: 2106–2129.

13 Lee, C.C., MacKay, J.A., Fréchet, J.M.J., and Szoka, F.C. (2005). Designing dendrimers for biological applications. *Nat. Biotechnol.* 23: 1517–1526.

14 Kitchens, K.M., El-Sayed, M.E.H., and Ghandehari, H. (2005). Transepithelial and endothelial transport of poly (amidoamine) dendrimers. *Adv. Drug Deliv. Rev.* 57: 2163–2176.

15 Burn, P.L., Lo, S.-C., and Samuel, I.D.W. (2007). The development of light-emitting dendrimers for displays. *Adv. Mater.* 19: 1675–1688.

16 Tomalia, D.A., Naylor, A.M., and Goddard, W.A. III, (1990). Starburst dendrimers: molecular-level control of size, shape, surface chemistry, topology, and flexibility from atoms to macroscopic matter. *Angew. Chem. Int. Ed. Engl.* 29: 138–175.

17 Moors, R. and Vögtle, F. (1993). Dendrimere Polymaine. *Chem. Ber.* 126: 2133–2135.

18 de Gennes, P.G. and Hervet, H.J. (1983). Statistics of "starburst" polymers. *Phys. Lett. (Paris)* 44: 351–360.

19 Hawker, C.J. and Frêchet, J.M.J. (1990). Preparation of polymers with controlled molecular architecture-a new convergent approach to dendritic macromolecules. *J. Am. Chem. Soc.* 112: 7638–7647.

20 Miller, T.M. and Neenan, T.X. (1990). Convergent synthesis of monodisperse dendrimers based upon 1,3,5 trisubstituted benzenes. *Chem. Mater.* 2: 346–349.

21 Frechet, J.M.J., Hawker, C.J., and Wooley, K.L. (1994). The convergent route to globular dendritic macromolecules: a versatile approach to precisely functionauzed three-dimensional polymers and novel block copolymers. *J. Macromol. Sci. Part A: Pure Appl. Chem.* 31: 1627–1645.

22 Lee, C.C., Mackay, J.A., Fréchet, J.M.J., and Szoka, F.C. (2005). Designing dendrimers for biological applications. *Nature Biotechnol.* 23: 1517–1526.

23 Kovvali, A.S., Chen, U., and Sirkar, K.K. (2000). Dendrimer membranes: a CO_2-selective molecular gate. *J. Am. Chem. Soc.* 122: 7594–7595.

24 Kovvali, A.S. and Sirkar, K.K. (2001). Dendrimer liquid membranes: CO_2 separation from gas mixtures. *Ind. Eng. Chem. Res.* 40: 2502–2511.

25 Kovvali, A.S. and Sirkar, K.K. (2002). Carbon dioxide separation with novel solvents as liquid membranes. *Ind. Eng. Chem. Res.* 41: 2287–2295.

26 Tomalia, D.A., Mardel, K., Henderson, S.A. et al. (2003). Dendrimers-an enabling synthetic science to controlled organic nanostructures. In: *Handbook of Nanoscience, Engineering and Technology* (eds. W.A. Goddard III,, D.W. Brenner, S.E. Lyshevski and G.J. Lafrate), 1–34. Boca Raton, FL: CRC Press.

27 Schults, J.S., Goddard, J.D., and Suchdeo, S.R. (1974). Facilitated transport via carrier-mediated diffusion in membranes part 1. Mechanistic aspects, experimental systems and characteristic regimes. *AIChE J.* 20: 417–445.

28 Pauly, S. (1999). Solid state properties, permeability and diffusion data. In: *Polymer Handbook*, 4e (eds. J. Brandrup, E.H. Immergut and E.A. Grulke), VI543–VI570. Hoboken, NJ: Wiley-Interscience.

29 Duan, S., Kouketsu, T., Kazama, S., and Yamada, K. (2006). Development of PAMAM dendrimer composite membranes for CO_2 separation. *J. Membr. Sci.* 283: 2–6.

30 Kouketsu, T., Duan, S., Kai, T. et al. (2007). PAMAM dendrimer composite membrane for CO_2 separation: formation of a chitosan gutter layer. *J. Membr. Sci.* 287: 51–59.

31 Zou, J. and Ho, W.S.W. (2006). CO_2-selective polymeric membranes containing amines in crosslinked poly(vinyl alcohol). *J. Membr. Sci.* 286: 310–321.

32 Huang, J., Zou, J., and Ho, W.S.W. (2008). Carbon dioxide capture using a CO_2-selective facilitated transport membrane. *Ind. Eng. Chem. Res.* 47: 1261–1267.

33 Xing, R. and Ho, W.S.W. (2011). Crosslinked polyvinylalcohol-polysiloxane/fumed silica mixed matrix membranes containing amines for CO_2/H_2 separation. *J. Membr. Sci.* 367: 91–102.

34 Bai, H. and Ho, W.S.W. (2011). Carbon dioxide-selective membranes for high-pressure synthesis gas purification. *Ind. Eng. Chem. Res.* 50: 12152–12161.

35 Zhao, Y. and Ho, W.S.W. (2012). Steric hindrance effect on amine demonstrated in solid polymer membranes for CO_2 transport. *J. Membr. Sci.* 415–416: 132–138.

36 Ramasubramanian, K., Verweij, H., and Ho, W.S.W. (2012). Membrane processes for carbon capture from coal-fired power plant flue gas: a modeling and cost study. *J. Membr. Sci.* 421: 299–310.

37 Zhao, Y. and Ho, W.S.W. (2013). Carbon dioxide-selective membranes for high-pressure synthesis gas purification, CO_2-selective membranes containing sterically hindered amines for CO_2/H_2 separation. *Ind. Eng. Chem. Res.* 52: 8774–8782.

38 Tong, Z., Vakharia, V.K., Gasda, M., and Ho, W.S.W. (2015). Water vapor and CO_2 transport through amine-containing facilitated transport membranes. *React. Funct. Polym.* 86: 111–116.

39 Vakharia, V.K., Ramasubramanian, K., and Ho, W.S.W. (2015). An experimental and modeling study of CO_2-selective membranes for IGCC syngas purification. *J. Membr. Sci.* 488: 56–66.

40 Duan, S., Taniguchi, I., Kai, T., and Kazama, S. (2012). Poly(amidoamine) dendrimer/poly(vinyl alcohol) hybrid membranes for CO_2 capture. *J. Membr. Sci.* 423–424: 107–112.

41 Duan, S., Kai, T., Taniguchi, I., and Kazama, S. (2013). Development of poly(amidoamine) dendrimer/poly(vinyl alcohol) hybrid membranes for CO_2 separation. *Desalin. Water Treat.* 51: 5337–5342.

42 Lin, H. and Freeman, B.D. (2004). Gas solubility, diffusivity and permeability in poly(ethylene oxide). *J. Membr. Sci.* 239: 105–117.

43 Lin, H. and Freeman, B.D. (2005). Materials selection guidelines for membranes that remove CO_2 from gas mixtures. *J. Mol. Struct.* 739: 57–74.

44 Liu, S.L., Shao, L., Chua, M.L. et al. (2013). Recent progress in the design of advanced PEO-containing membranes for CO_2 removal. *Prog. Polym. Sci.* 38: 1089–1120.

45 Lin, H.Q., Van wagner, E., Freeman, B.D. et al. (2006). Plasticization-enhanced hydrogen purification using polymeric membranes. *Science* 311: 639–642.

46 Taniguchi, I., Duan, S., Kazama, S., and Fujioka, Y. (2008). Facile fabrication of a novel high performance CO_2 separation membrane: immobilization of poly(amidoamine) dendrimers in poly(ethylene glycol) networks. *J. Membr. Sci.* 322: 277–280.

47 Jinnai, H., Nishikawa, Y., Koga, T., and Hashimoto, T. (1995). Direct observation of three-dimensional bicontinuous structure developed via spinodal decomposition. *Macromolecules* 28: 4782–4784.

48 Jinnai, H., Hishikawa, Y., Morimoto, H. et al. (2000). Geometrical properties and interface dynamics: time evolution of spinodal interface in a binary polymer mixture at the critical composition. *Langmuir* 16: 4380–4393.

49 Jinnai, H., Koga, T., Nishikawa, Y. et al. (1997). Curvature determination of spinodal interface in a condensed matter system. *Phys. Rev. Lett.* 78: 2248–2251.

50 Taniguchi, I., Kazama, S., and Jinnai, H. (2012). Structural analysis of poly(amidoamine) dendrimer immobilized in crosslinked poly(ethylene glycol). *J. Polym. Sci. B* 50: 1156–1164.

51 Andrews, H.C. (1976). Monochrome digital image enhancement. *Appl. Optics* 15: 495–503.

52 Harris, J.L. (1977). Constant variance enhancement: a digital processing technique. *Appl. Optics* 16: 1268–1271.

53 Taniguchi, I., Duan, S., Kai, T. et al. (2013). Effect of phase-separated structure on CO_2 separation performance of poly(amidoamine) dendrimer immobilized in a poly(ethylene glycol) network. *J. Mater. Chem. A* 1: 14514–14523.

54 Kesting, R.E. and Fritzsche, A.K. (1993). *Polymeric Gas Separation Membranes*. New York, NY: Wiley.

55 Taniguchi, I., Urai, H., Kai, T. et al. (2013). A CO_2-selective molecular gate of poly(amidoamine) dendrimer immobilized in a poly(ethylene glycol) network. *J. Membr. Sci.* 444: 96–100.

56 Casassa, E.Z., Sarquis, A.M., and van Dyke, C.H. (1986). The gelation of polyvinyl alcohol with borax: a novel class participation experiment involving the preparation and properties of a "slime". *J. Chem. Educ.* 63: 57–61.

57 Nagai, K., Masuda, T., Nakagawa, T. et al. (2001). Poly[1-(trimethylsilyl)-1-propyne] and related polymers: synthesis, properties and functions. *Prog. Polym. Sci.* 26: 721–798.

58 Côté, A.P., Benin, A.I., Ockwig, N.W. et al. (2005). Porous, crystalline, covalent organic frameworks. *Science* 310: 1166–1170.

59 McKeown, N.B., Gahnem, B., Msayib, K. et al. (2006). Towards polymer-based hydrogen storage materials: engineering ultramicroporous cavities within polymers of intrinsic microporosity. *Angew. Chem. Int. Ed.* 45: 1804–1807.

60 Budd, P.M., McKeown, N.B., and Fritsch, D. (2006). Polymers of intrinsic microporosity (PIMs): high free volume polymers for membrane applications. *Macromol. Symp.* 245: 403–405.

61 Powell, C.E. and Qiao, G.G. (2006). Polymeric CO_2/N_2 gas separation membranes for the capture of carbon dioxide from power plant flue gases. *J. Membr. Sci.* 279: 1–49.

62 Chung, T.-S., Jiang, L.Y., Li, Y., and Kulprathipanja, S. (2007). Mixed matrix membranes (MMMs) comprising organic polymers with dispersed inorganic fillers for gas separation. *Prog. Polym. Sci.* 32: 483–507.

63 Park, H.B., Jung, C.H., Lee, Y.M. et al. (2007). Polymers with cavities tuned for fast, selective transport of small molecules and ions. *Science* 318: 254–258.

64 El-Kaderi, H.M., Hunt, J.R., Mendoza-Cortés, J.L. et al. (2007). Designed synthesis of 3D covalent organic frameworks. *Science* 316: 268–272.

65 Park, H.B., Han, S.H., Jung, C.H. et al. (2010). Thermally rearranged (TR) polymer membranes for CO_2 separation. *J. Membr. Sci.* 359: 11–24.

66 Jeong, H.-K., Balbuena, P.B., Zhou, H.-C. et al. (2011). Carbon dioxide capture-related gas adsorption and separation in metal-organic frameworks. *Coord. Chem. Rev.* 255: 1791–1823.

67 Du, N., Park, H.B., Robertson, G.P. et al. (2011). Polymer nanosieve membranes for CO_2-capture applications. *Nat. Mater.* 10: 372–375.

68 Du, N., Park, H.B., Dal-Cin, M.M., and Guiver, M.D. (2012). Advances in high permeability polymeric membrane materials for CO_2 separations. *Energ. Environ. Sci.* 5: 7306–7322.

69 Luis, P., Gerven, T.V., and der Bruggen, B.V. (2012). Recent developments in membrane-based technologies for CO_2 capture. *Prog. Energy Combust. Sci.* 38: 419–448.

70 Merkel, T., Lin, H., Wei, X., and Baker, R. (2010). Power plant post-combustion carbon dioxide capture: an opportunity for membranes. *J. Membr. Sci.* 359: 126–139.

71 Nagumo, R., Kazama, S., and Fujioka, Y. (2009). Techno-economic evaluation of the coalbased integrated gasification combined cycle with CO_2 capture and storage technology. *Energy Proc.* 1: 4089–4093.

72 Taniguchi, I., Kai, T., Duan, S. et al. (2015). A compatible crosslinker for enhancement of CO_2 capture of poly(amidoamine) dendrimer-containing polymeric membranes. *J. Membr. Sci.* 475: 175–183.

73 Barillas, M.K., Enick, R.M., O'Brien, M. et al. (2011). The CO_2 permeability and mixed gas CO_2/H_2 selectivity of membranes composed of CO_2-philic polymers. *J. Membr. Sci.* 372: 29–39.

74 Xiao, Y., Low, B.T., Hosseini, S.S. et al. (2009). The strategies of molecular architecture and modification of polyimide-based membranes for CO_2 removal from natural gas-a review. *Prog. Polym. Sci.* 34: 561–580.

75 Scholes, C.A., Stevens, G.W., and Kentish, S.E. (2012). Membrane gas separation applications in natural gas processing. *Fuel* 96: 15–28.

76 Zhang, Y., Sunarso, J., Liu, S., and Wang, R. (2013). Current status and development of membranes for CO_2/CH_4 separation: a review. *Int. J. Greenhouse Gas Control* 12: 84–107.

77 Taniguchi, I., Wada, N., Kinugasa, K., and Higa, M. (2017). CO_2 capture by polymeric membranes composed of hyper-branched polymers with dense poly(oxytheylene) comb and poly(amidoamine). *Open Phys.* 15: 662–670.

78 Li, P., Wang, Z., Qiao, Z. et al. (2015). Recent developments in membranes for efficient hydrogen purification. *J. Membr. Sci.* 495: 130–168.

79 Savile, C.K. and Lalonde, J.J. (2011). Biotechnology for the acceleration of carbon dioxide capture and sequestration. *Curr. Opin. Biotechnol.* 22: 818–823.

80 Kimura, S., Honda, K., Kitamura, K. et al. (2014). Preliminary feasibility study for on-site hydrogen station with distributed CO_2 capture and storage system. *Energy Proc.* 63: 4575–4584.

11

Ionic Liquids for Chemisorption of CO$_2$

Mingguang Pan[1] and Congmin Wang[1,2]

[1]Department of Chemistry, ZJU-NHU United R&D Center, Zhejiang University, Hangzhou, China
[2]Key Laboratory of Biomass Chemical Engineering of Ministry of Education, Zhejiang University, Hangzhou, China

---

CHAPTER MENU

11.1 Introduction

Ionic liquids are low-temperature molten salts, usually with melting points below 100 °C, composed primarily of organic ions that may undergo almost countless structural variations. They have attracted increasing interest from chemists because of their unique properties, including negligible volatility, high thermal stability, non-flammability, low melting point, high ionic conductivity, and controlled miscibility. It is generally accepted that they play an important role in organic reactions as solvents [1], catalysts [2–4], and supports [5]. Substantial effort has been dedicated to ionic liquid-based materials, such as ionic liquid crystals [6–8], polyhedral oligomeric silsesquioxane (POSS) ionic liquids [9], ionic liquid propellants [10], magnetic and luminescent ionic liquids [11, 12], etc. The advent of ionic liquids (ILs) as eco-friendly and promising reaction media provides a new opportunity for the synthesis of functional advanced materials, such as energy storage materials

Materials for Carbon Capture, First Edition. Edited by De-en Jiang, Shannon M. Mahurin and Sheng Dai.
© 2020 John Wiley & Sons Ltd. Published 2020 by John Wiley & Sons Ltd.

[13–16], catalytic materials [17], and porous materials [18]. Some interesting properties and applications based on ILs, e.g. lower critical solution temperature or lower critical solution temperature (LCST)-type phase changes [19], dispersion of carbon nanotubes [20], and incorporation within the micropores of a metal-organic framework (MOF) [21], have been exploited to a great extent. Recently, significant effort has been devoted to the dissolution of natural biomaterials using polar ILs for the construction of biofuel cells [22]. It is doubtless that ILs also have gained in popularity in industry. One of the most famous industrial applications for ILs is the biphasic acid scavenging utilizing ionic liquids (BASIL) process of BASF [23, 24]. In this process, the IL performs not as a solvent but an acid-scavenging agent. In addition to these applications, ILs, as promising adsorbents, address the challenge of fossil-fuel-derived CO_2 accumulation in both the atmosphere and the ocean [25–28]. Key advantages that enable ILs to overcome the limits imposed by traditional sorption methods based on aqueous amine solutions for CO_2 capture are their negligible vapor pressures, high thermal stabilities, excellent CO_2 solubilities, and tunable properties [29–41].

CO$_2$ capture and storage (CCS) is among the most interesting and important emerging research areas where ILs play an essential role. A large number of experimental and theoretical studies have been presented related to understanding and increasing the physical solubility of CO_2 in ILs [42–54]. However, progress remains sluggish because the CO_2 capacity of ILs for physical dissolution at low partial pressures of CO_2 from post-combustion flue gas is too low to develop a reasonable separation process based on physical solubility. Thus, it is highly desirable to design functional ILs for chemical absorption of CO_2 to overcome this crucial problem. In 2002, the Davis group reported the first example of CO_2 chemisorption by an IL in which an amine group was attached to the cation; their work presented the capture of 0.5 mol of CO_2 per mole of IL under ambient pressure [55]. After this, other functionalized ILs such as amino acid–based ILs [56, 57], azole-based ILs [58], and phenol-based ILs [59] were developed for the chemical absorption of CO_2. In view of the rapid development in ILs for CO_2 capture, some recent representative works are presented as follows. Noble and coworkers projected room-temperature ILs (RTILs) onto absorptive and membrane technologies for CO_2 capture [60–62]. Brennecke et al. explored phase-change ILs (PCILs) that have the potential to reduce energy consumption during the regeneration for post-combustion CO_2 capture [63, 64]. Rogers demonstrated chemisorption of CO_2 in 1,3-dialkylimidazolium acetate ILs through the presence of an unstable N-heterocyclic carbene in a relatively stable IL based on single crystal X-ray diffraction analysis [65]. Wang et al. proposed a smart strategy to achieve an extremely high capacity of up to 1.60 mol of CO_2 per mol of IL in the presence of multiple-site cooperative interactions, which was originated from the π-electron delocalization in the pyridine ring of the anion-functionalized ILs [66, 67]. The unlimited structural variations make ILs key materials for the increasing emissions of CO_2. Therefore, one objective to enhance the absorption performance of CO_2 by tuning the structures of ILs has great significance for these urgent environmental concerns.

In this chapter, recent advances and technologies in CO_2 capture with functionalized ILs are discussed and evaluated. Functionalized ILs for chemical absorption of CO_2 play a major part. Desirable CO_2 capture for ILs usually includes efficient absorption capacity, rapid absorption kinetics, and energy-saving demand for regeneration. Herein, we focus on protic ionic liquids (PILs), aprotic ILs, metal chelate ILs, IL-based mixtures, and supported ILs for the enhancement of CO$_2$ absorption performance. With the aid of experimental results, spectroscopic methods, and theoretical calculations, CO$_2$ absorption by different functionalized ILs has been investigated and understood at the molecular level. Finally, we critically assess the current status of CO$_2$ capture and put forward our own perspectives on future directions and prospects for CO$_2$ absorption by the design of functionalized ILs.

11.2 PILs for Chemisorption of CO$_2$

PILs are formed through proton transfer from a Brønsted acid to a Brønsted base, where the proton is able to promote extensive hydrogen bonding [68]. The first reported IL is, in fact, a protic IL (ethanolammonium nitrate [EOAN]), which was reported by Gabriel in 1888 [69], followed by ethylammonium nitrate (EAN) in 1914 [70]. Considering the low reactivity of the conventional PILs toward CO$_2$ under ambient conditions [71, 72], Dai et al. [73] designed a series of superbase-derived PILs from strong organic bases and a wide variety of weak proton donors to achieve rapid and efficient CO$_2$ capture (Figure 11.1). Furthermore, these PILs were applied to organic separation as switchable solvents triggered by CO$_2$.

It was clear that the CO$_2$ absorption of [MTBDH$^+$][TFE$^-$] was almost complete in the first 5 minutes with a faster rate than that of [MTBDH$^+$][Im$^-$] (about 30 minutes), as shown in Figure 11.1. This rapid absorption is related to their low viscosities ([MTBDH$^+$][TFE$^-$], 8.63 cP; [MTBDH$^+$][Im$^-$], 31.85 cP). The superbase, as a strong proton acceptor, deprotonates the weak proton donors directly, providing a thermodynamic driving force for the chemical reaction of the PILs with CO$_2$. An equimolar absorption of CO$_2$ in this system is a result of the formation of a liquid carbonate, carbamate, or phenolate salt (Scheme 11.1), further evidenced by NMR and IR spectroscopy. For example, after CO$_2$ bubbling, a new band observed at 1696.4 cm^{-1} can be attributed to carbamate stretches [74, 75]. Furthermore, changes of energy at -116.8, -85.2, and -41.7 kJ mol^{-1} for TFE$^-$, Im$^-$, and PhO$^-$ from B3LYP/TZVP level of theory are in agreement with the variations of CO$_2$ absorption capacity (1.13, 1.03, and 0.49 mol of CO$_2$ per mole of IL, respectively).

Figure 11.1 CO$_2$ absorption by typical superbase-derived protic ionic liquids (PILs).

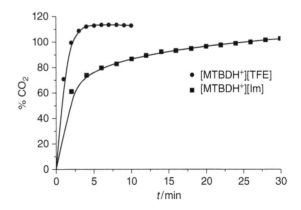

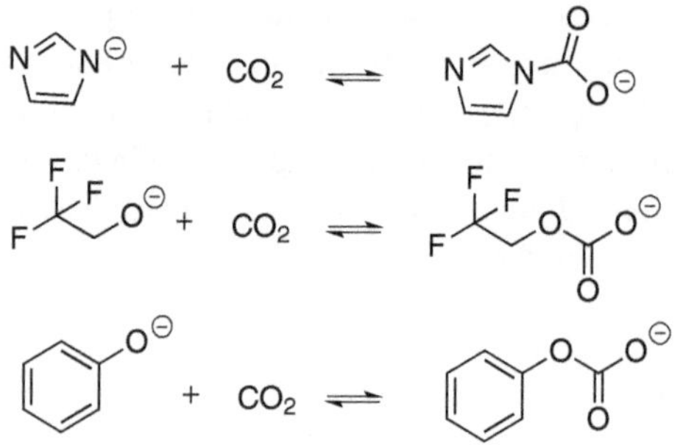

Scheme 11.1 CO$_2$ absorption by superbase-derived protic ionic liquids (PILs).

11.3 Aprotic Ionic Liquids for Chemisorption of CO$_2$

11.3.1 N as the Absorption Site

11.3.1.1 Amino-Containing Ionic Liquids

The first reported IL for chemisorption of CO$_2$ in 2002 was an amino-functionalized IL [55], which reacts with CO$_2$ in a carbamate mechanism. As witnessed by Scheme 11.2, this process is atom inefficient because one captured CO$_2$ molecule requires two amines. Subsequently, a great deal of effort was devoted to the capture efficiency of CO$_2$ based on amino-containing ILs [57, 76–84].

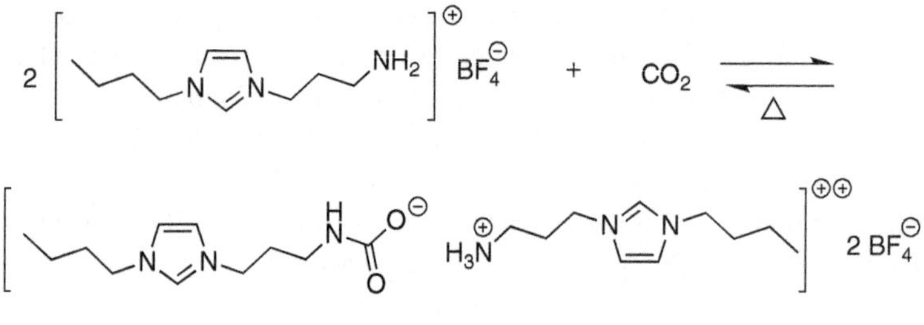

Scheme 11.2 CO$_2$ chemisorption by an amino-functionalized ionic liquid (IL).

As a typical example, Brennecke and co-workers [79] successfully achieved an equimolar absorption capacity of CO$_2$ by phosphonium-based amino acid ILs (Scheme 11.3). Theoretical calculations at the B3LYP/6-31G ++ (d, p) level also supported the experimental absorption results. The net energies for the prolinate and methioninate complexes are −71 and − 55 kJ mol^{-1}, respectively, in accordance with the experimental values at −80 kJ mol^{-1} and −64 kJ mol^{-1} measured using calorimetry. Ab initio calculations revealed that attaching the amine group to the cation favored the carbamate salt, reflecting the electrostatic stability of the zwitterions, resulting in inefficient capture capacity of CO$_2$. Tethering the amine group to the anion favored the carbamic acid because of the instability of the product dianion, leading to equimolar CO$_2$ capture. FT-IR spectroscopy is a powerful means to gain insight into the equimolar reaction mechanism of CO$_2$ with [P$_{66614}$][Pro] or [P$_{66614}$][Met]. For example, the prolinate N—H stretch at 3290 cm^{-1} disappears, no ammonium bands

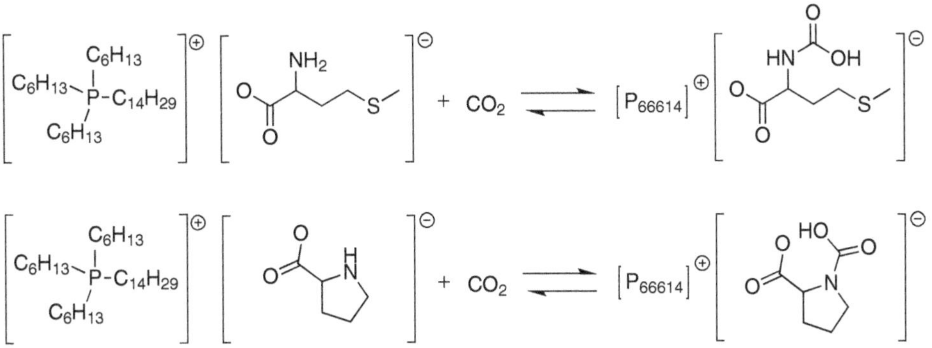

Scheme 11.3 Reaction schematics of CO$_2$ with [P$_{66614}$][Met] (top) and [P$_{66614}$][Pro] (bottom).

emerge, and a new peak at 1689 cm^{-1} reveals the formation of the COOH moiety from the reaction of [P$_{66614}$][Pro] and CO$_2$.

Subsequently, Brennecke [80] incorporated an additional amine group (e.g. lysine, asparaginate, glutaminate) into anions of naturally occurring amino acids to fabricate anion-functionalized ILs for high CO$_2$ capacity up to 1.4 mol mol^{-1}. Followed by this work, Riisager [82] investigated the effect of cations on the absorption capacity of CO$_2$ where the trihexyl(tetradecyl)phosphonium cation ([P$_{66614}$]) was replaced with the tri-hexyl(tetradecyl)ammonium cation ([N$_{66614}$]). An extremely high absorption capacity of up to 2.1 mol CO$_2$ per mol of IL by [N$_{66614}$][Lys] indicated that [P$_{66614}$][Lys]-CO$_2$ and [N$_{66614}$][Lys]-CO$_2$ adducts were composed of two different anion structures (Scheme 11.4). Recently, Wang [85] presented a new strategy for improving CO$_2$ capture through designing amino-functionalized ILs, where one amine could bind two CO$_2$. The results indicated that the basicity and steric hindrance of anions played a significant role in promoting amine group to capture two CO$_2$, where a high capacity of 1.96 mol mol^{-1} IL by [P$_{66614}$]$_2$[Asp] at 30 °C and 1 atm as well as excellent reversibility was achieved. However, the dramatic increase in viscosity of an amino-functionalized IL because of the extensive intermolecular hydrogen-bonding networks would make practical application for post-combustion CO$_2$ capture quite challenging.

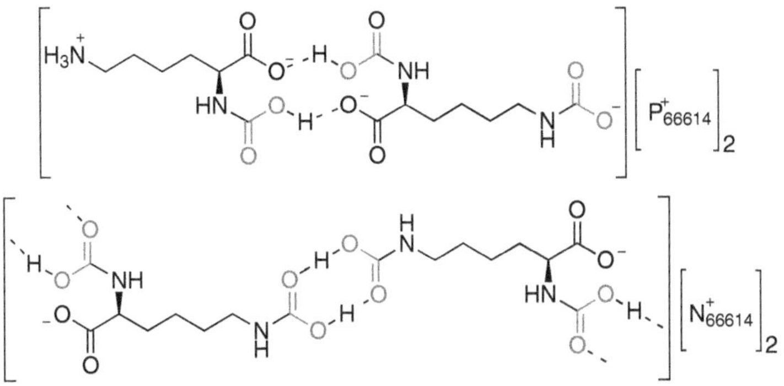

Scheme 11.4 Proposed structures of the anions in the [P$_{66614}$][Lys]–CO$_2$ and [N$_{66614}$][Lys]–CO$_2$ adducts (gray color, CO$_2$ groups).

11.3.1.2 Azolide Ionic Liquids

With regard to the high viscosity observed in CO_2 absorption in the presence of strong hydrogen-bonding networks, Brennecke selected suitably substituted azolides (aprotic heterocyclic anions [AHAs]) as the counter anions to react stoichiometrically and reversibly with CO_2 and to not suffer a large viscosity increase [56, 86]. The reaction enthalpy of CO_2 absorption was tailored by altering the substituent of the anion, such as [2-CNpyr]⁻ or [3-CF₃pyra]⁻, indicating energy demands during the desorption process could be properly controlled and modulated. More recently, PCILs, which are solid salts at normal flue gas processing temperatures and liquid PCIL-CO_2 complexes after reaction with CO_2, were discovered by Brennecke [63] based on AHAs. It can be anticipated that post-combustion CO_2 capture with PCILs has great potential to require less parasitic energy, because the phase-change that the liquid complex undergoes when it returns to a solid releases heat, reducing the added energy during the regeneration.

Wang et al. [58] investigated the tunability of the absorption enthalpy in azolide ILs (Scheme 11.5) for both efficient CO_2 capture and energy-saving release. When the pKa value of the anion in DMSO decreased from 19.8 to 8.2, correspondingly, the enthalpy of CO_2 absorption decreased from 91.0 to 19.1 kJ mol⁻¹. A quantitative relationship between the enthalpy of CO_2 absorption and the pKa value indicated that the enthalpy of CO_2 absorption can be quantified by tuning the basicity of the ILs (Figure 11.2), which offers an opportunity to achieve high CO_2 capacity with low energy demand. In contrast to amino-functionalized ILs, the rapid absorption kinetics by these non-amino basic ILs is due to a relatively low viscosity in the absence of strong hydrogen-bonding networks.

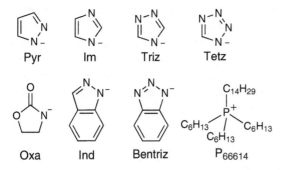

Scheme 11.5 Structure of anion and cation in basic azolide ILs.

The chemisorption of CO_2 in these azole-based IL systems was confirmed by TGA, NMR, and IR spectroscopy. The desorption of CO_2 is more facile when the basicity of the IL decreases, as indicated by TGA results. Reaction of aprotic heterocyclic anions with CO_2 to form a liquid carbamate resulted in an equimolar absorption of CO_2. For example, compared with fresh IL [P₆₆₆₁₄][Triz], a new peak at 1736 cm⁻¹ upon the uptake of CO_2 is assigned to a carbamate (C=O) stretch. Similarly, after the absorption of CO_2, a peak at 160.7 ppm appears in the ¹³C NMR spectrum, attributable to carbamate carbonyl carbon.

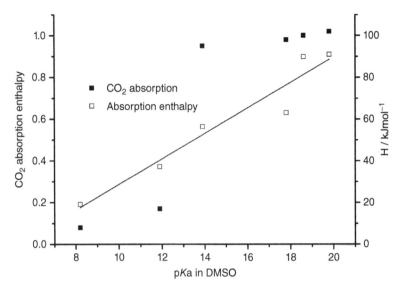

Figure 11.2 The relationship between CO_2 absorption capacity (■), absorption enthalpy (ΔH, □), and the pKa value of the anion in DMSO. The linear relationship between ΔH and the pKa of anion in DMSO is shown: $R^2 = 0.930$.

11.3.2 O as the Absorption Site

In recent years, phenolic ILs with facile structural variations, where the negative charge of the O atoms are the absorption sites, were reported by Wang et al. [59] for efficient and reversible capture of CO_2. Diverse phenolic ILs were obtained by modification of the electron-withdrawing or electron-donating ability, and the position and number of the substituents on the anion moieties. As shown in Table 11.11, substituent effects of the phenolic anions play a vital role in absorption capacity, absorption enthalpy, and absorption rate of CO_2.

Interestingly, when a carbonyl group was introduced to the *para*-position of a phenolic anion, a significant improvement in absorption capacity and cyclic reversibility was achieved due to the cooperative C—H···O hydrogen-bonding interaction between the carbonyl group and CO_2 [87].

11.3.3 Both N, O as Absorption Sites

Equimolar CO_2 absorption seems to be the ultimate capacity for non-amino-functionalized ILs in the past few decades. To go beyond equimolar capture, Wang and co-workers [67] reported a unique strategy to achieve an extremely high capacity of up to 1.60 mol CO_2 per mol IL and excellent reversibility with pyridine-containing anion-functionalized ILs through multi-site cooperative interactions. In their work, two site interactions between the electronic negative nitrogen and oxygen atoms in this anion and CO_2 contribute to such high capacities for CO_2 capture (Scheme 11.6).

Table 11.1 CO$_2$ chemisorption by phenolic ionic liquids.

Entry	Ionic liquid	CO$_2$ absorption[a, b]	T (°C)	η (cPa)[c]
1	[P$_{66614}$][4-Me-PhO]	0.91	30	392.7
2	[P$_{66614}$][4-MeO-PhO]	0.92	30	253.4
3	[P$_{66614}$][4-H-PhO]	0.85	30	246.7
4	[P$_{66614}$][4-Cl-PhO]	0.82	30	376.5
5	[P$_{66614}$][4-CF$_3$-PhO]	0.61	30	286.4
6	[P$_{66614}$][4-NO$_2$-PhO]	0.30	30	984.3
7	[P$_{66614}$][3-Cl-PhO]	0.72	30	223.2
8	[P$_{66614}$][2-Cl-PhO]	0.67	30	378.3
9	[P$_{66614}$][2,4-Cl-PhO]	0.48	30	472.5
10	[P$_{66614}$][2,4,6-Cl-PhO]	0.07	30	672.1
11	[P$_{66614}$][3-NMe$_2$-PhO]	0.94	30	512.1
12	[P$_{66614}$][1-Naph]	0.89	30	1077
13	[P$_{66614}$][2-Naph]	0.86	30	878.4

a) Mole of CO$_2$ per mol ionic liquid.
b) Determined for 30 minutes.
c) Determined at 23 °C.

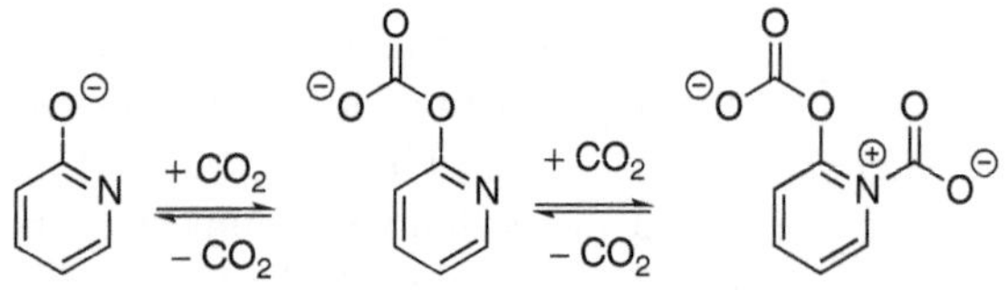

Scheme 11.6 The plausible mechanism of CO$_2$ absorption by [P$_{66614}$][2-Op] through multiple-site cooperative interactions.

The existence of multiple-site cooperative interactions was fully evidenced by absorption results, quantum-chemical calculations, spectroscopic investigations, and calorimetric data. CO$_2$ capacities of hydroxypyridine anion-containing ILs, such as [P$_{66614}$][2-Op] (1.58 mol mol^{-1}), are significantly higher than the combination of that from phenolic IL (0.85 mol mol^{-1}) and that from pyridine (0.013 mol mol^{-1}). The negative charge delocalization from the O atom to the N atom leads to an increase of the Mulliken atomic charges of nitrogen atoms in [P$_{66614}$][2-Op], [P$_{66614}$][3-Op], and [P$_{66614}$][4-Op] at −0.323, −0.235, and −0.285, respectively, far superior to that in pyridine (−0.161), indicating that it is possible for the N atom in these hydroxypyridine-containing ILs to act as the second reactive site. The multiple-site interactions between [P$_{66614}$][2-Op] with CO$_2$ were further confirmed by ^{13}C NMR and FTIR spectroscopy (Figure 11.3). The N-CO$_2$ and O-CO$_2$ interactions can be assigned at 1670 and 1650 cm^{-1}, respectively, in the IR spectra, and at 159.3 and 166.6 ppm, respectively, in the ^{13}NMR spectra after the uptake of CO$_2$. Furthermore, the appearance of two exothermic peaks in the calorimetry after the capture of CO$_2$ by [2-Op] also confirmed the presence of multiple-site cooperative interactions (N-CO$_2$ interaction and O-CO$_2$

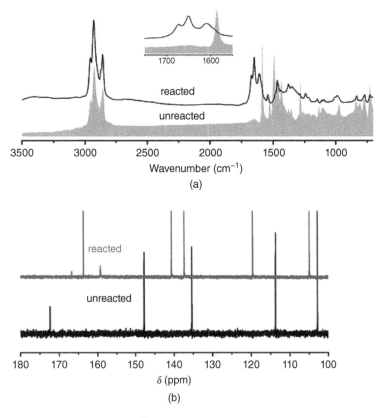

Figure 11.3 (a) IR and (b) ^{13}C NMR spectra of hydroxypyridine-functionalized IL [P$_{66614}$][2-Op] before and after CO$_2$ capture.

interaction) in the hydroxypyridine anion. In situ IR spectroscopy with two-dimensional correlation analysis revealed that the change of the peak at 1586 and 1650 cm^{-1} preceded 1670 cm^{-1}, indicating that the O atom is superior to the N atom and binds CO$_2$ first.

Furthermore, Wang et al. [87] presented a new strategy for improving CO$_2$ capture through the enhanced Lewis acid–base and cooperative C—H⋯O hydrogen bonding interactions by incorporating a carbonyl species into the imidazolate anions. As seen in Figure 11.4, [P$_{66614}$][4-CHO-Im] exhibits high absorption capacity and excellent desorption of CO$_2$, superior to that of [P$_{66614}$][Im]. In other words, the CO$_2$ absorption amount of [P$_{66614}$][4-CHO-Im] (1.24 mol mol^{-1}) exceeds that of [P$_{66614}$][Im] (0.98 mol mol^{-1}) with a large gap of 0.26; the release of CO$_2$ by [P$_{66614}$][4-CHO-Im] is complete and reversible, while 0.25 mol CO$_2$ per mole IL remained for [P$_{66614}$][Im] under the same desorption condition. These results demonstrate that the incorporation of the CHO group apparently improves the absorption performance of CO$_2$. The FTIR spectra further confirm the presence of a C—H⋯O hydrogen bonding interaction between the anion [4-CHO-Im] and CO$_2$ because the stretching vibration peaks at 2733 and 1640 cm^{-1} moved to 2743 and 1615 cm^{-1}, respectively, after the capture of CO$_2$ by [P$_{66614}$][4-CHO-Im]. On the other hand, Dai and co-workers [88] developed another method to improve CO$_2$ capture through introducing a

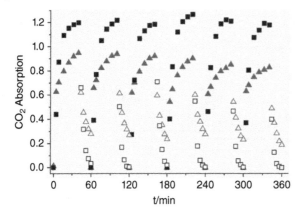

Figure 11.4 Six cycles of CO_2 absorption and desorption using $[P_{66614}][4\text{-CHO-Im}]$ and $[P_{66614}][Im]$. CO_2 absorption was carried out at $30\,^{\circ}C$ and 1 bar, and desorption was performed at $80\,^{\circ}C$ under N_2. For $[P_{66614}][4\text{-CHO-Im}]$: ■, absorption; □, desorption. For $[P_{66614}][Im]$: ▲, absorption; △, desorption.

carboxylate group in amino acid ILs, where the carboxylate group was activated, leading to high absorption capacity of CO_2 up to $1.69\ mol\ mol^{-1}$ by aminopolycarboxylate-based ILs. Cui et al [89] presented a new strategy for enhancing CO_2 capture through the concept of preorganization and cooperation. An extremely high gravimetric CO_2 capacity of up to 22 wt% ($1.65\ mol\ mol^{-1}$) and excellent reversibility (16 cycles) were achieved from 10 vol.% of CO_2 in N_2 when using an IL having a preorganized anion with multiple active sites, such as N and O.

11.3.4 C as the Absorption Site

The proton at the C(2) position of 1,3-dialkylimidazolium rings can be extracted by a basic enough anion because of its weak acidity, providing an idea for imidazolium ILs to capture CO_2 based on the C absorption site. Maginn [90] used NMR results to explain the absorption of CO_2 in 1-butyl-3-methylimidazolium acetate and proposed the abstraction of the proton at the C(2) position by the basic acetate anion formed via reaction of CO_2 with the carbene species. Rogers [65] demonstrated the chemisorption of CO_2 in 1,3-dialkylimidazolium acetate ILs (Scheme 11.7) and realized that complex anion

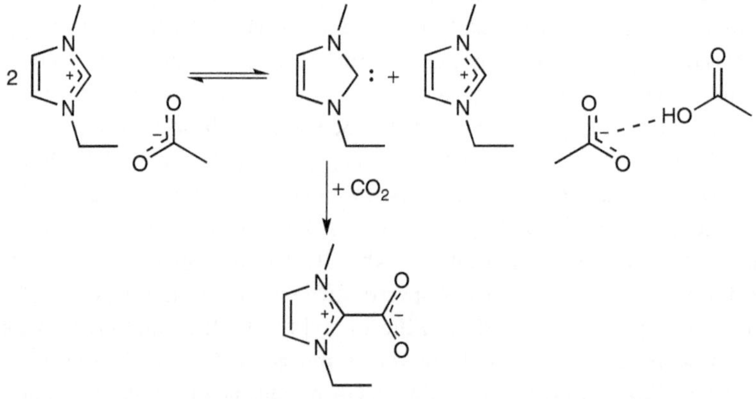

Scheme 11.7 Proposed reaction of CO_2 and $[C_2mim][OAc]$.

formation played an important role in stabilization of the volatile acetic acid, prevented further decomposition reactions, and allowed imidazolium ILs to act as stable reservoirs of carbenes based on single-crystal X-ray diffraction analysis. Thus, they concluded that an unstable N-heterocyclic carbene could exist in a relatively stable IL. To gain insight into the CO$_2$ reaction pathway of 1-ethyl-3-methylimidazolium ILs, Brennecke [91] selected basic aprotic heterocyclic anions that are capable of reacting stoichiometrically with CO$_2$ as the anion, and discovered that the carbene-CO$_2$ reaction is stronger than the anion-CO$_2$ reaction from the CO$_2$ uptake isotherm, as seen via ^{1}H and ^{13}C NMR spectroscopy.

11.4 Metal Chelate ILs for Chemisorption of CO$_2$

Metal chelate ILs are ILs that consist of a metallic coordination center and surrounding ligands. Aware of the intrinsic drawbacks of aqueous alkanolamine systems, including solvent loss, degradation, and high energy consumption for regeneration, Wang et al. [92] presented a method using tunable alkanolamine-based ILs with a multidentate coordinated cation to achieve high stability, good CO$_2$ absorption capacity, as well as excellent reversibility. The cations of this class of ILs are derived from complexation of a metal ion with neutral ligands [34, 93]. For example, mixing of 2,2'-(ethylenedioxy) bis(ethylamine) (DOBA) and LiTf$_2$N affords the IL [Li(DOBA)][Tf$_2$N]. The coordination sites in the DOBA ligand with Li$^+$ were N, O, O, and N. After the absorption of CO$_2$, one N site was replaced with an oxygen atom of CO$_2$, resulting in an equimolar capacity of CO$_2$ absorption (Scheme 11.8).

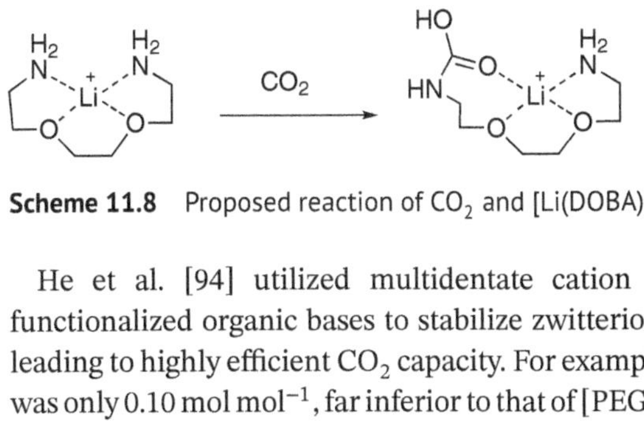

Scheme 11.8 Proposed reaction of CO$_2$ and [Li(DOBA)][Tf$_2$N].

He et al. [94] utilized multidentate cation coordination between Li$^+$ and PEG-functionalized organic bases to stabilize zwitterionic complexes after the capture of CO$_2$, leading to highly efficient CO$_2$ capacity. For example, the CO$_2$ capacity of PEG150-MeBu$_2$N was only 0.10 mol mol^{-1}, far inferior to that of [PEG150-MeBu$_2$NLi][NTf$_2$] (0.66 mol mol^{-1}).

11.5 IL-Based Mixtures for Chemisorption of CO$_2$

To overcome the disadvantages of amine-functionalized ILs, including high viscosity, multi-step synthesis, and no cost-competition with commodity chemicals, Noble et al. [95] used organic amine/RTIL solutions for the rapid and reversible capture of 1 mol CO$_2$ per 2 mol amine. This method is industrially attractive because RTILs with desired properties, i.e. nonvolatility, enhanced CO$_2$ solubility, and lower heat capacities, capture CO$_2$ with significant advantages, including increased energy efficiency, when mixed with commercial amines. The RTILs with Tf$_2$N anions have relatively low viscosities, and the insoluble amine-carbamate precipitate helps to drive the capture reaction, resulting in rapid and complete absorption in 25 minutes. In addition, Chen et al. [96] believe that hydrogen bonding between protonated monoethanolamine and the chloride ion in hydroxyl-imidazolium-based ILs and monoethanolamine mixture can benefit the capture and thermal stabilization of CO_2.

Mixtures of an amidine superbase with an alcohol, alkylamine, or amino alcohol have good reactivity and high absorption capacity for CO_2 capture. Unfortunately, the volatilization of alcohol, as well as the recombination of CO_2 with volatilized species (i.e. alcohols and/or base), led to the loss of organic solvents and high energy demand for regeneration. To address the loss of volatiles, Dai and coworkers [97] proposed an integrated strategy consisting of 1 : 1 mixtures of an alcohol-containing TSIL as a proton donor and an appropriate superbase as a proton acceptor to achieve rapid, reversible, and equimolar capture of CO_2. In further work [98], they selected imidazolium-based ILs as proton donors due to their weak acidities in the C-2 proton positions, where the protons were taken away by a superbase. As a result, the rapid and equimolar absorption of CO_2 in these systems was successfully achieved.

In consideration of the associated issues including high viscosity by amino-functionalized ILs for the capture of CO_2, the Han group [99] employed another strategy by mixing an amino-functionalized IL (2-hydroxyethyl-trimethyl-ammonium 2-pyrrolidine carboxylic acid salt, [Choline][Pro]) and polyethylene (PEG 200). Addition of PEG 200 in the IL decreased the viscosity of the system and thus enhanced the absorption rate of CO_2.

11.6 Supported ILs for Chemisorption of CO_2

Supported ILs provide new opportunities for CO_2 capture and separation because of their dual functions derived from both ILs and supports. Zhang et al. [76, 78] proposed a strategy using supported amino-functionalized ILs on porous silica gel to mitigate the high viscosity of amino-functionalized ILs, and thus fast and reversible CO_2 absorption was achieved, superior to that from bubbling CO_2 into the bulk of the IL. The large surface of the silica gel helps with this rapid absorption rate of CO_2. In a recent paper, Dai and coworkers [100] fabricated a porous liquid containing empty cavities by surface engineering of hollow silica spheres with suitable corona and canopy species (i.e. IL species), which was fully characterized by FT-IR spectra, TGA and DSC, SEM and TEM, small-angle X-ray scattering (SAXS), and N_2-sorption isotherms. The ether groups in the canopy chains enhance the gas solubility and selectivity toward CO_2 (CO_2/N_2 selectivity, c. 10) via Lewis acid/base interactions, and the presence of the empty cavities offers accelerated gas diffusivity through the porous liquid. Such a facile synthetic strategy to fabricate nano-structure-based porous liquids provides limitless potential for the promotion of gas separation.

Supported IL membranes attract growing interest in gas separation by taking advantage of their high permeability, selectivity, and stability [101–105]. Hanioka [101] first used a task-specific IL to selectively transport CO_2 in a supported liquid membrane (SLM). In their strategy, the amine moiety of the IL facilitated the chemical transport of CO_2, while CH_4 only relied on physical permeation, resulting in selective separation of CO_2 through the SLM. Subsequently, Brennecke and coworkers [102] separated CO_2 from H_2 with high permeability and selectivity even at elevated temperatures, with a supported IL membrane (the amino-functionalized ILs [$H_2NC_3H_6$mim][Tf_2N] in a cross-linked nylon-66 polymeric support). Matsuyama et al. [103] reported that SLMs incorporating amino acid ILs remarkably facilitated CO_2 permeation under dry and low-humidity conditions. In further work, they made use of tetrabutylphosphonium amine-functionalized glycinate or 2-cyanopyrrolide ILs to prepare task-specific IL-based facilitated transport membranes and evaluated CO_2 permeabilities and viscosity and CO_2 absorbance of ILs [104].

11.7   Conclusion and Perspectives

In summary, ILs address the grand challenge for CCS by taking advantage of their desired properties, including negligible vapor pressures, high thermal stabilities, excellent CO_2 solubilities, and tunable properties. ILs for chemisorption of CO_2 play an increasingly important role in CO_2 capture and separation, especially after Davis first reported a task-specific IL for chemical absorption of CO_2. Actually, the most attractive features of any CO_2 absorbent are high absorption capacity, energy-saving demand for regeneration, and rapid absorption rate. Thus, enormous efforts have been made to achieve these objectives through the design of functionalized ILs [106], including protic ILs, aprotic ILs, metal chelate ILs, IL-based mixtures, and supported ILs. Though great progress has been achieved in improving the chemisorption of CO_2 by amino-functionalized ILs, the dramatic increase in the viscosities upon CO_2 uptake became an inevitable problem in these systems. Therefore, non-amino functionalized ILs including azolate and phenolate ILs have been developed for enhancing the absorption performance of CO_2 in an atom-efficient manner. Furthermore, multiple-site interactions have been pursued to improve the capture of CO_2.

Recently, the incorporation of weak interactions, such as hydrogen bonds, ionic bonds (electrostatic interactions), π–π interactions, and Van der Waals forces have vitalized the field of CO_2 capture. Opportunities and challenges coexist in proper adjustment of chemical interactions and weak interactions with CO_2 by functional ILs for both efficient absorption capacity and energy-saving release. It is inevitable that anomalous or unique absorption phenomena or behavior will be encountered with any new system, but it is both challenging and meaningful to discover the reasons behind the phenomena, thus enabling the modulation and in-depth understanding of CO_2 capture. Furthermore, the physicochemical property changes (e.g. polarity, basicity, viscosity, and even phase change) of ILs upon CO_2 absorption present promising potential to construct CO_2-stimulus responsive materials based on ILs. The combination of ILs and other functional materials would provide a fantastic strategy for efficient CO_2 capture, with abundant fundamental science to be explored. We anticipate that there is considerable promise for ILs to realize new advances and innovations for CO_2 capture, especially as new concepts and strategies are introduced.

Acknowledgments

The National Key Basic Research Program of China (2015CB251401), the National Natural Science Foundation of China (No.21776239, No.21322602), the Zhejiang Provincial Natural Science Foundation of China (LZ17B060001), the Program for Zhejiang Leading Team of S&T Innovation (2011R50007), and the Fundamental Research Funds of the Central Universities are greatly acknowledged for their generous financial support.

References

1 Wasserscheid, P. and Welton, T. (2008). *Ionic Liquids in Synthesis*. Weinheim: Wiley-VCH.

2 Cole, A.C., Jensen, J.L., Ntai, I. et al. (2002). Novel Brønsted acidic ionic liquids and their use as dual solvent–catalysts. *J. Am. Chem. Soc.* 124: 5962–5963.

3 Xiao, J.C., Twamley, B., and Shreeve, J.M. (2004). An ionic liquid-coordinated palladium complex: a highly efficient and recyclable catalyst for the heck reaction. *Org. Lett.* 6: 3845–3847.

4 Kim, H.S., Kim, K.Y., Lee, C., and Chin, C.S. (2002). Ionic liquids containing anionic selenium species: applications for the oxidative Carbonylation of aniline. *Angew. Chem. Int. Ed.* 41: 4300.

5 Miao, W. and Chan, T.H. (2006). Ionic-liquid-supported synthesis: a novel liquid-phase strategy for organic synthesis. *Acc. Chem. Res.* 39: 897.

6 Binnemans, K. (2005). Ionic liquid crystals. *Chem. Rev.* 105: 4148–4204.

7 Ishiba, K., Morikawa, M., Chikara, C. et al. (2015). Photoliquefiable ionic crystals: a phase crossover approach for photon energy storage materials with functional multiplicity. *Angew. Chem. Int. Ed.* 54: 1532–1536.

8 Kouwer, P.H.J. and Swager, T.M. (2007). Synthesis and Mesomorphic properties of rigid-core ionic liquid crystals. *J. Am. Chem. Soc.* 129: 14042–14052.

9 Tanaka, K., Ishiguro, F., and Chujo, Y. (2010). POSS Ionic Li quid. *J. Am. Chem. Soc.* 132: 17649–17651.

10 Zhang, Q. and Shreeve, J.M. (2013). Ionic liquid propellants: future fuels for space propulsion. *Chem.-Eur. J.* 19: 15446–15451.

11 Okuno, M., Hamaguchi, H., and Hayashi, S. (2006). Magnetic manipulation of materials in a magnetic ionic liquid. *Appl. Phys. Lett.* 89: 132506.

12 Arenz, S., Babai, A., Binnemans, K. et al. (2005). Intense near-infrared luminescence of anhydrous lanthanide(III) iodides in an imidazolium ionic liquid. *Chem. Phys. Lett.* 402: 75–79.

13 Barpanda, P., Chotard, J.-N., Delacourt, C. et al. (2011). LiZnSO₄F made in an ionic liquid: a ceramic electrolyte composite for solid-state lithium batteries. *Angew. Chem. Int. Ed.* 50: 2526–2531.

14 MacFarlane, D.R., Tachikawa, N., Forsyth, M. et al. (2014). Energy applications of ionic liquids. *Energy Environ. Sci.* 7: 232–250.

15 Eshetu, G.G., Armand, M., Scrosati, B., and Passerini, S. (2014). Energy storage materials synthesized from ionic liquids. *Angew. Chem. Int. Ed.* 53: 13342–13359.

16 Armand, M., Endres, F., MacFarlane, D.R. et al. (2009). Ionic-liquid materials for the electrochemical challenges of the future. *Nat. Mater.* 8: 621–629.

17 Zhang, P., Wu, T.B., and Han, B.X. (2014). Preparation of catalytic materials using ionic liquids as the media and functional components. *Adv. Mater.* 26: 6810–6827.

18 Mahurin, S.M., Fulvio, P.F., Hillesheim, P.C. et al. (2014). Directed synthesis of nanoporous carbons from task-specific ionic liquid precursors for the adsorption of CO₂. *ChemSusChem* 7: 3284–3289.

19 Fukumoto, K. and Ohno, H. (2007). LCST-type phase changes of a mixture of water and ionic liquids derived from amino acids. *Angew. Chem. Int. Ed.* 46: 1852–1855.

20 Wang, J.Y., Chu, H.B., and Li, Y. (2008). Why single-walled carbon nanotubes can be dispersed in imidazolium-based ionic liquids. *ACS Nano* 2: 2540–2546.

21 Fujie, K., Yamada, T., Ikeda, R., and Kitagawa, H. (2014). Introduction of an ionic liquid into the micropores of a metal–organic framework and its anomalous phase behavior. *Angew. Chem. Int. Ed.* 53: 11302–11305.

22 Swatloski, R.P., Spear, S.K., Holbrey, J.D., and Rogers, R.D. (2002). Dissolution of cellulose with ionic liquids. *J. Am. Chem. Soc.* 124: 4974–4975.

23 Giernoth, R. (2010). Task-specific ionic liquids. *Angew. Chem. Int. Ed.* 49: 2834–2839.

24 Plechkova, N.V. and Seddon, K.R. (2008). Applications of ionic liquids in the chemical industry. *Chem. Soc. Rev.* 37: 123–150.

25 Pandolfi, J.M., Connolly, S.R., Marshall, D.J., and Cohen, A.L. (2011). Projecting coral reef futures under global warming and ocean acidification. *Science* 333: 418–422.

26 Canadell, J.G., Quéré, C.L., Paupach, M.R. et al. (2007). Contributions to accelerating atmospheric CO_2 growth from economic activity, carbon intensity, and efficiency of natural sinks. *Proc. Natl. Acad. Sci. U. S. A.* 104: 18866–18870.

27 Hasib-ur-Rahman, M., Siaj, M., and Larachi, F. (2010). Ionic liquids for CO_2 capture – development and progress. *Chem. Eng. Process.* 49: 313–322.

28 Yu, K.M.K., Curcic, I., Gabriel, J., and Tsang, S.C.E. (2008). Recent advances in CO_2 capture and utilization. *ChemSusChem* 1: 893–899.

29 Rochelle, G.T. (2009). Amine scrubbing for CO_2 capture. *Science* 325: 1652–1654.

30 Han, B., Zhou, C.G., Wu, J.P. et al. (2011). Understanding CO_2 capture mechanisms in aqueous monoethanolamine via first principles simulations. *J. Phys. Chem. Lett.* 2: 522–526.

31 Dupont, J., de Souza, R.F., and Suarez, P.A.Z. (2002). Ionic liquid (molten salt) phase organometallic catalysis. *Chem. Rev.* 102: 3667–3691.

32 Wasserscheid, P. and Keim, W. (2000). Ionic liquids – new "solutions" for transition metal catalysis. *Angew. Chem. Int. Ed.* 39: 3772–3789.

33 Huang, J.F., Luo, H.M., Liang, C.D. et al. (2005). Hydrophobic Brønsted acid–base ionic liquids based on PAMAM dendrimers with high proton conductivity and blue photoluminescence. *J. Am. Chem. Soc.* 127: 12784–12785.

34 Huang, J.F., Luo, H.M., and Dai, S. (2006). A new strategy for synthesis of novel classes of room-temperature ionic liquids based on complexation reaction of cations. *J. Electrochem. Soc.* 153: J9–J13.

35 Greaves, T.L. and Drummond, C.J. (2008). Protic ionic liquids: properties and applications. *Chem. Rev.* 108: 206–237.

36 Wu, W.Z., Han, B.X., Gao, H.X. et al. (2004). Desulfurization of flue gas: SO_2 absorption by an ionic liquid. *Angew. Chem. Int. Ed.* 43: 2415–2417.

37 Tempel, D.J., Henderson, P.B., Brzozowski, J.R. et al. (2008). High gas storage capacities for ionic liquids through chemical complexation. *J. Am. Chem. Soc.* 130: 400–401.

38 Fukumoto, K., Yoshizawa, M., and Ohno, H. (2005). Room temperature ionic liquids from 20 natural amino acids. *J. Am. Chem. Soc.* 127: 2398–2399.

39 Earle, M.J., Esperanca, J., Gilea, M.A. et al. (2006). The distillation and volatility of ionic liquids. *Nature* 439: 831–834.

40 Wang, C.M., Guo, L.P., Li, H.R. et al. (2006). Preparation of simple ammonium ionic liquids and their application in the cracking of dialkoxypropanes. *Green Chem.* 8: 603–607.

41 Wang, C.M., Zhao, W.J., Li, H.R., and Guo, L.P. (2009). Solvent-free synthesis of unsaturated ketones by the Saucy–Marbet reaction using simple ammonium ionic liquid as a catalyst. *Green Chem.* 11: 843–847.

42 Anderson, J.L., Dixon, J.K., and Brennecke, J.F. (2007). Solubility of CO_2, CH_4, C_2H_6, C_2H_4, O_2, and N_2 in 1-hexyl-3-methylpyridinium Bis(trifluoromethylsulfonyl)imide: comparison to other ionic liquids. *Acc. Chem. Res.* 40: 1208–1216.

43 Baltus, R.E., Culbertson, B.H., Dai, S. et al. (2004). Low-pressure solubility of carbon dioxide in room-temperature ionic liquids measured with a quartz crystal microbalance. *J. Phys. Chem. B* 108: 721–727.

44 Bara, J.E., Gabriel, C.J., Carlisle, T.K. et al. (2009). Gas separations in fluoroalkyl-functionalized room-temperature ionic liquids using supported liquid membranes. *Chem. Eng. J.* 147: 43–50.

45 Perez-Blanco, M.E. and Maginn, E.J. (2011). Molecular dynamics simulations of carbon dioxide and water at an ionic liquid interface. *J. Phys. Chem. B* 115: 10488–10499.

46 Zhang, X.C., Liu, Z.P., and Wang, W.C. (2008). Screening of ionic liquids to capture CO_2 by COSMO-RS and experiments. *AIChE J.* 54: 2717–2728.

47 Huang, X.H., Margulis, C.J., Li, Y.H., and Berne, B.J. (2005). Why is the partial molar volume of CO_2 so small when dissolved in a room temperature ionic liquid? Structure and dynamics of CO_2 dissolved in [Bmim⁺] [PF₆⁻]. *J. Am. Chem. Soc.* 127: 17842–17851.

48 Zhang, X.C., Huo, F., Liu, Z.P. et al. (2009). Absorption of CO_2 in the ionic liquid 1-n-hexyl-3-methylimidazolium Tris(pentafluoroethyl)trifluorophosphate ([hmim][FEP]): a molecular view by computer simulations. *J. Phys. Chem. B* 113: 7591–7598.

49 Finotello, A., Bara, J.E., Narayan, S. et al. (2008). Ideal gas Solubilities and solubility Selectivities in a binary mixture of room-temperature ionic liquids. *J. Phys. Chem. B* 112: 2335–2339.

50 Carlisle, T.K., Bara, J.E., Gabriel, C.J. et al. (2008). Interpretation of CO_2 solubility and selectivity in nitrile-functionalized room-temperature ionic liquids using a group contribution approach. *Ind. Eng. Chem. Res.* 47: 7005–7012.

51 Muldoon, M.J., Aki, S., Anderson, J.L. et al. (2007). Improving carbon dioxide solubility in ionic liquids. *J. Phys. Chem. B* 111: 9001–9009.

52 Bara, J.E., Gabriel, C.J., Lessmann, S. et al. (2007). Enhanced CO_2 separation selectivity in oligo(ethylene glycol) functionalized room-temperature ionic liquids. *Ind. Eng. Chem. Res.* 46: 5380–5386.

53 Cadena, C., Anthony, J.L., Shah, J.K. et al. (2004). Why is CO_2 so soluble in imidazolium-based ionic liquids? *J. Am. Chem. Soc.* 126: 5300–5308.

54 Wang, Y., Wang, C.M., Zhang, L.Q., and Li, H.R. (2008). Difference for SO_2 and CO_2 in TGML ionic liquids: a theoretical investigation. *Phys. Chem. Chem. Phys.* 10: 5976–5982.

55 Bates, E.D., Mayton, R.D., Ntai, I., and Davis, J.H. (2002). CO_2 capture by a task-specific ionic liquid. *J. Am. Chem. Soc.* 124: 926–927.

56 Gurkan, B., Goodrich, B.F., Mindrup, E.M. et al. (2010). Molecular Design of High Capacity, low viscosity, chemically tunable ionic liquids for CO_2 capture. *J. Phys. Chem. Lett.* 1: 3494–3499.

57 Liu, A.H., Ma, R., Song, C. et al. (2012). Equimolar CO_2 capture by N-substituted amino acid salts and subsequent conversion. *Angew. Chem. Int. Ed.* 51 (45): 11306–11310.

58 Wang, C.M., Luo, X.Y., Luo, H.M. et al. (2011). Tuning the basicity of ionic liquids for equimolar CO_2 capture. *Angew. Chem. Int. Ed.* 50: 4918–4922.

59 Wang, C.M., Luo, H.M., Li, H.R. et al. (2012). Tuning the physicochemical properties of diverse phenolic ionic liquids for equimolar CO_2 capture by the substituent on the anion. *Chem.-Eur. J.* 18: 2153–2160.

60 Bara, J.E., Camper, D.E., Gin, D.L., and Noble, R.D. (2010). Room-temperature ionic liquids and composite materials: platform technologies for CO_2 capture. *Acc. Chem. Res.* 43: 152–159.

61 Bara, J.E., Carlisle, T.K., Gabriel, C.J. et al. (2009). Guide to CO_2 separations in imidazolium-based room-temperature ionic liquids. *Ind. Eng. Chem. Res.* 48: 2739–2751.

62 Noble, R.D. and Gin, D.L. (2011). Perspective on ionic liquids and ionic liquid membranes. *J. Membr. Sci.* 369: 1–4.

63 Seo, S., Simoni, L.D., Ma, M. et al. (2014). Phase-change ionic liquids for postcombustion CO_2 capture. *Energy Fuel* 28: 5968–5877.

64 Seo, S., Guzman, M.Q., DeSilva, M.A. et al. (2014). Chemically tunable ionic liquids with aprotic heterocyclic anion (AHA) for CO_2 capture. *J. Phys. Chem. B* 118: 5740–5751.

65 Gurau, G., Rodríguez, H., Kelley, S.P. et al. (2011). Demonstration of chemisorption of carbon dioxide in 1,3-dialkylimidazolium acetate ionic liquids. *Angew. Chem. Int. Ed.* 50: 12024–12026.

66 Wang, C.M., Cui, G.K., Luo, X.Y. et al. (2012). Highly efficient and reversible SO_2 capture by tunable azole-based ionic liquids through multiple-site chemical absorption. *J. Am. Chem. Soc.* 133: 11916–11919.

67 Luo, X.Y., Guo, Y., Ding, F. et al. (2014). Significant improvements in CO_2 capture by pyridine-containing anion-functionalized ionic liquids through multiple-site cooperative interactions. *Angew. Chem. Int. Ed.* 53: 7053–7057.

68 Kennedy, D.F. and Drummond, C.J. (2009). Large aggregated ions found in some Protic ionic liquids. *J. Phys. Chem. B* 113: 5690–5693.

69 Gabriel, S. and Weiner, J. (1888). Ueber einige Abkömmlinge des Propylamins. *Ber. Dtsch. Chem. Ges.* 21: 2669–2679.

70 Walden, P. (1914). Ueber die Molekulargrösse und elektrische Leitfähigkeit einiger geschmolzenen Salze. *Bull. Acad. Imp. Sci. St.-Petersbourg* 8: 405.

71 Yuan, X.L., Zhang, S.J., Liu, J., and Lu, X.M. (2007). Solubilities of CO_2 in hydroxyl ammonium ionic liquids at elevated pressures. *Fluid Phase Equilib.* 257: 195–200.

72 Mattedi, S., Carvalho, P.J., Coutinho, J.A.P. et al. (2011). High pressure CO_2 solubility in N-methyl-2-hydroxyethylammonium Protic ionic liquids. *J. Supercrit. Fluids* 56: 224–230.

73 Wang, C.M., Luo, H.M., Jiang, D. et al. (2010). Carbon dioxide capture by Superbase-derived Protic ionic liquids. *Angew. Chem. Int. Ed.* 49: 5978–5981.

74 Jessop, P.G., Heldebrant, D.J., Li, X.W. et al. (2005). Green chemistry: reversible nonpolar-to-polar solvent. *Nature* 436: 1102–1102.

75 Phan, L., Chiu, D., Heldebrant, D.J. et al. (2008). Switchable solvents consisting of amidine/alcohol or guanidine/alcohol mixtures. *Ind. Eng. Chem. Res.* 47: 539–545.

76 Zhang, J.M., Zhang, S.J., Dong, K. et al. (2006). Supported absorption of CO$_2$ by Tetrabutylphosphonium amino acid ionic liquids. *Chem.-Eur. J.* 12: 4021–4026.

77 Soutullo, M.D., Odom, C.I., Wicker, B.F. et al. (2007). Reversible CO$_2$ capture by unexpected plastic-, resin-, and gel-like ionic soft materials discovered during the Combi-click generation of a TSIL library. *Chem. Mater.* 19: 3581–3583.

78 Zhang, Y.Q., Zhang, S.J., Lu, X.M. et al. (2009). Dual amino-functionalised phosphonium ionic liquids for CO$_2$ capture. *Chem.-Eur. J.* 15: 3003–3011.

79 Gurkan, B.E., de la Fuente, J.C., Mindrup, E.M. et al. (2010). Equimolar CO$_2$ absorption by anion-functionalized ionic liquids. *J. Am. Chem. Soc.* 132: 2116–2117.

80 Goodrich, B.F., de la Fuente, J.C., Gurkan, B.E. et al. (2011). Experimental measurements of amine-functionalized anion-tethered ionic liquids with carbon dioxide. *Ind. Eng. Chem. Res.* 50: 111–118.

81 Xue, Z.M., Zhang, Z.F., Han, J. et al. (2011). Carbon dioxide capture by a dual amino ionic liquid with amino-functionalized imidazolium cation and taurine anion. *Int. J. Greenhouse Gas Control* 5: 628–633.

82 Saravanamurugan, S., Kunov-Kruse, A.J., Fehrmann, R., and Riisager, A. (2014). Amine-functionalized amino acid-based ionic liquids as efficient and high-capacity absorbents for CO$_2$. *ChemSusChem* 7: 897–902.

83 Luo, X.Y., Ding, F., Lin, W.J. et al. (2014). Efficient and energy-saving CO$_2$ capture through the entropic effect induced by the intermolecular hydrogen bonding in anion-functionalized ionic liquids. *J. Phys. Chem. Lett.* 5: 381–386.

84 Gutowski, K.E. and Maginn, E.J. (2008). Amine-functionalized task-specific ionic liquids: a mechanistic explanation for the dramatic increase in viscosity upon complexation with CO$_2$ from molecular simulation. *J. Am. Chem. Soc.* 130: 14690–14704.

85 Luo, X.Y., Lv, X.Y., Shi, G.L. et al. (2019). Designing amino-functionalized ionic liquids for improved carbon capture: one amine binds two CO$_2$. *AIChE J.* 65: 230–238.

86 Gohndrone, T.R., Lee, T.B., DeSilva, A. et al. (2014). Competing reactions of CO$_2$ with cations and anions in azolide ionic liquids. *ChemSusChem* 7: 1970–1975.

87 Ding, F., He, X., Luo, X.Y. et al. (2014). Highly efficient CO$_2$ capture by Carbonyl-containing ionic liquids through Lewis acid–base and cooperative C-H···O hydrogen bonding interaction strengthened by the anion. *Chem. Commun.* 50: 15041–15044.

88 Chen, F.F., Huang, K., Zhou, Y. et al. (2016). Multi-molar absorption of CO$_2$ by the activation of carboxylate groups in amino acid ionic liquids. *Angew. Chem. Int. Ed.* 55: 7166–7170.

89 Huang, Y.J., Cui, G.K., Zhao, Y.L. et al. (2017). Preorganization and cooperation for highly efficient and reversible capture of low-concentration CO$_2$ by ionic liquids. *Angew. Chem. Int. Ed.* 56: 13293–13297.

90 E.J. Marginn. 2007. Design and evaluation of ionic liquids as novel CO$_2$ absorbents. Quarterly Technical Report to the DOE. DOE Office of Scientific and Technical Information.

91 Seo, S., DeSilva, M.A., and Brennecke, J.F. (2014). Physical properties and CO$_2$ reaction pathway of 1-Ethyl-3-Methylimidazolium ionic liquids with aprotic heterocyclic anions. *J. Phys. Chem. B* 118: 14870–14879.

92 Wang, C.M., Guo, Y., Zhu, X. et al. (2012). Highly efficient CO_2 capture by tunable Alkanolamine-based ionic liquids with multidentate cation coordination. *Chem. Commun.* 48: 6526–6528.

93 Tamura, T., Hachida, T., Yoshida, K. et al. (2010). New Glyme–cyclic imide lithium salt complexes as thermally stable electrolytes for lithium batteries. *J. Power Sources* 195: 6095–6100.

94 Yang, Z.-Z. and He, L.-N. (2014). Efficient CO_2 capture by tertiary amine-functionalized ionic liquids through Li^+-stabilized Zwitterionic adduct formation. *Beilstein J. Org. Chem.* 10: 1959–1966.

95 Camper, D., Bara, J.E., Gin, D.L., and Noble, R.D. (2008). Room-temperature ionic liquid–amine solutions: tunable solvents for efficient and reversible capture of CO_2. *Ind. Eng. Chem. Res.* 47: 8496–8498.

96 Huang, Q., Li, Y., Jin, X.B. et al. (2011). Chloride ion enhanced thermal stability of carbon dioxide captured by Monoethanolamine in hydroxyl imidazolium based ionic liquids. *Energy Environ. Sci.* 4: 2125–2133.

97 Wang, C.M., Mahurin, S.M., Luo, H.M. et al. (2010). Reversible and robust CO_2 capture by equimolar task-specific ionic liquid–superbase mixtures. *Green Chem.* 12: 870–874.

98 Wang, C.M., Luo, H.M., Luo, X.Y. et al. (2010). Equimolar CO_2 capture by imidazolium-based ionic liquids and Superbase systems. *GreenChem.* 12: 2019–2023.

99 Li, X.Y., Hou, M.Q., Zhang, Z.F. et al. (2008). Absorption of CO_2 by ionic liquid/polyethylene glycol mixture and the thermodynamic parameters. *Green Chem.* 10: 879–884.

100 Zhang, J.S., Chai, S.-H., Qiao, Z.-A. et al. (2015). Porous liquids: a promising class of media for gas separation. *Angew. Chem. Int. Ed.* 54: 932–936.

101 Hanioka, S., Maruyama, T., Sotani, T. et al. (2008). CO_2 separation facilitated by task-specific ionic liquids using a supported liquid membrane. *J. Membr. Sci.* 314: 1–4.

102 Myers, C., Pennline, H., Luebke, D. et al. (2008). High temperature separation of carbon dioxide/hydrogen mixtures using facilitated supported ionic liquid membranes. *J. Membr. Sci.* 322: 28–31.

103 Kasahara, S., Kamio, E.J., Ishigami, T., and Matsuyama, H. (2012). Amino acid ionic liquid-based facilitated transport membranes for CO_2 separation. *Chem. Commun.* 48: 6903–6905.

104 Kasahara, S., Kamio, E.J., Otani, A., and Matsuyama, H. (2014). Fundamental investigation of the factors controlling the CO_2 permeability of facilitated transport membranes containing amine-functionalized task-specific ionic liquids. *Ind. Eng. Chem. Res.* 53: 2422–2431.

105 Wang, C.M., Luo, X.Y., Zhu, X. et al. (2013). The strategies for improving carbon dioxide chemisorption by functionalized ionic liquids. *RSC Adv.* 3: 15518–15527.

106 Zeng, S.J., Zhang, X.P., Bai, L. et al. (2017). Ionic-liquid-based CO_2 capture systems: structure, interaction, Process. *Chem. Rev.* 117: 9625–9673.

12

Ionic Liquid-Based Membranes

Chi-Linh Do-Thanh[1], Jennifer Schott[1], Sheng Dai[1,2] and Shannon M. Mahurin[2]

[1]*Department of Chemistry, University of Tennessee, Knoxville, TN*
[2]*Chemical Sciences Division, Oak Ridge National Laboratory, Oak Ridge, TN*

CHAPTER MENU

12.1 Introduction

The removal of carbon dioxide from flue gas is a significant challenge that has become more pressing as the concentration of atmospheric CO_2 has increased and its effects have become more widely established. Future solutions could include increased efficiency in the generation and use of energy, replacing fossil-fuel-based energy sources with more benign sources, and a shift to a greater use of renewable, carbon-free energy sources. However, fossil fuels are currently the most economical method of energy generation and will continue to be used for the foreseeable future. In the interim, carbon capture and storage offers a potential short-term solution to reduce the emissions of CO_2 and mitigate their environmental effects.

A number of technologies have been used to separate CO_2 from a complex gas stream [1] including absorption, adsorption, cryogenic distillation, and membranes [2]. Post-combustion CO_2 absorption by chemical solvents such as aqueous alkanolamines has long been studied in separations applications. In particular, amine scrubbing has generally been used to remove CO_2 from natural gas streams during the "sweetening" process and as such is a mature technology with a relatively long history [3]. The use of

Materials for Carbon Capture, First Edition. Edited by De-en Jiang, Shannon M. Mahurin and Sheng Dai.
© 2020 John Wiley & Sons Ltd. Published 2020 by John Wiley & Sons Ltd.

amine scrubbing has also been explored for CO_2 capture, though this process is energy intensive and can significantly increase the cost of energy because of the high regeneration energy and because these solvents are volatile and corrosive. As a result, there has been considerable current interest in developing alternative sorbents for CO_2 capture including chilled ammonia as well as solid sorbents such as metal organic frameworks (MOFs) [4, 5], covalent organic frameworks [6], zeolitic imidazolate frameworks [7], and porous carbon [8, 9]. Membranes have also received considerable attention for CO_2 capture because membrane-based methods generally require less energy than other methods [10–13]. In addition, they have lower capital cost because of the smaller footprint required to accomplish the separation, simpler operation, and lower environmental impact because no solvents are used during the separation process.

One of the challenges in membrane-based separation is developing membrane materials that show optimal separation performance while retaining easy processability and cost-effective implementation. Consequently, a significant amount of effort has been devoted to polymeric membranes, which is the subject of a separate chapter. In addition to polymers, room-temperature ionic liquids (RTILs) have also received considerable attention as a novel material with applications in gas separations [14–43]. RTILs are room-temperature salts typically composed of an organic cation and an inorganic or organic anion, and these materials are unique in that they are composed completely of ions and have a melting point below $100\,°C$. RTILs have a number of physicochemical properties such as negligible vapor pressure, high thermal stability, and broad liquid range. In addition, RTIL properties such as CO_2 solubility, viscosity, stability, and molecular reactivity can be readily tuned through appropriate selection of cation and anion, thus introducing the idea of designer materials to describe RTILs. The ability to tune the interaction of RTILs with solute molecules is particularly important because it enables the interaction between the RTIL and CO_2 to be tuned and the separations properties to be modulated depending on the RTIL used [44, 45]. This leads to the design of RTILs through the selection of cation/anion pairs or through the introduction of functional groups with CO_2 interactions that can enhance the solubility in the material. Moreover, the strength of the CO_2/RTIL interaction can be uniformly varied and can range from a purely physical interaction that is relatively weak to a chemical interaction in which the CO_2 becomes chemically bound and the RTIL functions as an absorbent. For membrane-based separation, it is important to tune the properties appropriately so that there is good interaction with CO_2 leading to high solubility but not so strong that there is a chemical bond that would reduce the gas transport through the membrane and make it difficult to desorb on the low-pressure side. Much of the interest in RTILs for the separation of CO_2 began with the report that supercritical CO_2 could dissolve into RTILs [46–48]. Figure 12.1 shows a series of RTILs that have typically been used for the separation of CO_2.

Despite the great potential of RTILs to be used in membrane-based separations, a number of challenges remain before these promising materials will be widely implemented in applications. In general, RTILs have high viscosity and a relatively high production cost, both of which hinder wide acceptance. Additionally, RTILs must be supported in some way in order to function as a membrane, and there has been some work exploring various methodologies for fashioning RTIL-based membranes. In this chapter, we will explore the mechanisms of gas transport through RTIL membranes and present work that has sought to address

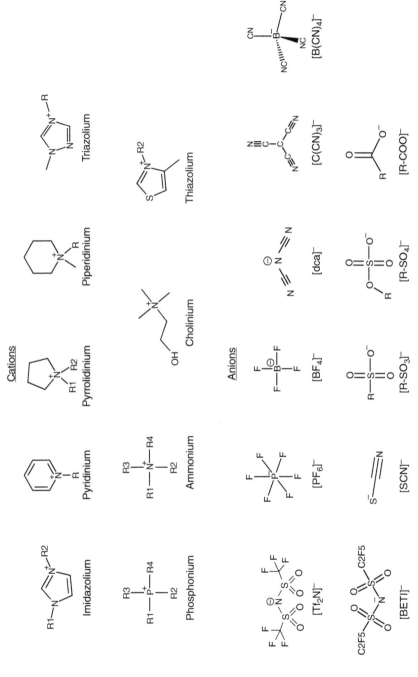

Figure 12.1 Structures of RTILs that have been used for membrane-based separations.

the challenges associated with implementation of these unique materials in separations applications.

12.1.1 Transport in Ionic Liquids

The separations performance of a membrane regardless of the material is generally defined in terms of two parameters: *permeability*, which is a measure of the intrinsic membrane productivity; and *selectivity*, which is a measure of the membrane separation efficiency [49, 50]. The permeability of a gas through a membrane is essentially the pressure- and thickness-normalized flux, as follows:

$$P_i = \frac{n_i \cdot l}{\Delta p_i}$$

where n_i is the flux of the gas molecule through the membrane, l is the thickness of the membrane, and Δp_i is the transmembrane pressure, which serves as the driving force for gas transport through the membrane. Though a number of units are used for gas permeability, the most common is the barrer which is defined as follows:

$$1\ barrer = 10^{-10} \frac{cm^3 \cdot cm}{cm^2 \cdot s \cdot cmHg}$$

The permeability of a gas through a material is an intrinsic property of the material and is independent of the thickness. In contrast, permeance, which is often used in applications, increases as the membrane thickness decreases.

The permeability of a gas in an RTIL-based membrane is generally governed by the product of a thermodynamic component (the gas solubility) and a kinetic component (the diffusivity). This so-called *solution/diffusion mechanism* in RTILs is the same mechanism that drives gas transport in polymeric membranes. In this pressure-driven process, gas is absorbed by the RTIL on the high-pressure side, diffuses through the ionic liquid membrane, and is then desorbed on the low-pressure side (see Figure 12.2). As mentioned, the two key properties that define the permeability in an RTIL are the solubility and diffusivity, according to the following equation:

$$P_{ij} = S_i \cdot D_j$$

where P is the permeability, S is the gas solubility, and D is the diffusivity. The ideal selectivity between a gas pair is then defined as the ratio of the individual permeabilities:

$$\alpha_{ij} = \frac{P_i}{P_j}$$

The ideal selectivity can then be rewritten as the product of solubility selectivity and diffusivity selectivity using the definition of permeability as follows:

$$\alpha_{ij} = \frac{S_i}{S_j} \cdot \frac{D_i}{D_j}$$

In polymer membranes and molecular sieves, differences in the diffusivity between penetrant molecules such as CO_2 and N_2 drive the overall selectivity of the membrane since the solubility selectivity is approximately one. In contrast, the selectivity in RTILs is generally due to differences in the solubility of the gases rather than diffusivity, e.g. the solubility of

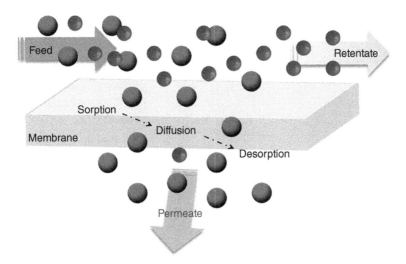

Figure 12.2 Illustration showing the solution/diffusion mechanism. Source: reprinted from Ref. [2] with permission from The Royal Society of Chemistry.

CO_2 in RTILs is higher than the solubility of N_2, resulting in a focus on enhancing CO_2 solubility by adding specific functional groups.

12.1.2 Facilitated Transport

In addition to traditional solution/diffusion transport, facilitated transport in materials has long been known in various applications, particularly for transport in biological systems [51, 52]. There was subsequent interest in understanding the mechanisms of facilitated transport in membrane-based systems where the focus was on the parameters that dominated this type of transport and how these parameters affected separations [53–55]. In general, facilitated transport is a coupled transport mechanism involving a reversible interaction between the solute and a carrier in the material followed by diffusion through the membrane (see Figure 12.3). Solute molecules can also undergo both pure diffusion and facilitated transport if the carrier concentration is much less than the concentration of the solute (e.g. CO_2).

There are then two types of carrier-based membranes: mobile carriers and fixed-site carriers. For mobile carriers, the species can diffuse through the membrane, thus transporting the solute. In contrast, the species is stationary for fixed-site carriers, and the motion is more complex and can involve hopping of the solute from one carrier to the next. For facilitated transport in ionic liquid-based membranes, the predominant motion is carrier-mediated, where gas molecules such as CO_2 absorb into the liquid, react with the carrier to form a complex, diffuse through the membrane, and are released. The addition of the carrier generally serves to enhance the solubility of the gas of interest compared to other gases. The more inert gases then simply follow the typical solution–diffusion mechanism, thus improving the selectivity of CO_2 over N_2, for example.

For CO_2 separation, interest originally focused on membranes that incorporated some form of amine, because of the known interaction between CO_2 and amines [57–60]. In many

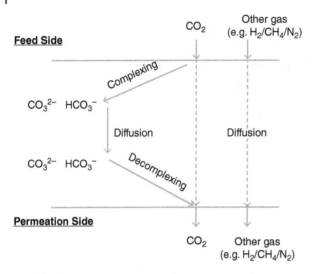

Figure 12.3 Illustration showing facilitated transport of CO_2 in a membrane. Source: reprinted from Ref. [56] with permission from Elsevier.

of the reports on facilitated transport in amine-based liquids, the presence of water was important to drive the interaction between CO_2 and the amine [61]. The development of task-specific ionic liquids [62], which incorporate certain functional groups such as amine moieties with increased reactivity, laid the groundwork for additional interest in facilitated transport where the presence of water was not necessary for the reaction to occur. For example, Myers et al. measured the CO_2 permeability and CO_2/H_2 selectivity for an imidazolium-based ionic liquid that incorporated a terminal amine on the alkyl chain of the cation [63]. The selectivity increased as the temperature was raised, reaching a peak value at approximately 85 °C, and then decreased with additional increases in temperature. This was compared to an imidazolium-based ionic liquid without an amine group, which showed a steadily decreasing selectivity with temperature to confirm the presence of facilitated transport. Similarly, Hanioka et al. measured the permeability and selectivity of CO_2 and CH_4 for two amine-containing alkylimidazolium ionic liquids, $[C_3NH_2mim][CF_3SO_3]$ and $[C_3NH_2mim][Tf_2N]$, and one traditional ionic liquid, $[bmim][Tf_2N]$ [64]. It should be noted that water vapor was used on both the feed and permeate sides during the permeability tests. The CO_2/CH_4 permselectivity of the amine-containing ionic liquids was higher than the traditional ionic liquid even at room temperature, and the membranes showed excellent stability. Moreover, the selectivity was measured as a function of transmembrane pressure, and the amine-containing ionic liquids showed an increase in selectivity as the CO_2 pressure was decreased, which is a typical method to confirm facilitated transport in membranes. This reflects the saturation of the carriers in the ionic liquid as the CO_2 partial pressure increases.

This same group continued their work by reporting separations performance of ionic liquids with a reactive amino acid anion coupled with both imidazolium and phosphonium cations [65]. They noted that while moisture was not necessary for CO_2 separation, low-moisture conditions did enhance CO_2 permeability slightly due to improved interaction with the amino acid. Additional work exploring the effect of water [66] and the impact of cation size on gas permselectivity [67] in the amino acid-based ionic liquids (ILs) has also been reported. Because the amino acid ILs still showed low CO_2 permeabilities at low temperature in dry conditions and in high CO_2 pressures, Kasahara et al. investigated the factors that determined CO_2 permeability in these materials and concluded that the viscosity of the complex and the amount of CO_2 absorbed were critical parameters controlling CO_2 permeability [68].

While significant work has focused on amine-containing ILs, carboxylate-based ILs also have good affinity for CO_2 [69, 70]. Wang et al. reported the CO_2 absorption of a series of 10 ionic analogs of triethylbutylammonium carboxylate and showed that the dicarboxylate-based ILs with fully deprotonated anions exhibited higher CO_2 affinity than monocarboxylate-based ILs [71]. Applying this to membranes, the same group found that membranes with the dicarboxylate-based ILs that are fully deprotonated were effective carriers in the selective separation of CO_2 [72]. The carboxylate-based membranes were also effective for the separation of SO_2, another important acid gas with relevance to separations from flue gas. Acetate-based ILs have been explored for the selective separation of CO_2 [73, 74]. Luis et al. also measured the facilitated transport of CO_2 and SO_2 using imidazolium cations with acetate anions supported in two different porous polymers and found that [butylimidazolium][acetate] in a hydrophilic poly(vinylidene fluoride) (PVDF) support had the highest CO_2 and SO_2 permeabilities while [methylimidazolium][acetate] in a hydrophilic PVDF support had the highest permselectivity [75]. Expanding further, Gouveia et al. explored equimolar binary mixtures of a cyano-based IL (imidazolium cation with tricyanomethanide anion) and a series of amino acid ILs (imidazolium cation with amino acid anions) for the facilitated transport of CO_2, confirming that these ILs could serve as carriers for CO_2 [76]. The mixtures exhibited increased CO_2 solubility compared to the pure cyano-based IL with the mixtures containing L-alaninate and taurinate anions showing large CO_2 permeability and high CO_2/N_2 selectivity. Facilitated transport in RTILs thus offers the potential to enhance the separations properties of these materials.

12.2 Supported IL Membranes

In order to create functional membranes from liquids, they must be contained or immobilized in a support, as illustrated in Figure 12.4 to prevent motion of the RTIL with the driving pressure across the membrane. While there has been interest in depositing RTILs between two dense films, the predominant approach to contain the RTILs is to incorporate them into the pores of a porous host. These supported ionic liquid membranes (SILMs) are generally commercially available membrane filters with a range of pore sizes and compositions. Though the most common supports are polymers such as polyethersulfone (PES) or PVDF, inorganic supports such as anodic alumina have also been used. In this architecture, the RTILs become immobilized within the pores of the support and are held in place via capillary forces where the strength is determined by the pore size and the RTIL/pore interaction.

Practically, the RTIL is loaded into the porous support by first immersing the support in the IL and allowing it to diffuse into the pores. The ability of the IL to wet the membrane

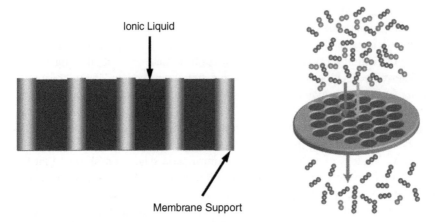

Figure 12.4 (Left) Graphic depicting the integrations of an ionic liquid (IL) in a supported ionic liquid membrane (SILM); (right) the selective transport of gas through the SILM. (*See color plate section for color representation of this figure*).

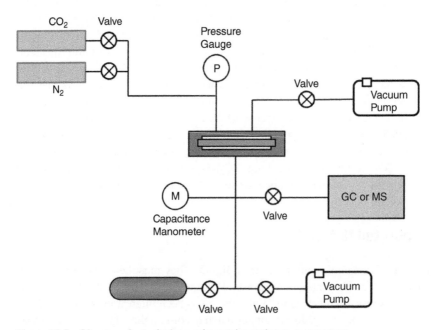

Figure 12.5 Diagram of a typical experimental membrane test system.

is important in obtaining good loading [77]. In addition, the immersed membrane must be placed in a vacuum oven or desiccator in order to remove air from the pores and allow complete filling. In some cases, some have reported using either vacuum or pressure to force the IL into the pores. For example, IL can be placed on one side of the membrane (on a suitable support) while pressure is applied (or vacuum is applied on the downstream side) to push the IL into the pores. Figure 12.5 shows a typical experimental setup used to measure the permeability and ideal selectivity of membranes. In much of the literature, the permeability of each individual gas is measured separately, and the ideal selectivity is presented. However, the addition of a gas chromatograph or mass spectrometer to the experimental setup along with appropriate flow controllers for each gas can enable the measurement of mixed gases, which is more representative of separations applications. SILMs have most commonly been reported using flat-sheet supports because a broad range of flat membranes are readily available [78]; however, hollow-fiber membranes offer greater interfacial area to pore volume and are more widely used in industrial separations applications, so there has been interest in using these fibers as supports. Despite the many advantages of SILMs, one of the challenges is the overall stability of the membrane due to the expulsion of the RTIL at transmembrane pressures that are high enough to overcome the capillary forces on the RTIL. This loss increases at higher transmembrane pressures, effectively limiting SILMs to low-pressure separations processes.

As mentioned, a variety of support compositions have been used in SILMs ranging from polymer supports such as PES, PVDF, and poly(dimethylsiloxane (PDMS) to ceramic supports and even carbon supports. In general, the support reduces the overall gas transport of the SILM because it is solid, i.e. the gas permeability through the SILM is lower than through the IL alone. This effect is often removed by loading the support with a material of well-known gas permeability, enabling the effect of the support to be removed and permitting measurement of the RTIL only. Some researchers, however, have reported that the support can play a larger role in modulating the gas permeability depending on the composition and pore size [79]. Close et al. reported CO_2 permeance values in ILs on 100 nm supports that were larger than those for the same ILs on 20 nm supports. In addition, the permeance of ILs in SILMs was measured to be larger than the bulk, which was attributed to the effect of interactions between the IL and the support [80]. Nanoporous carbon materials have also been used as supports for SILMs. Chai et al. used a soft-templated, nanoporous carbon membrane with controlled pore sizes up to 12 nm as the support and showed that the transmembrane pressure could be increased to 10 bar with minimal loss of the IL while maintaining good separations performance.

Since the first report of SILMs in 2002 [81], a wide variety of ILs have been explored for use in SILMs. Many of the early ILs focused on the common imidazolium-based ILs such as 1-ethyl-3-methylimidazolium bis(trifluoromethanesulfonyl)amide, or [emim][Tf$_2$N] [82]. The use of ILs for CO_2 separation began after Blanchard and coworkers suggested that supercritical CO_2 could be used to separate low-volatility solvents from ILs such as 1-butyl-3-methylimidazolium hexafluorophosphate, [bmim][PF$_6$] [48]. Afterward, the Brennecke group began exploring the phase behavior of ILs and found that CO_2 is remarkably soluble in imidazolium-based ILs, which set the stage for the use of RTILs in CO_2 separation [46, 47].

Scovazzo et al. incorporated the [bmim][PF$_6$] IL into a hydrophilic polyethersulfone membrane with a pore size of 0.1 µm and measured a CO_2 permeance of 4.6×10^{11} mol/(cm^2·kPa·s) [81]. This was the first demonstration that showed ILs could be supported in membranes and used to separate CO_2. In this case, the separation was CO_2 from air with a selectivity of 29. Soon after, Scovazzo expanded this work to include a variety of ILs, in particular, non-fluorinated ILs, partly because the [PF$_6$]$^-$ can readily decompose to HF if it is in water. An imidazolium cation, [emim], was coupled with a

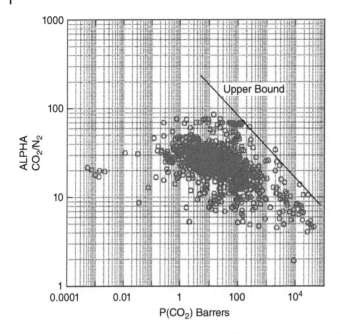

Figure 12.6 The upper bound correlation of CO_2/N_2 selectivity and CO_2 permeability. Source: reprinted from [84] with permission from Elsevier.

series of anions including bis(trifluoromethanesulfonyl)amide $[Tf_2N]^-$, trifluoromethane-sulfonate $[CF_3SO_3]^-$, chloride $[Cl]^-$, and dicyanamide $[dca]^-$, and the CO_2 separation properties were measured [83]. The CO_2 permeability values ranged from 350 barrers up to 1000 barrers. Ideal selectivity values ranged from 15 for the $[Cl]^-$ to 61 for $[dca]^-$. This work demonstrated the potential for ILs to be used for CO_2/N_2 separation with performance values that were near or above the Robeson upper bound for polymer membranes (Figure 12.6). The Robeson plot presents the ideal selectivity of a gas pair such as CO_2/N_2 versus the permeability of the more permeable gas, e.g. CO_2. In general, as the permeability of a membrane increases, the selectivity decreases, and the Robeson upper bound reflects this inherent trade-off between permeability and selectivity [84, 85].

A variety of conventional ILs have been explored in SILMs. In the early days of SILMs, most of the work was focused on the imidazolium-based ILs because they are readily available or easily synthesized [17]. However, some work has been reported on other IL systems, including phosphonium [23, 86], ammonium [87], and pyridinium-based ILs [88]. In addition to the cation, a variety of anions have also been explored, including tetrafluoroborate $([BF_4]^-)$, hexafluorophosphate $([PF6]^-)$, bis(trifluoromethylsulfonyl)imide $([Tf_2N]^-)$, and dicyanamide$([N(CN)_2]^-)$.

Because permeability in the SILM is based on the solution–diffusion mechanism, much effort was directed toward understanding the solubility and diffusivity of gases in these ILs. In general, the CO_2 diffusivity in ILs does not vary significantly for different cation/anion pairs [89]. As a result, selectivity is largely driven by the solubility selectivity. Consequently, various models have been developed to understand solubility in ILs and to predict the solubilities in new ILs. Camper and Noble developed a model based on regular solution theory that incorporates the molar volume of the IL to predict solubility [14]. This model was strictly applicable to the imidazolium-based ILs due to certain assumptions. This model was extended to a broader variety of ILs and incorporated both the molar volume and the IL viscosity [90].

Many inter-related factors influence CO_2 separation in ILs, such as the viscosity, cation, anion, and cation/anion interaction. Cadena et al. used a combination of experiment and molecular modeling to explore the effect of cation and anion on CO_2 solubility for a set of imidazolium-based ILs [82]. They found for these ILs that the anion had the greatest impact on CO_2 solubility. Molecular dynamics simulations indicated that there was stronger organization of the CO_2 around the anion compared to the cation, consistent with the experimental results in which the solubility did not vary significantly when two ILs with the same anion but different cations were measured. They expanded this work to additional gases, and the results further suggested the importance of the anion for gas solubility in ILs [91].

While significant effort has gone toward improving the permeability and selectivity of conventional IL membranes, it was recognized early on that the ability to readily functionalize ILs could offer a way to enhance their separations properties. Because of the effect of solubility selectivity on overall performance, the use of CO_2-philic functional groups was initially explored. A series of ILs with different nitrile groups in the anion were explored for CO_2 separation [92]. In this work, [emim][B(CN)$_4$] showed a high CO_2 permeability of ~2000 barrer and a high CO_2/N_2 selectivity of ~50. The high permeability was subsequently postulated to be the result of reduced cation–anion interactions and relatively strong CO_2–anion interaction [27, 29, 93].

12.2.1 Microporous Supports and Nanoconfinement

Cheng and co-workers created SILMs by depositing the IL [bmim][BF$_4$] into asymmetric and symmetric microporous poly(vinylidene fluoride) (PVDF) membranes [94]. They studied the gas permeation of these membrane structures and found the permeance to be in the following decreasing order: CO_2 > CO_2–air > N_2. While the asymmetric SILMs showed higher gas permeance, the symmetric SILMs possessed better CO_2/N_2 selectivity for single and mixed gases. Furthermore, the symmetric SILMs were capable of separating low concentrations of CO_2, even from ambient air. Therefore, these membranes may offer a method for capturing CO_2 directly from the atmosphere without requiring CO_2 transportation.

The transport properties of ILs can be considerably altered by confinement in porous materials. Along with the rise of new materials such as SILMs, studies of confinement effects have become more important. In order to predict the gas permeability of SILMs, Labropoulos et al. developed a method employing ILs [bmim][C(CN)$_3$] and [emim][C(CN)$_3$] as pore modifiers to study the effect of IL nanoconfinement on the CO_2 and N_2 permeability and selectivity of SILMs cast from the ILs into nanofiltration membranes [95]. The SILMs were developed on nanoporous (1 nm pore size) asymmetric ceramic membranes. The method's predicted permeability and selectivity deviated significantly from the experimental values for the SILMs. This difference is due to layering and crystallization of the IL phase that happen under extreme confinement into the nanopores. The bulkier [bmim] cation is tilted with respect to the pore surface, allowing the [C(CN)$_3$]$^-$ counteranion to occupy the center of the pore, which in turn forms a straight diffusion path, resulting in higher-than-predicted permeability values. In contrast, the less bulky [emim] cations are randomly placed in the pores, allowing the counteranions to interact with the pore surface, which leads to random diffusion paths similar to those in the unconfined IL phase. Moreover, [emim][C(CN)$_3$] solidified under extreme confinement into the nanopores, resulting in much slower CO_2 diffusion than predicted while maintaining high CO_2/N_2 separation. The authors concluded that for the confinement of ILs into nanopores, empirical models that predict the permeability performance of SILMs based on the physicochemical properties of the involved ILs cannot be applied.

The CO_2 separation characteristics of ILs [bmim][Tf$_2$N] and [hmim][Tf$_2$N] confined within a ceramic nanoporous film were determined by Banu and co-workers [96]. The solubility and diffusivity of CO_2 of the ILs were compared to those for bulk-phase ILs, with the confined ILs having enhanced values over those for the unconfined ILs. The researchers found that the physical properties of the confined ILs seemed to be changed by the porous support surface, possibly through reorganization of cations and anions at interface, which increased the available free volume for CO_2 absorption.

More recently, Hazelbaker et al. used proton and carbon-13 pulsed field gradient (PFG) NMR to examine the diffusion of a mixture of CO_2 and the IL [bmim][Tf$_2$N] under confinement in the mesopores of KIT-6 silica [97]. The results were compared to those for the corresponding bulk mixture of CO_2/IL, the unconfined pure IL, and the confined pure IL. Confinement was shown to display CO_2 diffusivity similar to or even larger than in the bulk CO_2/IL mixture. Diffusivities of the cations and CO_2 decreased with increasing diffusion time, consistent with diffusion slowdown as a result of transport resistances at the external surface of KIT-6 particles.

12.2.2 Hollow-Fiber Supports

There has been increased interest in hollow fibers (Figure 12.7) as support materials for SILMs because their high surface area-to-volume ratio would benefit permeation quantity by allowing a more efficient mass transfer than flat membranes. The first example of these fibers was published by Koros et al. where imidazolium-based RTILs were coupled with a superbase and incorporated into the hollow fiber (see Figure 12.8) [99, 100]. Lan and co-workers reported the use of polyethersulfone (PES) hollow fibers as support for RTILs

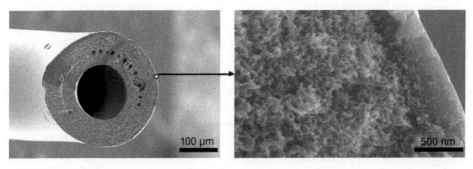

Figure 12.7 Image of a typical hollow-fiber membrane. Source: reprinted from Ref. [98] with permission from Elsevier.

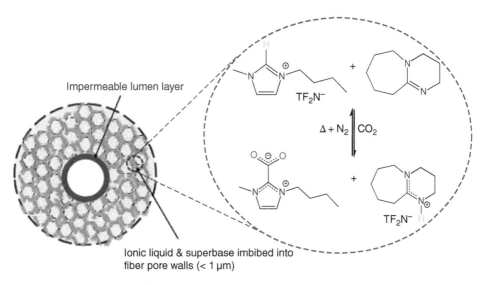

Figure 12.8 Illustration of RTILs imbibed into hollow-fiber membrane supports. Source: reprinted from Ref. [99] with permission from Elsevier.

to study CO_2/N_2 separation [101]. The fiber structures were adjusted by changing the dope solution composition to generate hollow fibers with and without finger-like macrovoids. The stability of SILMs was examined using ILs [bmim][BF$_4$] and [bmim][PF$_6$].

The authors also studied the effect of permeate direction and support structure on the pure gas permeability and ideal selectivity. CO_2 permeability and CO_2/N_2 selectivity were higher when the gas permeated from the shell side to the tube side (shell-feed permeation) compared to that when the gas permeated from the tube side to the shell side (bore-feed permeation). The presence of the finger-like macrovoids resulted in higher CO_2 permeability and lower CO_2/N_2 selectivity. Most of the results were close to the Robeson upper bound.

Another example of hollow-fiber membranes was reported by Wickramanayake et al., who used Matrimid® polymer and the IL [hmim][Tf$_2$N] as starting materials [102]. Four different types of membranes were produced using four morphologically different hollow-fiber supports to study the effect of fiber-wall morphology on fiber performance. Porosity influenced the transport and mechanical properties of the fiber. Fibers with high surface porosity also exhibited high permeability and selectivity. The mechanical properties are a function of the bulk porosity of the fiber. As bulk porosity increases, the volume fraction of the polymer decreases, resulting in less polymer matrix available to hold the fiber together. Tensile strength decreased significantly for fibers soaked in IL, most likely due to the plasticization effect that ILs can have on polymers. When ILs are mixed with polymer chains, they can disrupt intermolecular bonding and thus weaken the polymer matrix. Additionally, increasing fiber porosity and adding IL improve gas-transport properties while degrading the mechanical properties at the same time. Understanding this trade-off is key to a successful membrane system for applications as both gas separation and fiber durability are required.

The same research group later created another set of polymeric hollow-fiber SILMs using Matrimid and Torlon® polymers as supporting structures and the IL [hmim][Tf$_2$N] as the gas-transport media [103]. With the purpose of optimizing support structure, the polymer and fiber pore morphology were varied. The permeance and selectivity for CO_2/H_2 separation were similar for both Matrimid and Torlon supports when using a high enough fiber porosity, thereby maximizing the IL content of the membrane. Nevertheless, Matrimid supports displayed low mechanical strength when saturated with IL. Torlon supports were stronger but had a lower glass transition temperature due to a more flexible backbone chain. Molecular modeling revealed that the polymer chains in Torlon become more frequently interlocked with each other than in Matrimid, and [hmim][Tf_2N] interacts less with Torlon than with Matrimid.

More recently, Ge and Lee immobilized the ultra-hydrophobic IL [hmim] *tris*(pentafluoroethyl)trifluorophosphate (FAP) in the pores of a polypropylene hollow fiber for the liquid–liquid–liquid microextraction (HF-LLLME) of chlorophenols [104]. For HF-LLLME, a hollow fiber is used to support organic phase, which is held within the wall pores, and acceptor phase, which is confined within the channel of the membrane. The [hmim][FAP], used in this work as a wall pore-impregnated solvent in HF-LLLME, is far more hydrophobic than other common ILs containing PF_6^- and PO_4^{3-}, making it more suitable for extraction from aqueous samples. The extraction method was utilized in the analysis of chlorophenols in canal water samples, showing effective cleanup and high sensitivity.

12.3 Polymerizable ILs

More widespread adoption of IL membranes for gas separations will require improvements in the mechanical stability of the membrane. In SILMs, there is a tendency for the IL to be pushed out of the support at transmembrane pressures higher than a few bar. One approach to improve the mechanical stability of IL-based membranes while simultaneously enhancing the separations performance compared to dense polymer membranes is to use polymerizable ILs, or poly(ionic liquid)s, which can be prepared through either direct polymerization of IL monomers or by modification of existing polymers [56, 105, 106]. Polymerizable IL units are used to create these novel materials, and various polymerization techniques can be utilized depending on the strategy, i.e. direct polymerization or polymer modification. The advantage of the poly(ionic liquid) system lies in the fact that the macroscopic membrane is mechanically stable while it is largely liquid on the microscopic length scale, which provides enhanced gas transport compared to purely solid membranes, thus combining the advantages of the IL with the processability and stability of the polymer.

The application of poly(ionic liquid)s to gas separation was first reported by Tang et al., where IL monomers such as poly[p-vinylbenzyltrimethyl ammonium tetrafluoroborate] and poly[2-(methacryloyloxy)ethyltrimethylammonium tetrafluoroborate] were synthesized by anion exchange reactions [107]. The poly(ionic liquid)s were subsequently prepared by free radical polymerization and characterized for CO_2 sorption. The CO_2 sorption capacity for the poly(ionic liquid)s was 7.6 and 6.0 times higher, respectively, than an analogous RTIL, [bmim][BF_4], with fast and reversible sorption and desorption kinetics. Interestingly, the IL monomers showed no CO_2 uptake, indicating that polymerization of IL monomers could serve as a general strategy for preparing novel gas-separation materials. This was subsequently extended to imidazolium-based IL monomers where the poly(ionic liquid)s were prepared by free radical polymerization using an initiator [108]. These materials exhibited highly selective CO_2 sorption compared to N_2, with the poly(ionic liquid)s showing higher sorption capacity than the IL monomers. Because of the very low surface area of the poly(ionic liquid) materials, CO_2 absorption into the bulk was assumed to play a major role in defining the relatively high capacity. Additional work on imidazolium-based poly(ionic liquid)s explored the effect of the anion, cation, and backbone [109]. A series of ammonium-based poly(ionic liquid)s with a polystyrene backbone were then synthesized and showed selective CO_2 sorption up to 10.67 mol% [110]. These results, particularly the enhanced CO_2 sorption of the poly(ionic liquid)s compared to the corresponding monomer, sparked significant interest in these materials for CO_2 separations [111, 112] and laid the foundation for their use as membranes [113].

Bara et al. first reported a series of mechanically stable imidazolium-based poly(ionic liquid) membranes synthesized from simple IL monomers with varying alkyl chain lengths using either an acrylate or styrene polymerizable backbone (see Figure 12.9) [114].

The monomers were converted into stable membranes using photopolymerization and then tested for permeability and selectivity using CO_2, N_2, and CH_4. The permeability values were more dependent on the alkyl chain length rather than the particular polymerizable backbone. With CO_2 permeabilities ranging from 10 to 30 barrer and CO_2/N_2 selectivities of approximately 30, the performance of these membranes was near the Robeson upper bound, comparable to traditional polymer membranes. This same group followed up on this first work by exploring poly(ionic liquid)s with various polymerizable backbones such as polystyrene-, polyvinyl-, and polyacrylate-based groups, and with different compositions and substituents, though these were primarily based on imidazolium ILs [115–118].

Among the advantages of ILs are the broad range of possible cation/anion combinations, various backbone choices, and potential to add functional groups to improve gas permeability and selectivity. Bhavsar et al. examined the effect of anions on the CO_2 sorption of poly(ionic liquid)s and found that the carboxylate anion yielded materials with high CO_2 sorption and high selectivity over N_2 and H_2 [119]. Much of the work, particularly in the

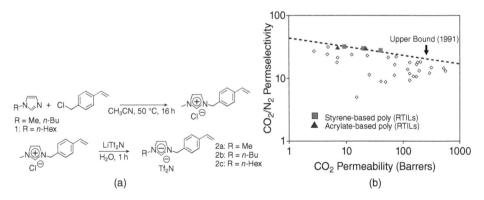

Figure 12.9 (a) Synthesis of styrene-based monomers; (b) performance of styrene- and acrylate-based poly(ionic liquids). Source: reprinted with permission from [114]. Copyright (2007) American Chemical Society.

beginning, was focused on imidazolium-based ILs as the monomers. More recently, this has expanded to include different IL monomers. For example, Tomé et al. reported the synthesis and characterization of pyrrolidinium-based poly(ionic liquid) membranes for CO_2 separation [120]. Though the CO_2 permeability was comparable to imidazolium-based poly(ionic liquid)s, the synthesis was much simpler and involved a direct anion exchange in a commercial poly(diallyldimethylammonium) chloride followed by solvent casting to form the membrane. Imidazolium-based poly(ionic liquid)s involve multiple organic syntheses and purification steps. Pyridinium-based membranes synthesized by converting aromatic polyethers with main chain pyridine groups followed by anion exchange metathesis reactions were recently reported [121]. Though the permeabilities were comparable to polymer membranes, the poly(ionic liquid) membranes had high mechanical and thermal stability and were relatively straightforward to synthesize. A series of polyurethane-based anionic poly(ionic liquid)s using two different polyols and a range of counter cations were synthesized, and the effect of the polyol structure and cation on CO_2 sorption capacity was measured [122]. Poly(ionic liquid)s have also been prepared from polybenzimidazole, where their physical properties have been investigated as well as gas-separations performance [123, 124]. A number of reports demonstrated the use of N-quaternization of the polybenzimidazole followed by metathesis to create different materials that were mechanically stable and could readily form membranes [125–127]. Many of these polybenzimidazole-based membranes exhibit high selectivity and high thermal stability and can also withstand high operating pressures. Moreover, the gas-separations properties could be tuned using various anions and substituents, making them potentially useful in separations applications despite modest permeabilities.

There has been significant interest in poly(ionic liquid)s over the past decade, and much progress has been made by varying the components and compositions as well as by adding functional groups. There remains, however, much work to be done to achieve the high permeabilities needed for gas-separations applications, which should inspire additional efforts in this area given the potential of these materials. In addition, tremendous knowledge has been gained, which can be used as a foundation for further study.

12.4 Mixed-Matrix ILs

Mixed-matrix membranes have historically been composed of porous inorganic filler particles dispersed in a polymer matrix, which generally provides improved performance compared to the polymer alone [128]. The filler particles used have included MOFs [129–131], zeolites [132–134], zeolitic imidazolate frameworks [135], carbon aerogels [136], porous silica [137, 138], carbon molecular sieves [139], and carbon nanotubes [140, 141]. One of the challenges with these types of membranes is that there must be good interfacial contact between the filler particle and the polymer. The presence of non-selective void space at the interface leads to higher permeability due to the large space but at the expense of reduced selectivity. A number of strategies have been developed to improve the compatibility between the polymer and the filler and thus address this challenge. One unique strategy is the incorporation of ILs into the mixed-matrix membranes to help improve the adhesion and interfacial interactions that serve to mitigate the formation of void space.

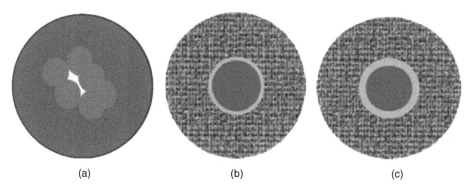

(a)   (b)   (c)

Figure 12.10 Illustration of room-temperature ionic liquid (RTIL) enhancing the adhesion of a mixed-matrix membrane. Source: reprinted from Ref. [142] with permission from Elsevier.

A number of approaches to incorporate ILs into mixed-matrix membranes have been pursued. One method is to simply use the IL as a means to improve the interfacial adhesion, while a second method is to directly mix ILs into polymers to improve performance. The first method is called a *three-component mixed-matrix membrane* as shown in Figure 12.10 and was first reported by Hudiono et al. [143]. In that work, the membrane contained a polymerizable IL, an IL, and a zeolite (SAPO-34). The SAPO-34, a microporous zeolite, was immersed in [emim][Tf_2N] in order to first coat the surface. A styrene-based imidazolium IL monomer was then added and crosslinked to form the three-component mixed-matrix membrane. The addition of the IL served to enhance the interaction between the zeolite filler particles and the poly(ionic liquid), which was evidenced by the improved CO_2/N_2 selectivity of the membrane compared to two-component systems and by the fact that the performance of the composite membrane better fit the Maxwell model equations. A subsequent report by the same group varied the poly(ionic liquid) while keeping the same SAPO-34 filler particle and [emim][Tf_2N] in order to optimize performance and better understand the structure of the hybrid system [142]. The combination of SAPO-34, [emim][Tf_2N], and a series of imidazolium-based poly(ionic liquids) showed high CO_2/CH_4 selectivity and CO_2 permeability with performance that exceeded the Robeson upper bound [144].

As shown in Figure 12.11, Casado-Coterillo incorporated a MOF, ZIF-8, into chitosan, which is an abundant and cheap polysaccharide, and 1-ethyl-3-methylimidazolium acetate, an IL with a high CO_2 solubility [145]. The three-component system with a ZIF-8 concentration of 10 wt% exhibited optimal performance, particularly with improved permselectivity, which was attributed to better adhesion between the ZIF-8 filler particles and the chitosan polymer. The free IL essentially wets the surface of the filler particle and promotes adhesion to the polymer matrix.

Extending this idea to materials that generally have weak interactions with polymers, a MOF, HKUST-1, was decorated with a thin layer of [emim][Tf_2N] followed by incorporation into a polyimide to generate a three-component mixed-matrix membrane with CO_2/CH_4 performance above the upper bound [146]. Electron microscopy showed the improved adhesion and minimization of non-selective voids at the interface between the MOF and the polymer, thus confirming that ILs can effectively extend the selection

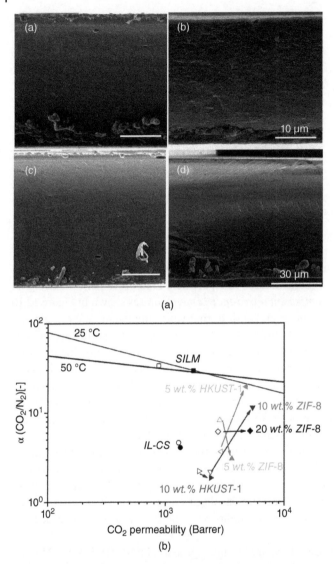

Figure 12.11 Scanning electron microscope images of the cross-sectional area of mixed-matrix membranes (MMMs) filled by (a) 5 wt% HKUST-1; (b) 20 wt% HKUST-1; (c) 10 wt% ZIF-8; (d) 20 wt% ZIF-8, respectively. (e) MMMs in comparison to the Robeson upper bound. Source: reprinted from Ref. [145] with permission from the Royal Society of Chemistry.

of porous materials that can be dispersed in the polymer due to enhanced interfacial interactions. Additional work has been reported on a variety of ILs, polymer matrices, and filler particles, creating three-component systems with improved separations performance resulting from enhanced polymer/matrix adhesion and reduced void space through the addition of the IL to the system [147–150].

Figure 12.12 Poly(ionic liquid)/ionic liquid composite. The red circles correspond to free [C$_2$mim] cations, the green circles are [Tf$_2$N] anions, and the blue circles are polymer bound cations. Source: reprinted from Ref. [56] with permission from Elsevier.

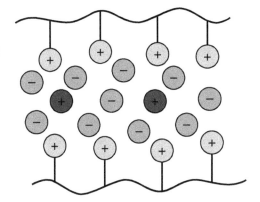

In most of these cases, the IL was first coated onto the filler particle, and then the surface-modified filler was incorporated into the polymer to form the membrane. In a slightly different approach, ILs can be mixed with polymers in a solvent compatible with both components followed by casting and solvent evaporation to create a composite IL/polymer membrane (see Figure 12.12). Abdollahi et al. created a two-component system where the IL, [emim][Tf$_2$N], was directly blended with Matrimid to form a hybrid membrane capable of withstanding high transmembrane pressure and exhibiting an increase in CO$_2$ permeability of more than 480% compared to the polymer alone and a significant enhancement in CO$_2$/CH$_4$ selectivity [151]. Regions of IL were formed during the synthesis process, leading to a heterogeneous structure in which the IL was essentially encapsulated within the polymer, significantly improving the stability of the membrane. This is an advantage compared to SILMs, which must be used at relatively low pressure.

Chen et al. blended PVDF with the IL [emim][B(CN)$_4$] to create a two-component mixed-matrix membrane that exhibited good mechanical strength, allowing high transmembrane pressure differences and CO$_2$/N$_2$ separation performance that equaled the Robeson upper bound line [152]. In addition to the improved separation performance and higher stability of the composite IL/polymer membrane, the impact of temperature on the stability of these membranes is less deleterious compared to SILMs. For example, blending a prototypical IL such as [bmim][Tf$_2$N] with thermally stable polymers such as polybenzimidazole yielded stable blended membranes with enhanced CO$_2$ permeability up to 200 °C [153]. Generally, blending polymers and ILs produces dense membranes with encapsulated IL. However, a recent report showed that a polymer/IL blend can also generate porous composite membranes with a thin separating layer [154]. By mixing polysulfone with different ILs, porous composite membranes were formed that showed improved CO$_2$ permeance and CO$_2$/N$_2$ selectivity compared to the dense polysulfone alone. The formation of pores was attributed to the presence of water in the IL, resulting in a phase inversion in the polysulfone. Composite polymer/IL membranes thus offer the potential to improve the permeability, selectivity, and stability of membranes compared to the polymer, which results, at least in part, from the heterogeneous nature of the blended membranes.

12.5 Conclusion and Outlook

Interest in RTILs for CO_2 separation has clearly grown in the last 15 years and represents an emerging field with significant potential to impact the growing area of carbon capture. The seemingly limitless possible ILs synthesized from a broad range of cation/anion pairs with tunable physicochemical properties including viscosity, chemical and thermal stability, and CO_2 reactivity is an important advantage with this class of material. Many of the RTILs have outstanding CO_2 permeability and selectivity, often surpassing the Robeson upper bound, which makes them valuable for membrane-based separations. In fact, when compared to the nearly infinite possible ILs, we have only just begun to explore these materials, which suggests that significant improvements in gas-separations properties are yet to be achieved. In addition, many of the structure/property relationships with the RTIL-based membranes are not well understood, providing the opportunity for significant advances in the future, particularly in the design of task-specific ILs for membranes. Another important area of research in RTIL membranes is the stability of the membrane at realistic operating pressures. A number of advances have occurred in this area, particularly in the area of polymerizable RTILs and polymer/RTIL composites. However, there is still room for improvement before these materials can be more widely applied. Despite these challenges, many novel RTIL membrane materials with outstanding CO_2 separation performance remain to be discovered and could tremendously impact the field of carbon capture.

References

1 Boot-Handford, M.E., Abanades, J.C., Anthony, E.J. et al. (2014). Carbon capture and storage update. *Energy Environ. Sci.* 7: 130–189.

2 Tome, L.C. and Marrucho, I.M. (2016). Ionic liquid-based materials: a platform to design engineered CO_2 separation membranes. *Chem. Soc. Rev.* 45: 2785–2824.

3 Rochelle, G.T. (2009). Amine scrubbing for CO_2 capture. *Science* 325: 1652–1654.

4 McDonald, T.M., Mason, J.A., Kong, X.Q. et al. (2015). Cooperative insertion of CO_2 in diamine-appended metal-organic frameworks. *Nature* 519: 303–308.

5 D'Alessandro, D.M., Smit, B., and Long, J.R. (2010). Carbon dioxide capture: prospects for new materials. *Angew. Chem. Int. Ed.* 49: 6058–6082.

6 Ding, S.Y. and Wang, W. (2013). Covalent organic frameworks (COFs): from design to applications. *Chem. Soc. Rev.* 42: 548–568.

7 Zhang, Z.J., Yao, Z.Z., Xiang, S.C., and Chen, B.L. (2014). Perspective of microporous metal-organic frameworks for CO_2 capture and separation. *Energy Environ. Sci.* 7: 2868–2899.

8 Liu, J., Wickramaratne, N.P., Qiao, S.Z., and Jaroniec, M. (2015). Molecular-based design and emerging applications of nanoporous carbon spheres. *Nat. Mater.* 14: 763–774.

9 Saufi, S.M. and Ismail, A.F. (2004). Fabrication of carbon membranes for gas separation – a review. *Carbon* 42: 241–259.

10 Koros, W.J. and Mahajan, R. (2000). Pushing the limits on possibilities for large scale gas separation: which strategies? *J. Membr. Sci.* 175: 181–196.

11 Luis, P. and Van der Bruggen, B. (2013). The role of membranes in post-combustion CO_2 capture. *Greenhouse Gases Sci. Technol.* 3: 318–337.

12 Park, H.B., Kamcev, J., Robeson, L.M. et al. (2017). Maximizing the right stuff: the trade-off between membrane permeability and selectivity. *Science* 356: 10.

13 Merkel, T.C., Lin, H.Q., Wei, X.T., and Baker, R. (2010). Power plant post-combustion carbon dioxide capture: an opportunity for membranes. *J. Membr. Sci.* 359: 126–139.

14 Camper, D., Bara, J., Koval, C., and Noble, R. (2006). Bulk-fluid solubility and membrane feasibility of Rmim-based room-temperature ionic liquids. *Ind. Eng. Chem. Res.* 45: 6279–6283.

15 Camper, D., Becker, C., Koval, C., and Noble, R. (2006). Diffusion and solubility measurements in room temperature ionic liquids. *Ind. Eng. Chem. Res.* 45: 445–450.

16 Camper, D., Scovazzo, P., Koval, C., and Noble, R. (2004). Gas solubilities in room-temperature ionic liquids. *Ind. Eng. Chem. Res.* 43: 3049–3054.

17 Baltus, R.E., Counce, R.M., Culbertson, B.H. et al. (2005). Examination of the potential of ionic liquids for gas separations. *Sep. Sci. Technol.* 40: 525–541.

18 Baltus, R.E., Culbertson, B.H., Dai, S. et al. (2004). Low-pressure solubility of carbon dioxide in room-temperature ionic liquids measured with a quartz crystal microbalance. *J. Phys. Chem. B* 108: 721–727.

19 Hou, Y. and Baltus, R.E. (2007). Experimental measurement of the solubility and diffusivity of CO_2 in room-temperature ionic liquids using a transient thin-liquid-film method. *Ind. Eng. Chem. Res.* 46: 8166–8175.

20 Kilaru, P.K., Condemarin, P.A., and Scovazzo, P. (2008). Correlations of low-pressure carbon dioxide and hydrocarbon solubilities in imidazolium-, phosphonium-, and ammonium-based room-temperature ionic liquids. Part 1. Using surface tension. *Ind. Eng. Chem. Res.* 47: 900–909.

21 Muldoon, M.J., Aki, S., Anderson, J.L. et al. (2007). Improving carbon dioxide solubility in ionic liquids. *J. Phys. Chem. B* 111: 9001–9009.

22 Iarikov, D.D., Hacarlioglu, P., and Oyama, S.T. (2011). Supported room temperature ionic liquid membranes for CO_2/CH_4 separation. *Chem. Eng. J.* 166: 401–406.

23 Ferguson, L. and Scovazzo, P. (2007). Solubility, diffusivity, and permeability of gases in phosphonium-based room temperature ionic liquids: data and correlations. *Ind. Eng. Chem. Res.* 46: 1369–1374.

24 Finotello, A., Bara, J.E., Camper, D., and Noble, R.D. (2008). Room-temperature ionic liquids: temperature dependence of gas solubility selectivity. *Ind. Eng. Chem. Res.* 47: 3453–3459.

25 Hojniak, S.D., Khan, A.L., Holloczki, O. et al. (2013). Separation of carbon dioxide from nitrogen or methane by supported ionic liquid membranes (SILMs): influence of the cation charge of the ionic liquid. *J. Phys. Chem. B* 117: 15131–15140.

26 Lee, B.-S. and Lin, S.-T. (2015). Screening of ionic liquids for CO_2 capture using the cosmo-sac model. *Chem. Eng. Sci.* 121: 157–168.

27 Liu, H., Dai, S., and Jiang, D.E. (2014). Molecular dynamics simulation of anion effect on solubility, diffusivity, and permeability of carbon dioxide in ionic liquids. *Ind. Eng. Chem. Res.* 53: 10485–10490.

28 Mahurin, S.M., Dai, T., Yeary, J.S. et al. (2011). Benzyl-functionalized room temperature ionic liquids for CO_2/N_2 separation. *Ind. Eng. Chem. Res.* 50: 14061–14069.

29 Mahurin, S.M., Hillesheim, P.C., Yeary, J.S. et al. (2012). High CO_2 solubility, permeability and selectivity in ionic liquids with the tetracyanoborate anion. *RSC Adv.* 2: 11813–11819.

30 Bara, J.E., Carlisle, T.K., Gabriel, C.J. et al. (2009). Guide to CO_2 separations in imidazolium-based room-temperature ionic liquids. *Ind. Eng. Chem. Res.* 48: 2739–2751.

31 Carvalho, P.J. and Coutinho, J.A.P. (2010). On the nonideality of CO_2 solutions in ionic liquids and other low volatile solvents. *J. Phys. Chem. Lett.* 1: 774–780.

32 Tome, L.C., Patinha, D.J.S., Freire, C.S.R. et al. (2013). CO_2 separation applying ionic liquid mixtures: the effect of mixing different anions on gas permeation through supported ionic liquid membranes. *RSC Adv.* 3: 12220–12229.

33 Tome, L.C., Florindo, C., Freire, C.S.R. et al. (2014). Playing with ionic liquid mixtures to design engineered CO_2 separation membranes. *Phys. Chem. Chem. Phys.* 16: 17172–17182.

34 Zhang, L., Chen, J., Lv, J.X. et al. (2013). Progress and development of capture for CO_2 by ionic liquids. *Asian J. Chem.* 25: 2355–2358.

35 Pereiro, A.B., Tome, L.C., Martinho, S. et al. (2013). Gas permeation properties of fluorinated ionic liquids. *Ind. Eng. Chem. Res.* 52: 4994–5001.

36 Fredlake, C.P., Crosthwaite, J.M., Hert, D.G. et al. (2004). Thermophysical properties of imidazolium-based ionic liquids. *J. Chem. Eng. Data* 49: 954–964.

37 Shannon, M.S., Tedstone, J.M., Danielsen, S.P.O. et al. (2012). Free volume as the basis of gas solubility and selectivity in imidazolium-based ionic liquids. *Ind. Eng. Chem. Res.* 51: 5565–5576.

38 Carvalho, P.J., Alvarez, V.H., Marrucho, I.M. et al. (2009). High pressure phase behavior of carbon dioxide in 1-butyl-3-methylimidazolium bis(trifluoromethylsulfonyl)imide and 1-butyl-3-methylimidazolium dicyanamide ionic liquids. *J. Supercrit. Fluids* 50: 105–111.

39 Carvalho, P.J., Alvarez, V.H., Marrucho, I.M. et al. (2010). High carbon dioxide solubilities in trihexyltetradecylphosphonium-based ionic liquids. *J. Supercrit. Fluids* 52: 258–265.

40 Carvalho, P.J., Alvarez, V.H., Schroder, B. et al. (2009). Specific solvation interactions of CO_2 on acetate and trifluoroacetate imidazolium based ionic liquids at high pressures. *J. Phys. Chem. B* 113: 6803–6812.

41 Hillesheim, P.C., Mahurin, S.M., Fulvio, P.F. et al. (2012). Synthesis and characterization of thiazolium-based room temperature ionic liquids for gas separations. *Ind. Eng. Chem. Res.* 51: 11530–11537.

42 Gurkan, B.E., de la Fuente, J.C., Mindrup, E.M. et al. (2010). Equimolar CO_2 absorption by anion-functionalized ionic liquids. *J. Am. Chem. Soc.* 132: 2116–2117.

43 Seo, S., DeSilva, M.A., Xia, H., and Brennecke, J.F. (2015). Effect of cation on physical properties and CO_2 solubility for phosphonium-based ionic liquids with 2-cyanopyrrolide anions. *J. Phys. Chem. B* 119: 11807–11814.

44 Teague, C.M., Dai, S., and Jiang, D.E. (2010). Computational investigation of reactive to nonreactive capture of carbon dioxide by oxygen-containing Lewis bases. *J. Phys. Chem. A* 114: 11761–11767.

45 Gurkan, B., Goodrich, B.F., Mindrup, E.M. et al. (2010). Molecular design of high capacity, low viscosity, chemically tunable ionic liquids for CO_2 capture. *J. Phys. Chem. Lett.* 1: 3494–3499.

46 Blanchard, L.A. and Brennecke, J.F. (2001). Recovery of organic products from ionic liquids using supercritical carbon dioxide. *Ind. Eng. Chem. Res.* 40: 287–292.

47 Blanchard, L.A., Gu, Z.Y., and Brennecke, J.F. (2001). High-pressure phase behavior of ionic liquid/CO_2 systems. *J. Phys. Chem. B* 105: 2437–2444.

48 Blanchard, L.A., Hancu, D., Beckman, E.J., and Brennecke, J.F. (1999). Green processing using ionic liquids and CO_2. *Nature* 399: 28–29.

49 Kiyono, M., Williams, P.J., and Koros, W.J. (2010). Effect of polymer precursors on carbon molecular sieve structure and separation performance properties. *Carbon* 48: 4432–4441.

50 Kiyono, M., Williams, P.J., and Koros, W.J. (2010). Effect of pyrolysis atmosphere on separation performance of carbon molecular sieve membranes. *J. Membr. Sci.* 359: 2–10.

51 Osterhout, W.J.V. (1940). Some models of protoplasmic surfaces. *Cold Spring Harb. Symp. Quant. Biol.* 8: 51–62.

52 Rea, R., De Angelis, M.G., and Baschetti, M.G. (2019). Models for facilitated transport membranes: a review. *Membranes* 9: 55.

53 Basaran, O.A., Burban, P.M., and Auvil, S.R. (1989). Facilitated transport with unequal carrier and complex diffusivities. *Ind. Eng. Chem. Res.* 28: 108–119.

54 Kemena, L.L., Noble, R.D., and Kemp, N.J. (1983). Optimal regimes of facilitated transport. *J. Membr. Sci.* 15: 259–274.

55 Noble, R.D., Koval, C.A., and Pellegrino, J.J. (1989). Facilitated transport membrane systems. *Chem. Eng. Prog.* 85: 58–70.

56 Dai, Z.D., Noble, R.D., Gin, D.L. et al. (2016). Combination of ionic liquids with membrane technology: a new approach for CO_2 separation. *J. Membr. Sci.* 497: 1–20.

57 Matsuyama, H., Terada, A., Nakagawara, T. et al. (1999). Facilitated transport of CO_2 through polyethylenimine/poly(vinyl alcohol) blend membrane. *J. Membr. Sci.* 163: 221–227.

58 Bao, L.H. and Trachtenberg, M.C. (2006). Facilitated transport of CO_2 across a liquid membrane: comparing enzyme, amine, and alkaline. *J. Membr. Sci.* 280: 330–334.

59 Deng, L.Y., Kim, T.J., and Hagg, M.B. (2009). Facilitated transport of CO_2 in novel PVAm/PVA blend membrane. *J. Membr. Sci.* 340: 154–163.

60 He, W., Zhang, F., Wang, Z. et al. (2016). Facilitated separation of CO_2 by liquid membranes and composite membranes with task-specific ionic liquids. *Ind. Eng. Chem. Res.* 55: 12616–12631.

61 Zhang, X.M., Tu, Z.H., Li, H. et al. (2017). Supported protic-ionic-liquid membranes with facilitated transport mechanism for the selective separation of CO_2. *J. Membr. Sci.* 527: 60–67.

62 Bates, E.D., Mayton, R.D., Ntai, I., and Davis, J.H. (2002). CO_2 capture by a task-specific ionic liquid. *J. Am. Chem. Soc.* 124: 926–927.

63 Myers, C., Pennline, H., Luebke, D. et al. (2008). High temperature separation of carbon dioxide/hydrogen mixtures using facilitated supported ionic liquid membranes. *J. Membr. Sci.* 322: 28–31.

64 Hanioka, S., Maruyama, T., Sotani, T. et al. (2008). CO_2 separation facilitated by task-specific ionic liquids using a supported liquid membrane. *J. Membr. Sci.* 314: 1–4.

65 Kasahara, S., Kamio, E., Ishigami, T., and Matsuyama, H. (2012). Amino acid ionic liquid-based facilitated transport membranes for CO_2 separation. *Chem. Commun.* 48: 6903–6905.

66 Kasahara, S., Kamio, E., Ishigami, T., and Matsuyama, H. (2012). Effect of water in ionic liquids on CO_2 permeability in amino acid ionic liquid-based facilitated transport membranes. *J. Membr. Sci.* 415: 168–175.

67 Kasahara, S., Kamio, E., and Matsuyama, H. (2014). Improvements in the CO_2 permeation selectivities of amino acid ionic liquid-based facilitated transport membranes by controlling their gas absorption properties. *J. Membr. Sci.* 454: 155–162.

68 Kasahara, S., Kamio, E., Otani, A., and Matsuyama, H. (2014). Fundamental investigation of the factors controlling the CO_2 permeability of facilitated transport membranes containing amine-functionalized task-specific ionic liquids. *Ind. Eng. Chem. Res.* 53: 2422–2431.

69 Shiflett, M.B. and Yokozeki, A. (2009). Phase behavior of carbon dioxide in ionic liquids: [emim][acetate], [emim][trifluoroacetate], and [emim][acetate] + [emim][trifluoroacetate] mixtures. *J. Chem. Eng. Data* 54: 108–114.

70 Yokozeki, A., Shiflett, M.B., Junk, C.P. et al. (2008). Physical and chemical absorptions of carbon dioxide in room-temperature ionic liquids. *J. Phys. Chem. B* 112: 16654–16663.

71 Wang, G.N., Dai, Y., Hu, X.B. et al. (2012). Novel ionic liquid analogs formed by triethylbutylammonium carboxylate-water mixtures for CO_2 absorption. *J. Mol. Liq.* 168: 17–20.

72 Huang, K., Zhang, X.-M., Li, Y.-X. et al. (2014). Facilitated separation of CO_2 and SO_2 through supported liquid membranes using carboxylate-based ionic liquids. *J. Membr. Sci.* 471: 227–236.

73 Santos, E., Albo, J., and Irabien, A. (2014). Acetate based supported ionic liquid membranes (SILMs) for CO_2 separation: influence of the temperature. *J. Membr. Sci.* 452: 277–283.

74 Shi, W., Myers, C.R., Luebke, D.R. et al. (2012). Theoretical and experimental studies of CO_2 and H_2 separation using the 1-ethyl-3-methylimidazolium acetate ([emim][CH_3COO]) ionic liquid. *J. Phys. Chem. B* 116: 283–295.

75 Luis, P., Neves, L.A., Afonso, C.A.M. et al. (2009). Facilitated transport of CO_2 and SO_2 through supported ionic liquid membranes (SILMs). *Desal.* 245: 485–493.

76 Gouveia, A.S.L., Tome, L.C., and Marrucho, I.M. (2016). Towards the potential of cyano and amino acid-based ionic liquid mixtures for facilitated CO_2 transport membranes. *J. Membr. Sci.* 510: 174–181.

77 Cichowska-Kopczynska, I., Joskowska, M., and Aranowski, R. (2014). Wetting processes in supported ionic liquid membranes technology. *Physicochem. Probl. Miner. Process* 50: 373–386.

78 Hopkinson, D., Zeh, M., and Luebke, D. (2014). The bubble point of supported ionic liquid membranes using flat sheet supports. *J. Membr. Sci.* 468: 155–162.

79 Neves, L.A., Crespo, J.G., and Coelhoso, I.M. (2010). Gas permeation studies in supported ionic liquid membranes. *J. Membr. Sci.* 357: 160–170.

80 Close, J.J., Farmer, K., Moganty, S.S., and Baltus, R.E. (2012). CO_2/N_2 separations using nanoporous alumina-supported ionic liquid membranes: effect of the support on separation performance. *J. Membr. Sci.* 390: 201–210.

81 Scovazzo, P., Visser, A.E., Davis, J.H. et al. (2002). Supported ionic liquid membranes and facilitated ionic liquid membranes. In: *Ionic Liquids: Industrial Applications for Green Chemistry, Vol. 818* (eds. R.D. Rogers and K.R. Seddon), 69–87. Washington: American Chemical Society.

82 Cadena, C., Anthony, J.L., Shah, J.K. et al. (2004). Why is CO_2 so soluble in imidazolium-based ionic liquids? *J. Am. Chem. Soc.* 126: 5300–5308.

83 Scovazzo, P., Kieft, J., Finan, D.A. et al. (2004). Gas separations using non-hexafluorophosphate $[PF_6]^-$ anion supported ionic liquid membranes. *J. Membr. Sci.* 238: 57–63.

84 Robeson, L.M. (2008). The upper bound revisited. *J. Membr. Sci.* 320: 390–400.

85 Freeman, B.D. (1999). Basis of permeability/selectivity tradeoff relations in polymeric gas separation membranes. *Macromolecules* 32: 375–380.

86 Morgan, D., Ferguson, L., and Scovazzo, P. (2005). Diffusivities of gases in room-temperature ionic liquids: data and correlations obtained using a lag-time technique. *Ind. Eng. Chem. Res.* 44: 4815–4823.

87 Condemarin, R. and Scovazzo, P. (2009). Gas permeabilities, solubilities, diffusivities, and diffusivity correlations for ammonium-based room temperature ionic liquids with comparison to imidazolium and phosphonium RTIL data. *Chem. Eng. J.* 147: 51–57.

88 Anderson, J.L., Dixon, J.K., and Brennecke, J.F. (2007). Solubility of CO_2, CH_4, C_2H_6, C_2H_4, O_2, and N_2 in 1-hexyl-3-methylpyridinium bis(trifluoromethylsulfonyl)imide: comparison to other ionic liquids. *Acc. Chem. Res.* 40: 1208–1216.

89 Scovazzo, P. (2009). Determination of the upper limits, benchmarks, and critical properties for gas separations using stabilized room temperature ionic liquid membranes (SILMs) for the purpose of guiding future research. *J. Membr. Sci.* 343: 199–211.

90 Ki99u, P.K. and Scovazzo, P. (2008). Correlations of low-pressure carbon dioxide and hydrocarbon solubilities in imidazolium-, phosphonium-, and ammonium-based room-temperatuire ionic liquids. Part 2. Using activation energy of viscosity. *Ind. Eng. Chem. Res.* 47: 910–919.

91 Anthony, J.L., Anderson, J.L., Maginn, E.J., and Brennecke, J.F. (2005). Anion effects on gas solubility in ionic liquids. *J. Phys. Chem. B* 109: 6366–6374.

92 Mahurin, S.M., Lee, J.S., Baker, G.A. et al. (2010). Performance of nitrile-containing anions in task-specific ionic liquids for improved CO_2/N_2 separation. *J. Membr. Sci.* 353: 177–183.

93 Babarao, R., Dai, S., and Jiang, D.E. (2011). Understanding the high solubility of CO_2 in an ionic liquid with the tetracyanoborate anion. *J. Phys. Chem. B* 115: 9789–9794.

94 Cheng, L.-H., Rahaman, M.S.A., Yao, R. et al. (2014). Study on microporous supported ionic liquid membranes for carbon dioxide capture. *Int. J. Greenhouse Gas Control* 21: 82–90.

95 Labropoulos, A.I., Romanos, G.E., Kouvelos, E. et al. (2013). Alkyl-methylimidazolium tricyanomethanide ionic liquids under extreme confinement onto nanoporous ceramic membranes. *J. Phys. Chem. C* 117: 10114–10127.

96 Banu, L.A., Wang, D., and Baltus, R.E. (2013). Effect of ionic liquid confinement on gas separation characteristics. *Energy Fuel* 27: 4161–4166.

97 Hazelbaker, E.D., Guillet-Nicolas, R., Thommes, M. et al. (2015). Influence of confinement in mesoporous silica on diffusion of a mixture of carbon dioxide and an imidazolium-based ionic liquid by high field diffusion nmr. *Microporous Mesoporous Mater.* 206: 177–183.

98 Sanders, D.F., Smith, Z.P., Guo, R. et al. (2013). Energy-efficient polymeric gas separation membranes for a sustainable future: a review. *Polymer* 54: 4729–4761.

99 Lee, J.S., Hillesheim, P.C., Huang, D.K. et al. (2012). Hollow fiber-supported designer ionic liquid sponges for post-combustion CO_2 scrubbing. *Polymer* 53: 5806–5815.

100 Lee, J.S., Lively, R.P., Huang, D.K. et al. (2012). A new approach of ionic liquid containing polymer sorbents for post-combustion CO_2 scrubbing. *Polymer* 53: 891–894.

101 Lan, W., Li, S., Xu, J., and Luo, G. (2013). Preparation and carbon dioxide separation performance of a hollow fiber supported ionic liquid membrane. *Ind. Eng. Chem. Res.* 52: 6770–6777.

102 Wickramanayake, S., Hopkinson, D., Myers, C. et al. (2013). Investigation of transport and mechanical properties of hollow fiber membranes containing ionic liquids for pre-combustion carbon dioxide capture. *J. Membr. Sci.* 439: 58–67.

103 Wickramanayake, S., Hopkinson, D., Myers, C. et al. (2014). Mechanically robust hollow fiber supported ionic liquid membranes for CO_2 separation applications. *J. Membr. Sci.* 470: 52–59.

104 Ge, D. and Lee, H.K. (2015). Ultra-hydrophobic ionic liquid 1-hexyl-3-methylimidazolium tris(pentafluoroethyl)trifluorophosphate supported hollow-fiber membrane liquid-liquid-liquid microextraction of chlorophenols. *Talanta* 132: 132–136.

105 Adzima, B.J., Venna, S.R., Klara, S.S. et al. (2014). Modular polymerized ionic liquid block copolymer membranes for CO_2/N_2 separation. *J. Mater. Chem. A* 2: 7967–7972.

106 Cheng, H., Wang, P., Luo, J. et al. (2015). Poly(ionic liquid)-based nanocomposites and their performance in CO_2 capture. *Ind. Eng. Chem. Res.* 54: 3107–3115.

107 Tang, J.B., Tang, H.D., Sun, W.L. et al. (2005). Poly(ionic liquid)s: a new material with enhanced and fast CO_2 absorption. *Chem. Commun.*: 3325–3327.

108 Tang, J.B., Sun, W.L., Tang, H.D. et al. (2005). Enhanced CO_2 absorption of poly(ionic liquid)s. *Macromolecules* 38: 2037–2039.

109 Tang, J.B., Tang, H.D., Sun, W.L. et al. (2005). Poly(ionic liquid)s as new materials for CO_2 absorption. *J. Polym. Sci. A Polym. Chem.* 43: 5477–5489.

110 Tang, J.B., Tang, H.D., Sun, W.L. et al. (2005). Low-pressure CO_2 sorption in ammonium-based poly(ionic liquid)s. *Polymer* 46: 12460–12467.

111 Wilke, A., Yuan, J.Y., Antonietti, M., and Weber, J. (2012). Enhanced carbon dioxide adsorption by a mesoporous poly(ionic liquid). *ACS Macro Lett.* 1: 1028–1031.

112 Tang, H.D., Tang, J.B., Ding, S.J. et al. (2005). Atom transfer radical polymerization of styrenic ionic liquid monomers and carbon dioxide absorption of the polymerized ionic liquids. *J. Polym. Sci. A Polym. Chem.* 43: 1432–1443.

113 Hu, X.D., Tang, J.B., Blasig, A. et al. (2006). CO_2 permeability, diffusivity and solubility in polyethylene glycol-grafted polyionic membranes and their CO_2 selectivity relative to methane and nitrogen. *J. Membr. Sci.* 281: 130–138.

114 Bara, J.E., Lessmann, S., Gabriel, C.J. et al. (2007). Synthesis and performance of polymerizable room-temperature ionic liquids as gas separation membranes. *Ind. Eng. Chem. Res.* 46: 5397–5404.

115 Bara, J.E., Gabriel, C.J., Hatakeyama, E.S. et al. (2008). Improving CO_2 selectivity in polymerized room-temperature ionic liquid gas separation membranes through incorporation of polar substituents. *J. Membr. Sci.* 321: 3–7.

116 Carlisle, T.K., Wiesenauer, E.F., Nicodemus, G.D. et al. (2013). Ideal CO_2/light gas separation performance of poly(vinylimidazolium) membranes and poly(vinylimidazolium)-ionic liquid composite films. *Ind. Eng. Chem. Res.* 52: 1023–1032.

117 Bara, J.E., Hatakeyama, E.S., Gabriel, C.J. et al. (2008). Synthesis and light gas separations in cross-linked gemini room temperature ionic liquid polymer membranes. *J. Membr. Sci.* 316: 186–191.

118 Simons, K., Nijmeijer, K., Bara, J.E. et al. (2010). How do polymerized room-temperature ionic liquid membranes plasticize during high pressure CO_2 permeation? *J. Membr. Sci.* 360: 202–209.

119 Bhavsar, R.S., Kumbharkar, S.C., and Kharul, U.K. (2012). Polymeric ionic liquids (PILs): effect of anion variation on their CO_2 sorption. *J. Membr. Sci.* 389: 305–315.

120 Tome, L.C., Isik, M., Freire, C.S.R. et al. (2015). Novel pyrrolidinium-based polymeric ionic liquids with cyano counter-anions: high performance membrane materials for post-combustion CO_2 separation. *J. Membr. Sci.* 483: 155–165.

121 Vollas, A., Chouliaras, T., Deimede, V. et al. (2018). New pyridinium type poly(ionic liquids) as membranes for CO_2 separation. *Polymers* 10: 19.

122 Bernard, F.L., dos Santos, L.M., Schwab, M.B. et al. (2019). Polyurethane-based poly (ionic liquid)s for CO_2 removal from natural gas. *J. Appl. Polym. Sci.* 136: 8.

123 Bhavsar, R.S., Kumbharkar, S.C., and Kharul, U.K. (2014). Investigation of gas permeation properties of film forming polymeric ionic liquids (PILs) based on polybenzimidazoles. *J. Membr. Sci.* 470: 494–503.

124 Bhavsar, R.S., Kumbharkar, S.C., Rewar, A.S., and Kharul, U.K. (2014). Polybenzimidazole based film forming polymeric ionic liquids: synthesis and effects of cation-anion variation on their physical properties. *Polym. Chem.* 5: 4083–4096.

125 Rewar, A.S., Bhavsar, R.S., Sreekumar, K., and Kharul, U.K. (2015). Polybenzimidazole based polymeric ionic liquids (PILs): effects of controlled degree of n-quaternization on physical and gas permeation properties. *J. Membr. Sci.* 481: 19–27.

126 Kumbharkar, S.C., Bhavsar, R.S., and Kharul, U.K. (2014). Film forming polymeric ionic liquids (PILs) based on polybenzimidazoles for CO_2 separation. *RSC Adv.* 4: 4500–4503.

127 Shaligram, S.V., Wadgaonkar, P.P., and Kharul, U.K. (2015). Polybenzimidazole-based polymeric ionic liquids (PILs): effects of 'substitution asymmetry' on CO_2 permeation properties. *J. Membr. Sci.* 493: 403–413.

128 Rezakazemi, M., Amooghin, A.E., Montazer-Rahmati, M.M. et al. (2014). State-of-the-art membrane based CO_2 separation using mixed matrix membranes

(MMMs): an overview on current status and future directions. *Prog. Polym. Sci.* 39: 817–861.

129 Zornoza, B., Martinez-Joaristi, A., Serra-Crespo, P. et al. (2011). Functionalized flexible MOFs as fillers in mixed matrix membranes for highly selective separation of CO_2 from CH_4 at elevated pressures. *Chem. Commun.* 47: 9522–9524.

130 Zornoza, B., Seoane, B., Zamaro, J.M. et al. (2011). Combination of MOFs and zeolites for mixed-matrix membranes. *ChemPhysChem* 12: 2781–2785.

131 Perez, E.V., Balkus, K.J., Ferraris, J.P., and Musselman, I.H. (2009). Mixed-matrix membranes containing MOF-5 for gas separations. *J. Membr. Sci.* 328: 165–173.

132 Pechar, T.W., Kim, S., Vaughan, B. et al. (2006). Preparation and characterization of a poly(imide siloxane) and zeolite l mixed matrix membrane. *J. Membr. Sci.* 277: 210–218.

133 Pechar, T.W., Kim, S., Vaughan, B. et al. (2006). Fabrication and characterization of polyimide-zeolite l mixed matrix membranes for gas separations. *J. Membr. Sci.* 277: 195–202.

134 Pechar, T.W., Tsapatsis, M., Marand, E., and Davis, R. (2002). Preparation and characterization of a glassy fluorinated polyimide zeolite-mixed matrix membrane. *Desalination* 146: 3–9.

135 Ordonez, M.J.C., Balkus, K.J., Ferraris, J.P., and Musselman, I.H. (2010). Molecular sieving realized with ZIF-8/matrimid® mixed-matrix membranes. *J. Membr. Sci.* 361: 28–37.

136 Zhang, Y.F., Musselman, I.H., Ferraris, J.P., and Balkus, K.J. (2008). Gas permeability properties of mixed-matrix matrimid membranes containing a carbon aerogel: a material with both micropores and mesopores. *Ind. Eng. Chem. Res.* 47: 2794–2802.

137 Zornoza, B., Tellez, C., and Coronas, J. (2011). Mixed matrix membranes comprising glassy polymers and dispersed mesoporous silica spheres for gas separation. *J. Membr. Sci.* 368: 100–109.

138 Ahn, J., Chung, W.J., Pinnau, I. et al. (2010). Gas transport behavior of mixed-matrix membranes composed of silica nanoparticles in a polymer of intrinsic microporosity (PIM-1). *J. Membr. Sci.* 346: 280–287.

139 Vu, D.Q., Koros, W.J., and Miller, S.J. (2002). High pressure CO_2/CH_4 separation using carbon molecular sieve hollow fiber membranes. *Ind. Eng. Chem. Res.* 41: 367–380.

140 Kim, S., Chen, L., Johnson, J.K., and Marand, E. (2007). Polysulfone and functionalized carbon nanotube mixed matrix membranes for gas separation: theory and experiment. *J. Membr. Sci.* 294: 147–158.

141 Zhao, Y.A., Jung, B.T., Ansaloni, L., and Ho, W.S.W. (2014). Multiwalled carbon nanotube mixed matrix membranes containing amines for high pressure CO_2/H_2 separation. *J. Membr. Sci.* 459: 233–243.

142 Hudiono, Y.C., Carlisle, T.K., LaFrate, A.L. et al. (2011). Novel mixed matrix membranes based on polymerizable room-temperature ionic liquids and SAPO-34 particles to improve CO_2 separation. *J. Membr. Sci.* 370: 141–148.

143 Hudiono, Y.C., Carlisle, T.K., Bara, J.E. et al. (2010). A three-component mixed-matrix membrane with enhanced CO_2 separation properties based on zeolites and ionic liquid materials. *J. Membr. Sci.* 350: 117–123.

144 Singh, Z.V., Cowan, M.G., McDanel, W.M. et al. (2016). Determination and optimization of factors affecting CO_2/CH_4 separation performance in poly(ionic liquid)-ionic liquid-zeolite mixed-matrix membranes. *J. Membr. Sci.* 509: 149–155.

145 Casado-Coterillo, C., Fernandez-Barquin, A., Zornoza, B. et al. (2015). Synthesis and characterisation of MOF/ionic liquid/chitosan mixed matrix membranes for CO_2/N_2 separation. *RSC Adv.* 5: 102350–102361.

146 Lin, R.J., Ge, L., Diao, H. et al. (2016). Ionic liquids as the MOFs/polymer interfacial binder for efficient membrane separation. *ACS Appl. Mater. Interfaces* 8: 32041–32049.

147 Mohshim, D.F., Mukhtar, H., and Man, Z. (2018). A study on carbon dioxide removal by blending the ionic liquid in membrane synthesis. *Sep. Purif. Technol.* 196: 20–26.

148 Ahmad, N.N.R., Leo, C.P., Mohammad, A.W., and Ahmad, A.L. (2017). Modification of gas selective SAPO zeolites using imidazolium ionic liquid to develop polysulfone mixed matrix membrane for CO_2 gas separation. *Microporous Mesoporous Mater.* 244: 21–30.

149 Huang, G.J., Isfahani, A.P., Muchtar, A. et al. (2018). Pebax/ionic liquid modified graphene oxide mixed matrix membranes for enhanced co2 capture. *J. Membr. Sci.* 565: 370–379.

150 Ilyas, A., Muhammad, N., Gilani, M.A. et al. (2018). Effect of zeolite surface modification with ionic liquid aptms ac on gas separation performance of mixed matrix membranes. *Sep. Purif. Technol.* 205: 176–183.

151 Abdollahi, S., Mortaheb, H.R., Ghadimi, A., and Esmaeili, M. (2018). Improvement in separation performance of matrimid® 5218 with encapsulated [emim][Tf$_2$N] in a heterogeneous structure: CO_2/CH_4 separation. *J. Membr. Sci.* 557: 38–48.

152 Chen, H.Z., Li, P., and Chung, T.S. (2012). PVDF/ionic liquid polymer blends with superior separation performance for removing CO_2 from hydrogen and flue gas. *Int. J. Hydrog. Energy* 37: 11796–11804.

153 Liang, L., Gan, Q., and Nancarrow, P. (2014). Composite ionic liquid and polymer membranes for gas separation at elevated temperatures. *J. Membr. Sci.* 450: 407–417.

154 Lu, S.C., Khan, A.L., and Vankelecom, I.F.J. (2016). Polysulfone-ionic liquid based membranes for CO_2/N_2 separation with tunable porous surface features. *J. Membr. Sci.* 518: 10–20.

Index

Note: Page numbers in *italics* refer to Figures; those in **bold** to Tables.

Materials for Carbon Capture, First Edition. Edited by De-en Jiang, Shannon M. Mahurin and Sheng Dai.
© 2020 John Wiley & Sons Ltd. Published 2020 by John Wiley & Sons Ltd.